Vacuum Technique

Vacuum Technique

L.N. Rozanov

St. Petersburg Technical University, Russia

Edited by M.H.Hablanian

Translated from Russian by A.V. Pavlov

Routledge
Taylor & Francis Group

LONDON AND NEW YORK

First published 2002 by Taylor & Francis

2 Park Square, Milton Park, Abingdon, Oxfordshire OX14 4RN
52 Vanderbilt Avenue, New York, NY 10017

Routledge is an imprint of the Taylor & Francis Group, an informa business

First issued in paperback 2019

British Library Cataloguing in Publication Data
A catalogue record for this book is available from the British Library

Library of Congress Cataloging in Publication Data
A catalog record for this book has been requested.

ISBN 978-0-415-27351-0 (hbk)
ISBN 978-1-138-38130-8 (pbk)

Contents

Introduction

I.1   The Concept of Vacuum

The definition of vacuum currently has several meanings. The Latin word 'vacuum' means emptiness, and in this sense it has long been used in philosophy. Ancient Greek philosopher Democritus considered emptiness as an origin of the world.

Later on, Aristotle introduced the term of ether as an imperceptible medium that is capable of transmitting pressure. Up to now, the substantiality of vacuum remains a subject of discussions among philosophers.

The property of vacuum to transmit electromagnetic waves and the annihilation and random occurrence of microparticles and antiparticles therein are of interest for modern theoretical and experimental physics.

In engineering, vacuum is considered as the state of gas in which its pressure is lower that the atmospheric one. Close to the normal atmospheric pressure, vacuum can be quantitatively defined as the difference of the atmospheric pressure and the absolute pressure in a vacuum system. Differential manometers, i.e. the so-called vacuum gages directly measure this difference of pressures.

Vacuum was quantitatively defined for the first time in the beginning of the development of vacuum engineering when very low pressures could not be achieved.

At absolute pressures that differ from the atmospheric pressure by more than two orders of magnitude, this difference of pressures remains virtually constant and cannot characterize the state of rarefied gas. Under these conditions, the state of gas is quantitatively defined by its absolute pressure, and the term 'vacuum' is determined as the state of rarefied gas in which it obeys the regularities of basic physical phenomena, such as mass transport, gas flow, etc. Low, average, and high vacuum are distinguished using the dimensionless Knudsen criterion, which depends not only on pressure, but also on the size of vacuum chamber.

At low pressures that cannot be directly measured using conventional devices, the state of gas can be characterized by its molecular concentration, i.e., the number of molecules per unit volume. However, for the unification of measuring units, the devices are calibrated in the units of absolute pressure by recalculating them using the equation of gas state.

At even lower pressures, it becomes insufficient to quantitatively characterize the state of rarefied gas by the mean absolute pressure or the molecular concentration, because these quantities become strongly dependent on the coordinate and direction in a vacuum chamber. In this case, to describe the state of rarefied gas, vector characteristics of molecular flows should be used.

Ultrahigh vacuum is usually treated in literature as the state of gas that satisfies the conditions of high vacuum and low surface coverage with adsorbed gas molecules. The surface properties of pure materials can only be studied under ultrahigh vacuum. Any surface pollution, even within a monomolecular layer, strongly affects the electron emission, gas evolution, gas permeability, adherence, etc.

Advanced vacuum engineering allows achievement and measurement of absolute pressures of to 10^{-6} Pa for conventional industrial equipment and to 10^{-10} Pa for unique laboratory instruments. At room temperature, these pressures correspond to 10^8 and 10^4 molecules per one cubic centimeter, respectively. A pressure of 10^{-12} Pa was achieved in unique experiments. At this pressure one cubic centimeter contains several hundreds gas molecules. Therefore, although considerable advance has been made in vacuum engineering by reducing the absolute pressure by 10^{15} times, emptiness free from gas molecules has not yet been obtained.

Experiments conducted on board space vehicles prove that ionized hydrogen is the main residual gas that is present at large distances from the Earth, the least measured concentration of hydrogen corresponding to a pressure of 10^{-10} Pa. In the interstellar space, the concentration of residual gases is obviously quite nonuniform and its determination is a task of the future.

I.2   History of Vacuum Engineering

The development of vacuum engineering as a science has its roots in the research works of Italian scientist Galileo Galilei (1564–1642) who measured the atmospheric pressure. The impetus for these works was the impossibility to extract water from deep wells. In 1643, Galelei's disciple E. Torricelli (1608–1647) invented a mercury manometer. B. Pascal (1623–1662) determined atmospheric pressure as a function of height and suggested a design of membrane barometer.

In 1654 in Magdeburg, O. Guericke (1602–1686) conducted experiments with rarefied gases and developed a design of a vacuum piston-drive pump with water sealing. Further vacuum research dealt with the regularities of various physical and chemical processes and the effect of vacuum on living organisms. Experiments with electric charge in vacuum led to the discovery of X-ray radiation (1895) and the electron (1897). The heat-insulating properties of vacuum helped in the correct understanding of the heat transfer mechanisms and promoted the development of cryogenic engineering.

The successful study of the properties of rarefied gas allowed its general use in technology since 1872, when the first electric vacuum device, i.e., incandescent lamp with a coal electrode, was invented by Russian scientist A.N. Lodygin (1847–1923). Thermoelectronic emission was discovered in 1883 by American scientist and inventor T. Edison (1847–1931). From this point onwards, vacuum engineering became a technological basis for electric vacuum industry and other branches.

The expansion of the practical use of vacuum devices was accompanied by the development of techniques for obtaining vacuum. Over a short period of time in the early 20th century, presently widely used vacuum pumps were invented, i.e., rotary pumps (Gaede, 1905), adsorption pumps (D. Dewar, 1906), molecular pumps (Gaede, 1912), and diffusion pumps (Gaede, 1913). Success was also achieved in the development of vacuum measurement techniques, examples being the mercury compression manometer (G. McLeod, 1874), the thermal manometer (M. Pirani, 1909), and ionization manometers (O. Buckley, 1916).

The scientific fundamentals of vacuum engineering were also developed. In Russia, P.N. Lebedev (1901) in his experiments used the idea to remove residual gases using mercury vapor. That time, the fundamental properties of gases at low pressure were extensively investigated (M. Knudsen, M. Smoluchovsky, I. Langmuir, and S. Dushman).

In Russia, vacuum engineering was put forward by S.A. Vekshinski (1896–1974) who organized in 1928 a vacuum laboratory on the premises of the Svetlana factory in Leningrad and then headed the Research Institute of Vacuum Engineering in Moscow.

Up to the 1950s, pressures below 10^{-5} Pa were considered unachievable. The works of American scientists Nottingham (1948) and Alpert (1952) on the measurement of background currents of ionization gauge allowed one to expand the range of measured pressures by 3–4 orders of magnitude to ultrahigh vacuum.

For that purpose, new pumps were invented: the turbomolecular pump (Becker, 1958), the magnetic discharge pump (Jepsen and Holland, 1959), and the cryosorption pump (Lazarev and Fedorova, 1957); diffusion oil-vapor pumps were improved. The development of non-polluting pumping devices opened new application fields of vacuum engineering.

Partial pressure analyzers came into use to measure low pressures, such as static mass-spectrometers, omegatrons, chronotrons, farvitrons, topotrons, and electric mass filters.

Reliable assembly and operation of vacuum systems was ensured by the invention of leak detectors. In the 1960s, new sensitive techniques were developed to detect leakage in vacuum systems, i.e., mass-spectrometer, halide, katharometry, radio isotope and other techniques. To reduce gas evolution from vacuum chamber walls, vacuum units were subjected to high-temperature heating. The all-metal vacuum systems were manufactured, and ultrahigh vacuum sealing and vacuum electric contacts were designed. The technology of undetachable metal-glass connections and electron-beam and gas welding was developed.

Since the 1960s, new methods for vacuum system calculation were elaborated. Numerical computed-assisted methods have found general use. The method of static tests for vacuum system calculation was put forward by Davis (1960) to calculate the conductance of complex elements in vacuum systems. Space research gave impetus to a number of experiments with the dynamics of rarefied gases in free and limited space.

Vacuum engineering currently develops by increasing the efficiency of already existing devices and the improvement of new techniques for achievement and measurement of even lower gas pressures [1–9].

I.3 Applications of Vacuum Instruments

The use of vacuum began long before its properties have been understood. A few thousand years ago, vacuum was used in water-lifting and pneumatic mechanisms. The first applications of vacuum made use of its mechanical power. Water extraction from wells and the Magdeburg hemispheres were the first examples of this kind. At present, vacuum still finds general use in vacuum clamps, pneumatic mechanisms, and medical instruments. The absence of oxygen shows negative effect on all living organisms, with the exception of some viruses. However, controlled gas media can be used in barochambers in medicine and for food storage. In the absence of oxygen, active metals and other materials can be obtained, and these conditions are favorable for the operation of high-temperature heaters and cathodes.

In vacuum, the rate of evaporation of materials increases substantially. This phenomenon is used for drying food products, wood, and other commodities at room temperature. At pressures 40 times lower than the atmospheric one, water boils at room temperature. Vacuum evaporation of materials is used for producing thin coatings. The heat and sound isolating properties of vacuum are used in construction and for storage of heated or cooled materials. The property of vacuum to transmit electromagnetic radiation finds application in heating, spectral diagnostics, and temperature measurement in vacuum.

The application scope of vacuum broadens, but from the end of the 19th century till now its main application field remains electronics. Vacuum is a necessary condition for the operation of electronic tubes. Low and medium vacuum is used in illumination and gas discharge devices, and high vacuum in amplifying and generator tubes. The most strict requirements are imposed on vacuum in the production of CRTs and RF devices. Semiconductor devices do not require vacuum in operation, but their production widely uses vacuum techniques. This primarily concerns IC manufacturing, in which thin film deposition, ion etching, implantation, and electron lithography currently handle submicron device sizes.

In metallurgy, vacuum melting and remelting of metals eliminates detrimental dissolved gases and makes metals strong, ductile, and tough. This technology is used for the production of carbon-free iron electric motors, highly conducting copper, magnesium, calcium, tantalum, platinum, titanium, zirconium, beryllium, rare metals, alloys, and high-quality steels. Vacuum sintering of tungsten and molybdenum powders is a basic technological process of powder metallurgy. High-purity semiconductor and dielectric crystals are grown in vacuum. Diffusion vacuum welding produces joints of materials having much different melting points, e.g., ceramics with metals, steel with aluminum, etc.. Electron beam vacuum welding ensures high quality of joints of materials whose properties are similar.

In machine-building, vacuum is used in the study of the coalescence and unlubricated friction of metals, for deposition of strengthening coatings on tools and wear-resistant coatings on machine parts, and in clamping and transportation of worked parts in automated technological routes.

Chemical industry uses vacuum dryers for synthetic fibers, polyamides, aminoplastics, polyethylene, and organic solvents. Vacuum filters are used in the production of pulp, paper, and lubricant oils. Vacuum crystallizers are used in the production of paints and fertilizers.

Vacuum impregnation is the most cost-effective method for the manufacture of transformers, electric motors, capacitors, and cables in electrical engineering. The use of vacuum increases the durability and reliability of electric switches.

Optical industry has changed from chemical silver plating in mirror-making to vacuum aluminum plating. Antireflecting and protective coatings and interference filters are obtained by vacuum deposition.

In food industry, long-term storage and preservation are performed by freeze-out vacuum drying. Vacuum packaging of perishable products increases their storage periods. Sugar and salt production and freshening of sea water also involve vacuum processes.

In agriculture, vacuum is used in milking mashines. Vacuum cleaners have become indispensable in our homes.

In vehicles, vacuum is used for fuel feeding in carburetors and in vacuum braking systems. Simulation of space on Earth is necessary for testing of satellites and rockets.

In medicine, vacuum is used for the preservation of hormones, vaccines, and vitamins and for the preparation of antibiotics and anatomic and bacteriological samples.

Scientific studies of physical processes, such as evaporation, condensation, surface phenomena, and heat transfer, cryogenic experiments, and nuclear and thermonuclear reactions are carried out in vacuum instruments. The accelerator of charged particles, which is the main instrument of up-to-date nuclear physics, is another vacuum instrument.

Vacuum systems are used in chemistry for the study of pure substances, composition and separation of mixtures, and rates of chemical reactions. Such modern instruments for materials characterization as electron and ion microscopes and mass-spectrometers include vacuum chambers.

In materials science, the method of molecular beam vacuum epitaxy (MBVE) enables alloys of any compositions to be obtained. Ion implantation is a promising tool for the modification of surface properties of materials. Synthetic diamond, ruby, and sapphire are grown in vacuum.

The development of space investigations are much related to the simulation of space on Earth. Mechanisms and devices can be tested in vacuum chambers hundreds cubic meters in volume that produce very low pressures and simulate direct sun radiation.

Vacuum engineering finds new industrial applications that provide for the development of advanced technology.

Chapter 1

Properties of Gases at Low Pressures

1.1 Gas Pressure

The properties of gases at low pressures are studied by the physics of vacuum under the section of the molecular-kinetic theory of gases. The basic postulates of the physics of vacuum are as follows:

1. gas molecules move separately;
2. there is a permanent velocity distribution of gas molecules, i.e. the same number of molecules always has the same velocity;
3. there are no predominant directions in the transport of gas molecules, i.e., the space of gas molecules is isotropic;
4. the temperature of a gas is proportional to the mean kinetic energy of its molecules;
5. gases are adsorbed during the interaction with solid surfaces.

The state of gas in which its pressure is lower than the atmospheric one is referred to as vacuum. At pressures close to the atmospheric one, vacuum is quantitatively defined as the difference of the atmospheric pressure and the absolute pressure of a gas. At absolute pressures that are lower than the atmospheric one by more than two orders of magnitude, this difference remains virtually constant and cannot be accepted as a quantitative characteristic of rarefied gas.

Under these conditions, vacuum is quantitatively characterized by absolute gas pressure. At very low pressures that cannot be directly measured by existing devices, the state of gas can be characterized by the number of molecules in unit volume, i.e., the molecular concentration.

During the interaction of gas molecules with the surface of a solid body, the normal component of the change in the quantity of movement of a molecule is $mv\cos(\theta)$; where θ is the angle between the normal to the surface and the direction of molecule's movement, and m and v are the mass and velocity of the molecule, respectively. We will consider the case with a surface and a gas environment in the energetic and adsorption equilibrium. Each adsorbed molecule corresponds to one desorbed molecule with the opposite direction of the velocity vector. Thus, the total change in the impetus of adsorbed and desorbed molecules is $dK = mv\cos(\theta)$.

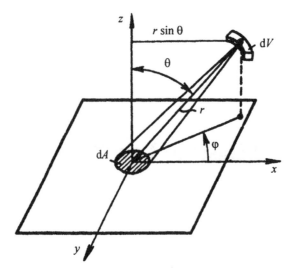

Figure 1.1 Scheme for calculation of gas pressure.

According to the second Newton law, the pressure of a molecule against the surface of a solid body is

$$p = \frac{dK}{dA\,dt} = \frac{2mv\cos(\theta)}{dA\,dt}, \tag{1.1}$$

where dA is the area of the surface and dt is the time of interaction of the molecule with the surface.

Taking into account the absence of predominant directions in a low-pressure gas, molecular concentration n – the number of molecules in unit volume, number of molecules in volume dV moving in the direction dA is proportional to the solid angle $d\omega$, at which the area dA is seen from the center of dV:

$$dN = n\frac{d\omega}{4\pi}dV. \tag{1.2}$$

The solid angle is

$$d\omega = \frac{\cos(\theta)dA}{r^2}, \tag{1.3}$$

where r is the distance between the surface and the separated volume (Fig. 1.1). For the volume dV in the spherical system of coordinates we may write the following formula:

$$dV = r\sin(\theta)d\varphi r d\theta dr. \tag{1.4}$$

We will determine the gas pressure at the surface of a solid body by integration over the volume of the hemisphere with radius $R = vdt$ from which molecules reach the surface in time dt. Taking into account Eq. (1.1),

$$p = \int_V \frac{2mv\cos(\theta)}{dAdt}dN. \tag{1.5}$$

Substituting Eqs. (1.2), (1.3), and (1.4) into Eq. (1.5), we obtain

$$p = \frac{nmv}{2\pi \cdot dt}\int_0^{2\pi}d\varphi \int_0^{\pi/2}\cos^2(\theta)\sin(\theta)d\theta \int_0^R dr = \frac{nmv^2}{3}. \tag{1.6}$$

According to the above postulates, there is a velocity distribution of gas molecules, and, therefore, we introduce the root-mean-square velocity of molecules $v_k^2 = (1/n)\sum_{i=1}^n v_i^2$ (1.6) instead of the constant. Then the equation of gas pressure is

$$p = \frac{nmv_k^2}{3}. \tag{1.7}$$

Because the density of gas is $\rho = nm$, Eq. (1.7) can be rewritten as

$$p = \frac{\rho v_k^2}{3}.$$

The equilibrium conditions used for the derivation of Eq. (1.7) may not be satisfied. An example is a condensing surface from which molecules are not desorbed because of a very long adsorption time. A body leaving Earth into space desorbs gas molecules from its surface. The number of molecules that collide with the surface of this body can be neglected. Then, the gas pressure on the surface at constant concentration, mass, and temperature decreases by a factor of two and can be calculated using the formula

$$p = \frac{nmv_k^2}{6}. \tag{1.8}$$

Real surfaces may be in such a state that Eqs. (1.7) and (1.8) are the extreme cases. To accurately calculate gas pressure, especially at very low pressures, one should know the ratio of the flows of gas molecules incident onto and evolved from the surface of a solid body.

If unit volume contain K gases that do not interact chemically, then to define the pressure of mixture p_m one should calculate the sum:

$$p_m = \sum_{i=1}^K \frac{1}{3}m_i n_i v_{ki}^2 = \sum_{i=1}^K p_i. \tag{1.9}$$

The latter expression is referred to as the Dalton law and is formulated as follows: the total pressure of gases that do not interact chemically is the sum of their partial pressures.

Using the definition of temperature as proportional to the mean kinetic energy of gas molecules, one may write $mv_k^2/2 = cT$, where c is the constant. Then Eq. (1.7) for the calculation of gas pressure can be presented in the form $p = (2/3)ncT$. Denominating $k = 2c/3$, we obtain

$$p = nkT. \qquad (1.10)$$

The mean kinetic energy of the molecules is

$$\frac{mv^2}{2} = \frac{3}{2}kT. \qquad (1.11)$$

Equation (1.10) describes the state of a gas and relates the three main parameters of the gas, i.e., its pressure, molecular concentration, and temperature.

The constant k is the Boltzmann constant: $k = 1.38 \times 10^{-23}$ J/ K.

Equation (1.10) can be presented in another form:

$$p = \frac{Nm}{VM}RT, \qquad (1.12)$$

where M is the molecular mass of the gas, V is the volume of the gas, R is the universal gas constant, $R = kN_A = 8.31 \times 10^3$ J/(K kmol), and N_A is the Avogadro number,

$$N_A = \frac{N}{m} = 6.02 \times 10^{26} \ (\text{kmol})^{-1}.$$

As follows from Eqs. (1.10) and (1.12), given constant mass and pressure of a gas, its volume is proportional to its absolute temperature (the Gay–Lussac law). Alternatively, if the mass and volume of a gas are constant, its pressure is proportional to the absolute temperature (the Charles law). The Boyle–Mariotte law postulates that at constant mass and temperature the product of gas pressure and volume is constant. According to the Avogadro law, the molecular concentration of a gas does not depend on the type of its molecules if its pressure and temperature are constant.

The CI unit of pressure is Pa (Pascal), which is equal to 1 N/m^2. Literature on vacuum engineering may use also other units. The most popular out-of-system unit of pressure is millimeter of mercury column (mm Hg, or torr). A gas pressure of 1 mm Hg is equal to the pressure produced by a 1 mm high mercury column, provided the density of mercury is 13595.1 kg/m^3 (at 0°C), and g is 9.80665 m/s^2 at the 45 deg lattitude. The pressure of a liquid column is determined as $p = \rho gh$, whence

$$1 \text{ mm Hg} = 13595.1 \times 9.80665 \times 10^{-3} = 133.32239 \text{ N/m}^2.$$

In meteorology, pressure is often measured in bars (1 bar = 10^5 Pa) or millibars (1 mbar = 100 Pa). Relationships between various units of pressure are given in Table C.1.

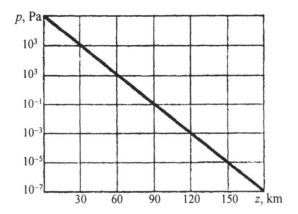

Figure 1.2 Atmospheric pressure at different altitudes above sea level.

Atmospheric air is the main gas mixture dealt with in vacuum engineering (Table C.2). Its major components are nitrogen, oxygen, and water vapor. At 25°C and 50% humidity, the partial pressure of water vapor is 1.2×10^3 Pa.

The reference conditions are accepted as 10^5 Pa and 273 K. Then, according to Eq. (1.12), the volume of 1 kmol of any gas is 22.4 m³. Under the reference conditions, the molecular concentration of air is 2.7×10^{25} m⁻³.

Most of air components are gaseous, only H_2O, CO_2, and Xe being vapors.

The pressure of atmospheric air depends on altitude above sea level. For altitude changing by dz, the pressure changes by d$p = -\rho g dz$, where ρ is the density of the gas, $\rho = nm = pm/(kT)$, and g is the acceleration of gravity. Separating variables, we obtain

$$\frac{dp}{p} = -\frac{mg}{kT}dz. \tag{1.13}$$

Integration yields

$$\ln p = -\frac{mg}{kT}z + C \text{ or } p = C\exp\left(-\frac{mg}{kT}z\right). \tag{1.14}$$

At $z = 0$, the integration coefficient C is equal to gas pressure p_0 at the Earth's surface, whence

$$p = p_0\exp\left(-\frac{mg}{kT}z\right). \tag{1.15}$$

This latter expression is referred to as the Boltzmann equation, which suggests that, elevating 15 km at a time, one measures the pressure of air to decrease by approximately one order of magnitude. Figure 1.2 shows change in the pressure of air with altitude.

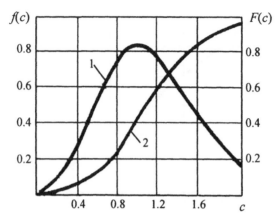

Figure 1.3 (*1*) differential *f(c)* and (*2*) integral *F(c)* velocity distributions of gas molecules.

1.2   Velocity Distribution of Gas Molecules

Colliding with one another or with chamber walls, gas molecules change their velocity and direction of movement. On the basis of the steady-state velocity distribution of gas molecules and the isotropy of gas volume and taking into account that, according to Eq. (1.11), the mean-root-square velocity of gas molecules is $v_s = \sqrt{3kT/m}$, Maxwell obtained the velocity distribution of gas molecules in the form

$$dn_v = 4n\pi v^2 \left(\frac{m}{2\pi kT}\right)^{3/2} \exp\left(-\frac{(mv)^2}{2kT}\right)dv, \tag{1.16}$$

where dn_v is the number of molecules whose velocities are within the range from v to $v+dv$. Equation (1.16) is derived in detail in Appendix A.1.

The peak velocity of the distribution is called the most probable velocity. Equalizing the derivative of the function in Eq. (1.16) to zero, we obtain

$$v_p = \sqrt{2kT/m}. \tag{1.17}$$

Denoting $c = v/v_p$, we rewrite Eq. (1.16) as

$$dn_v = \frac{4n}{\sqrt{\pi}}c^2\exp(-c^2)dc.$$

The dimensionless functions $f(c) = dn_v/(ndc)$ and $F(c) = \int_0^c f(c)dc$ are represented in Fig. 1.3. The function $F(c)$ describes the total number of molecules whose velocities are less than c.

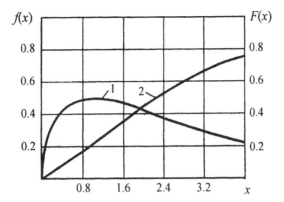

Figure 1.4 (*1*) differential *f*(*x*) and (*2*) integral *F*(*x*) energy distributions of gas molecules.

Vacuum engineering calculations often use the mean parameters, i.e., the arithmetical mean velocity

$$v_a = \frac{1}{n}\int_0^\infty v^2 dn_v = \sqrt{\frac{8kT}{\pi m}} \qquad (1.18)$$

and the mean-root-square velocity

$$v_s = \sqrt{\frac{1}{n}\int_0^\infty v^2 dn_v} = \sqrt{\frac{3kT}{m}}. \qquad (1.19)$$

The ratio of the velocities v_p, v_a, and v_s is 1:1.128:1.225.

For nitrogen at 0°C, these velocities are 402, 453, and 492 m/s, respectively. Table C.2 lists arithmetical mean velocities of molecules of some gases at different temperatures.

Transforming Eq. (1.16) we obtain the energy distribution function of gas molecules:

$$dn_E = \frac{2n}{\sqrt{\pi}}(kT)^{-3/2}\exp\left(-\frac{E}{kT}\right)\sqrt{E}dE.$$

Here $E = mv_s^2/2$ is the progressive motion energy of a molecule and dn_E is the number of molecules whose energies are within the range from E to $E+ dE$.

The peak energy or most probable energy of gas molecules is E_p. Introducing the variable $x = E/E_p$, we obtain

$$f(x) = \frac{d(n_E/n)}{dx} = 4\sqrt{\frac{2x}{\pi}}\exp\left(-\frac{x}{2}\right). \qquad (1.20)$$

Table 1.1 Dimensionless distribution functions of gas molecules.

c	$f(c)$	$F(c)$	x	$f(x)$	$F(x)$
0.1	0.0223	0.0008	0.1	0.2401	0.0082
0.2	0.0867	0.0059	0.2	0.3229	0.0024
0.3	0.1856	0.0193	0.4	0.4131	0.0598
0.4	0.3077	0.0438	0.6	0.4578	0.1036
0.5	0.4393	0.0812	0.8	0.4785	0.1505
0.6	0.5668	0.1316	1.0	0.4839	0.1987
0.7	0.6775	0.1939	1.2	0.4797	0.2470
0.8	0.7613	0.2663	1.4	0.4688	0.2945
0.9	0.8129	0.3453	1.6	0.4535	0.3406
1.0	0.8302	0.4276	1.8	0.4352	0.3851
1.2	0.7697	0.5896	2.0	0.4152	0.4276
1.4	0.6232	0.7286	2.8	0.3294	0.5765
1.6	0.4464	0.8369	3.6	0.2502	0.6920
1.8	0.2862	0.9096	4.0	0.2160	0.7385
2.0	0.1652	0.9540	6.0	0.0973	0.8884
2.5	0.0272	0.9941	10.0	0.0170	0.9814

The integral curve representing the fraction of gas molecules whose energy is less than E is

$$F(x) = \int_0^x f(x)\mathrm{d}x. \tag{1.21}$$

Figure 1.4 shows the dimensionless functions (1.20) and (1.21). The maximum of the differential curve corresponds to the most probable energy $E_p = 0.5kT$. The calculated arithmetical mean energy of gas molecules is $E_a = 1.5kT$. The most important values of the dimensionless functions are given in Table 1.1.

1.3   Mean Free Path Length

Due to the collisions, at frequency K, with randomly moving molecules, the initial number of molecules N_0 in a directional gas flow decreases in time $\mathrm{d}t$ by $\mathrm{d}N = -KN\mathrm{d}t$. Integration of this formula yields

$$N = N_0\exp(-Kt) = N_0\exp\left(-\frac{l}{L}\right), \tag{1.22}$$

where $L = v/K$ is the mean free path length of the molecules, which is determined by the velocity of molecules related to the frequency of collisions in unit time, and $l = vt$ is the path passed by a molecule over time t.

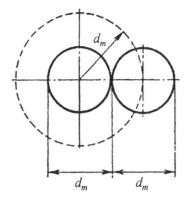

Figure 1.5 Collision of two similar molecules.

The molecules collide if the distance between their centers is equal to the diameter of one molecule d_m (Fig. 1.5). We assume that one of the molecules has radius d_m, the others being mathematical points with zero radii. Moving with velocity v in a gas with molecular concentration n, this molecule will cover in 1 s the volume $V = \pi d_m^2 v$ and undergo $K = n d_m^2 v$ collisions. The mean free path length is then

$$L = \frac{v}{K} = \frac{1}{\pi d_m^2 n}. \tag{1.23}$$

Taking into account the relative velocities of gas molecules, which were disregarded in deriving Eq. (1.23), the more accurate equation of mean free path length is

$$L = \frac{1}{\sqrt{2}\pi d_m^2 n} \tag{1.24}$$

It can be seen from Eq. (1.24) that at a constant molecular concentration the mean free path length does not depend on temperature. However, experiments suggest an increase in the mean free path length with heating. This temperature dependence can be allowed for by the experimental coefficient

$$L = \frac{1}{\sqrt{2}\pi d_m^2 n(1 + C/T)}. \tag{1.25}$$

Here C is the Sutherland constant, which is equal to the temperature, at which, given constant molecular concentration, the mean free path length is half the value at infinitely high temperature. The coefficients C of various gases are given in Table C.4.

To allow for the interaction (attraction) of gas molecules, the effective molecular diameter d_T is introduced:

$$d_T^2 = d_m^2\left(1 + \frac{C}{T}\right).$$ (1.26)

The effective diameter of a molecule decreases with an increase in temperature. Equation (1.25) can be written, taking into account Eq. (1.26), as

$$L = \frac{1}{\sqrt{2}\pi d_T^2 n}.$$ (1.27)

Using the equation of gas state (Eq. (1.10)) and Eq. (1.26), one may transform Eq. (1.27) to

$$L = \frac{kT}{\sqrt{2}\pi p d_m^2 (1 + C/T)} = \frac{kT^2}{\sqrt{2}\pi p d_m^2 (T + C)}.$$ (1.28)

For air at $T = 293$ K and 1 Pa, Eq. (1.28) suggests that $L_1 = 0.7 \times 10^3$ m Pa. At any other pressures,

$$L = \frac{L_1}{p} \cong \frac{0.7 \times 10^{-3}}{p},$$ (1.29)

where p is in Pa and L is in meters.

Calculating the mean free path length of gas molecules at various temperatures and constant pressure on the basis of Eq. (1.28), we obtain

$$L_T = L_0 \frac{T^2(T_0 + C)}{T_0^2(T + C)} = bL_0,$$ (1.30)

where

$$b = \frac{T^2(T_0 + C)}{T_0^2(T + C)}.$$

Table 1.2 shows mean free path lengths of molecules of different gases at unit pressure. For a mixture of two gases with molecular masses m_1 and m_2, the mean free path length of particles of mass m_1 is calculated as

$$L_1 = \frac{1}{\sqrt{2}\pi n_1 d_{T_1}^2 + \pi n_2 d_{12}^2 \sqrt{1 + m_1/m_2}},$$ (1.31)

where d_{T_1} is the effective diameter of molecules of mass m_1 and concentration n_1, $d_{12} = 0.5(d_{T_1} + d_{T_2})$, and d_{T_2} is the effective diameter of molecules of mass m_2 and concentration n_2.

Table 1.2 Mean free path length of molecules of different gases at 1 Pa.

Gas	$L_1 \times 10^3$, m Pa at T, K				Gas	$L_1 \times 10^3$, m Pa at T, K			
	600	293	77	4.2		600	293	77	4.2
N_2	20.8	8.67	1.26	0.0061	H_2	28.2	12.2	0.197	0.0108
O_2	16.9	7.02	1.00	0.0047	Xe	10.5	3.93	0.448	0.0017
Ar	16.7	6.79	0.933	0.0042	H_2O	13.9	4.38	0.391	0.0013
CO_2	11.6	4.32	0.492	0.0019	Air	16.0	6.72	0.995	0.0048
Ne	30.7	13.9	2.50	0.0165	He	43.6	19.1	3.13	0.0174
Kr	14.1	5.52	0.691	0.0029					

The first term in the denominator of Eq. (1.31) is determined by the collisions of similar molecules m_1, and the second term, by the collisions of different molecules, m_1 and m_2. If $n_1 \ll n_2$, the formula simplifies:

$$L_1 = \frac{1}{\pi n_2 d_{12}^2 \sqrt{1 + m_1/m_2}} . \tag{1.32}$$

1.4 Interaction of Gas Molecules with Surfaces

The number of molecules incident on unit area of a surface in unit time is the molecular collision frequency:

$$N_q = \int_V \frac{dN}{dAdt} = \frac{n}{4\pi} \int_0^{2\pi} d\varphi \int_0^{\pi/2} \sin(\theta)\cos(\theta)d\theta \int_0^R \frac{dr}{dt} = \frac{nv}{4} , \tag{1.33}$$

where dN is taken as in Eq. (1.2). Taking into account the velocity distribution function of the molecules, we have

$$N_q = \frac{nv_a}{4} , \tag{1.34}$$

where v_a is the arithmetical mean velocity of gas molecules.

The volume of gas interacting with unit area of a surface in unit time can be expressed through the collision frequency and molecular concentration:

$$V_q = \frac{N_q}{n} = \frac{v_a}{4} . \tag{1.35}$$

L.N. ROZANOV

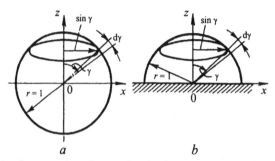

Figure 1.6 Schematic of evaporation of gas molecules from (*a*) point and (*b*) planar source.

The latter formula does not include pressure and determines the maximum speed of a perfect vacuum pump, i.e., that which pumps off all gas molecules ending up in its working volume. For air at atmospheric pressure and $T = 300$ K, $M = 29$ kg/kmole, the frequency of molecular collisions with a surface is $N_q = 2.9 \times 10^{23}$ s^{-1}cm^{-2}, and the volume of such molecules in unit time is $V_q = 11.6 \times 10^{-3}$ m^3/(s cm^2).

The flow of molecules evaporated from a solid surface in high vacuum, for which no molecular collisions occur in the gas phase, is determined by the directions of the evaporating molecules.

Evaporation from a point source is described by an isotropic distribution, for which the probabilities of all the directions within solid angle $d\omega$ are equal.

$$dP = \frac{d\omega}{4\pi}. \tag{1.36}$$

Using the definition of solid angle

$$d\omega = 2\pi \sin(\gamma)d\gamma,$$

(see Fig. 1.6, *a*), we have $dP = \sin(\gamma)d\gamma/2$, and the total probability of molecular flow over a sphere is

$$P = \int_0^\pi \frac{\sin(\gamma)}{2} d\gamma = 1.$$

Experiments showed that the probability that a molecule departs from a surface is proportional to the cosine of the angle between the normal direction and the molecule's path (Fig. 1.6, *b*):

$$dP = A\frac{d\omega}{2\pi}\cos(\gamma), \tag{1.37}$$

where A is the coefficient (the normalization condition $\int_0^{\pi/2} dP = 1$ suggests $A = 2$).

Thus, the probability of molecule's depart from a surface is proportional to the double product of the relative solid angle and the cosine of the angle between the normal direction and the molecule's path (the cosine law).

Integration of Eq. (1.37) in the range from 0 to γ at $A = 2$ and

$$d\omega = 2\pi\sin(\gamma)d\gamma,$$

suggests the fraction of molecules leaving the surface within angle γ be

$$\varepsilon = \int_0^\gamma \sin(2\gamma)d\gamma = \sin^2(\gamma). \tag{1.38}$$

Hence we derive the equation of the angle γ corresponding to the fraction of molecular flow ε:

$$\gamma = \arcsin(\sqrt{\varepsilon}). \tag{1.39}$$

This formula is generally used for mathematical modeling of molecular flows in high-vacuum engineering.

1.5 Adsorption Time

The molecules of gases interact with surfaces during the time of adsorption, which depends on surface properties and the type of gas. Gas adsorption by solids is used in vacuum engineering for producing and measuring vacuum. The role of surface processes increases as pressure goes down.

Gas or vapor sorption by solids, either on the surface or in the bulk, is referred to as sorption, and the particular case of sorption on the surface is called adsorption. Physical adsorption and chemisorption are distinguished. Sorption of gases in the bulk is called absorption. This latter process involves dissolution of gases in solids.

Materials that sorb gases are called sorbents (adsorbents and absorbents), and sorbed substances are called sorbates (adsorbates and absorbates). Gas evolution from solids is referred to as desorption.

Sorption is an exothermic process. The energy released during gas sorption has a physical and chemical nature. The physical component of this interaction energy is determined by several effects responsible for the attraction and repulsion of molecules.

The induction-caused attraction during the interaction of two dipoles occurs if at least one of the interacting molecules has a permanent dipole moment; the same moment will be induced in the other molecule. The dispersion effect of attraction is accounted for by the interaction of fluctuating dipoles that are produced by electrons spinning about atomic nuclei.

The energy of physical attraction can be calculated as

$$Q_{ph} = \frac{2\mu_0^2\alpha_0}{r^6} + \frac{2}{3}\frac{\mu_0^4}{kTr^6} + \frac{3}{4}\frac{\alpha_0^2 J}{r^6}, \tag{1.40}$$

where r is the distance between the molecules, μ_0 is the dipole moment of one molecule, α_0 is the polarizability, and J is the ionization potential.

The terms in Eq. (1.40) correspond to the energy of induction, orientation, and dispersion attraction, respectively. The magnitudes of these energies for various pairs of similar molecules are listed in Table C.5.

For polar molecules H_2O and NH_3, substantial part of the interaction energy is due to the orientation effect. Unpolar molecules interact only because of the dispersion effect, which increases with atomic number. The induction effect is small for any considered molecules.

The energy of physical attraction is inversely proportional to the sixth power of the distance between the interacting molecules.

Under chemical interaction, the attraction energy Q_{ch} is accounted for by the formation of covalent or ionic bonds.

Covalent bonds are formed by two electrons belonging to different atoms and having opposite spins. Ionic bonds are due to the electrostatic attraction of opposite-charged ions. The type of chemical bond may be combined.

The repulsion of molecules is accounted for by the interaction of positively charged nuclei of close molecules. The energy of repulsion is inversely proportional to the twelfth power of the distance between the molecules:

$$Q_0 = \frac{B}{r^{12}}.$$

Taking into account all the abovementioned effects of interaction between two molecules, we may write the interaction energy of molecules as $\Delta Q = Q_0 - Q_{ph} - Q_{ch}$. At $\Delta Q = 0$, two molecules are in the state of equilibrium, under which the repulsion and attraction energies are equal.

To evaluate the energy of interaction of a molecule with a solid surface, one should sum the energies of interaction of this molecule with each atom of the solid matrix. If the distance between a gas molecule and an adsorbent surface is large as compared to the distance between the adsorbent atoms, then integration over volume can be used. The interaction of a molecule with a solid surface has the energy that is equal to

$$\varphi = \int_V Q n_a dV, \tag{1.41}$$

where n_a and V are the concentration and the volume of the adsorbent atoms.

After integration, the attraction component will be proportional to the third, and the repulsion component, to the ninth power of the distance between the surface and the molecule.

The energy of physical adsorption is, as a rule, not higher than 100 MJ/kmol. The heats of adsorption of some gases by carbon are listed in Table C.6. The energy of chemisorption

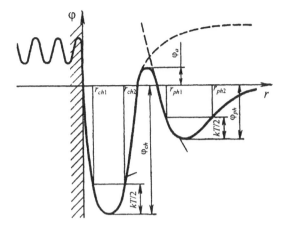

Figure 1.7 Potential interaction energy φ of multiatomic molecules with surfaces: (*1*) chemisorption and (*2*) physical adsorption.

is greater and usually ranges from 100 to 400 MJ/kmol. The heats of adsorption of some gases by materials are listed in Table C.7.

Graphic representation of Eq. (1.41) by potential curves is shown in Fig. 1.7. Approaching the surface, the molecule ends up in the first potential well and releases energy corresponding to physical adsorption. The heat energy of progressive motion of the molecule $kT/2$ causes molecular oscillations in the potential well between r_{ph1} and r_{ph2}.

If the energy of a polyatomic molecule is greater than activation energy φ_a, this molecule will dissociate into atoms that may interact chemically with the surface. These atoms will then end up in the second potential well and oscillate in it between r_{ch1} and r_{ch2}.

At the next process stage, the chemisorbed molecules transfer to the lattice of the solid. This stage corresponds to the absorption or dissolution of gas by solid.

Desorption of gases proceeds in the inverse manner. Gas molecules transfer from the bulk to the chemisorbed state, and, if their energy is $kT/2 > (\varphi_{ch} + \varphi_a)$, they will leave the surface. For real surfaces with lattice defects, the adsorption heat is not constant and is determined by the energy distribution of adsorption centers.

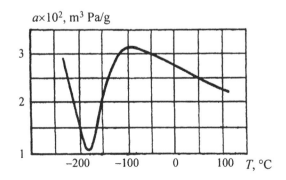

Figure 1.8 Amount of of hydrogen adsorption by nickel powder at 2.7×10^4 Pa.

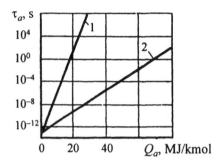

Figure 1.9 Adsorption time as a function of adsorption heat at (*1*) 77 and (*2*) 293 K.

Equilibrium sorption of hydrogen by nickel powder at different temperatures is shown in Fig. 1.8. With an increase in temperature, the amount of sorbed gas first decreases, this stage corresponding to physical adsorption. After the minimum, the sorption increases due to chemisorption.

Adsorption heat changes across the surface of a solid. For perfect surfaces, this change is caused by the periodicity of their lattice. The potential barrier of molecular migration over a surface is the migration heat, which is lower than the adsorption heat and determined by the relation $Q_m = zQ_a$, where z is the coefficient. For cubic lattices, $z = 1/2$, and for hexagonal ones, $z = 2/3$. The real ratios of the adsorption heat and the migration heat may be considerably different from the above perfect values because of lattice defects.

The minimum adsorption time is the oscillation period of a molecule in a potential well. This time is approximately considered independent of temperature and similar for all gas molecules: $\tau_0 = 10^{-13}$ s. The total adsorption time increases with adsorption heat and decreases with an increase in temperature; it can be calculated using the Frenkel equation

$$\tau_a = \tau_0 \exp\left(\frac{Q_a}{RT}\right), \tag{1.42}$$

where Q_a is the adsorption heat.

The adsorption time of the main components of air with an adsorption heat of 20 MJ/kmol is 10^{-10} s at room temperature and 1 s at the LN temperature (Fig. 1.9). For water and oil vapors with an adsorption heat of 80 MJ/kmol the adsorption time is 100 s at 293 K and 10^{43} s at 77 K. Helium has an adsorption heat of 2 MJ/кmole; hence, its adsorption time is as small as 10^{-13} s even at 77 K. Helium adsorption time on smooth surfaces becomes notable only below 4 K.

For chemisorption at adsorption heats of above 100 MJ/kmol the adsorption time is so large that adsorption at room temperature becomes irreversible if the pressure of gas over the sorbent surface changes, and the system can be brought to its initial state only by increasing temperature. The time of physical adsorption is small, and the dependence of the amount of sorbed gas on gas pressure is reversible.

The time in which molecules transfer from one to the other potential well on a surface is referred to as the migration time and determined, by analogy with adsorption time, as

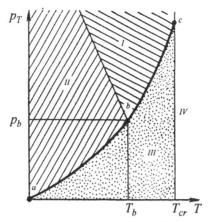

Figure 1.10 Diagramm of state of aggregation of substances: (*I*) luquids, (*II*) solid, (*III*) vapor, and (*IV*) gas.

$$\tau_m = \tau_0 \exp\left(\frac{Q_m}{RT}\right), \tag{1.43}$$

where Q_m is the migration heat.

At large migration times gas molecules are localized and do not migrate.

1.6 Saturation Pressure

Depending on temperature and pressure, substances may be in various states of aggregation. Vacuum engineering is mostly concerned in the region of low pressures, in which substances may transform from liquid to vapor state (evaporation) and *vice versa* (condensation), and from solid to vapor state (sublimation) and *vice versa* (desublimation). Figure 1.10 shows diagram of the state of aggregation for some substances, and the critical temperature T_c and the parameters of the ternary point T_b and pressures p_T are listed in table C8.

Most of air components are gaseous under the reference conditions, and only H_2O, CO_2, and Xe are vapors. At 77 K, which is the temperature of LN-cooled vacuum traps, most air components are vapors, and only He, H_2 and Ne remain gaseous.

The pressure p_T at which a substance changes state of aggregation is its saturation pressure at a particular temperature. Curve *abc* in Fig. 1.10, which determines the saturation pressure at pressures below 100 Pa, can be approximated as

$$\log p_T = M - \frac{N}{T}, \tag{1.44}$$

where M and N are specific constants of substances (Table 1.3). More exact experimental saturation pressures of various substances are shown in Fig. 1.11. The room temperature

Table 1.3 Coefficients of Eq. (1.44).

	Cu	Al	Zn	Ni	Fe	Cr	N_2
M	11.08	10.91	10.00	11.87	11.56	12.06	24.1
$N \times 10^3$	16.98	15.94	14.87	20.96	19.97	20.00	8.9×10^5

saturation pressures of organic and polymer materials used in vacuum engineering as sealers are shown in Table C.9.

The saturation pressure of alloys is approximately

$$\frac{p_{AS}}{p_A} = \frac{n_A}{n_A + n_B} = X_A,$$
(1.45)

where p_A is the saturation pressure of substance A, p_{AS} is the saturation pressure of substance A dissolved with substance B, and n_A and n_B are the molar quantities of substances A and B in the solution.

The concentration of A in the solution can be expressed in weight percents:

$$q_A = \frac{100 X_A M_A}{X_A M_A + (1 - X_A) M_B},$$

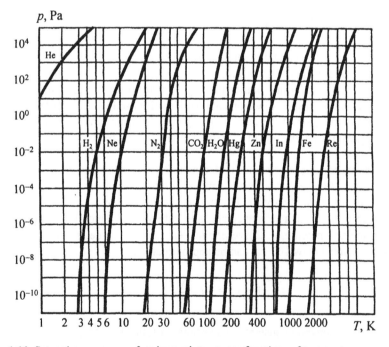

Figure 1.11 Saturation pressures of various substances as functions of temperature.

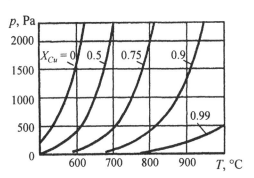

Figure 1.12 Saturation pressure of zinc alloyed with copper as a function of temperature.

where X_A is the molar fraction of A and M_A and M_B are the molar masses of A and B.

By way of example, the temperature dependence of Zn saturation pressure in Zn–Cu alloy at various molar concentrations of copper X_{Cu} is shown in Fig. 1.12. As can be seen, an increase in the molar fraction of copper at constant temperature reduces the saturation pressure of zinc.

Colliding with a solid surface, a molecule may condense or be reflected to the gas. The probability of condensation increases with increasing evaporation heat and with decreasing gas temperature. If the evaporation heat is uniform across a surface, the probability of condensation can be defined as the fraction of molecules whose energy is lower than the condensation heat. Using the integral energy distribution of gas molecules (Eq. (1.21)), we write the equation of the probability of condensation for a free surface:

$$\gamma = 4 \int_{0}^{x} \sqrt{\frac{2x}{\pi}} e^{-x/2} dx, \tag{1.46}$$

where $x = E/(kT/2)$ and E is the evaporation heat. The parameter $\gamma(x)$ can be found as $F(x)$ in Table 1.1.

The mass of the molecular flow incident on unit surface area in unit time is the product of molecular mass and the collision frequency of the molecules with the surface N_q (Eq. (1.34)). Thus, the condensation rate is

$$G_c = \gamma m N_q = \gamma p_g \sqrt{\frac{M}{2\pi R T_g}}, \tag{1.47}$$

where p_g and T_g are the gas pressure and temperature. In the CI system, M is measured in kg/kmol and $R = 8.31 \times 10^3$ J/(kmol K), whence G_c is measured in kg/(m^2 s).

At the saturation pressure of a substance, its condensation and evaporation are in equilibrium. Thus, the evaporation rate can be determined from the condensation rate:

$$G_e = p_T \gamma \sqrt{\frac{M}{2\pi R T}}, \tag{1.48}$$

where p_T is the saturation pressure at temperature T.

The mass exchange rate on a surface is

$$G = G_c - G_e = 4.38 \times 10^{-3}(p - p_T)\gamma \sqrt{\frac{M}{T}}. \tag{1.49}$$

At $p > p_T$, the substance is condensed on the surface, and at $p < p_T$ it is evaporated. The sublimation-desublimation processes can be described as above. Table C.10 lists evaporation rates of various materials at a saturation pressure of 1.33 Pa.

The condensation of substances is an exothermic process. The condensation heat can be determined by using the coefficient N of Eq. (1.44):

$$E = 2.3RN. \tag{1.50}$$

1.7   Surface Coverage with Gas Molecules

The coverage of solid surfaces by adsorbed gas molecules affects the properties and the surface processes of solids, such as electron emission, surface electric discharge, unlubricated friction and wear, adhesion of films, etc.

We denote the fraction of surface a/a_m covered by adsorbed molecules as θ and distinguish three degrees of coverage: high ($\theta \gg 1$), medium ($\theta = 1$) and low ($\theta \ll 1$).

At low surface coverage, the condensation rate is

$$\mu = \gamma N_q = \frac{\gamma n v_a}{4}, \tag{1.51}$$

where γ is the probability of condensation of molecules on free surface. This coefficient is also called capture or sticking probability.

The rate of evaporation from an adsorbed monomolecular layer at a constant adsorption heat can be calculated using the definition of adsorption time (Eq. (1.42)):

$$\vartheta = \frac{a_m}{\tau_a}, \tag{1.52}$$

where a_m is the number of molecules adsorbed on unit area of the monomolecular layer. In Eqs. (1.51) and (1.52), the condensation and evaporation rates are determined by the numbers of molecules interacting with unit surface area in unit time and have the dimension $(m^2 s)^{-1}$. Substituting Eqs. (1.51) and (1.52) into the condition of adsorption equilibrium

$$\frac{da}{dt} = \mu - \theta \vartheta = 0,$$

we obtain the degree of surface coverage

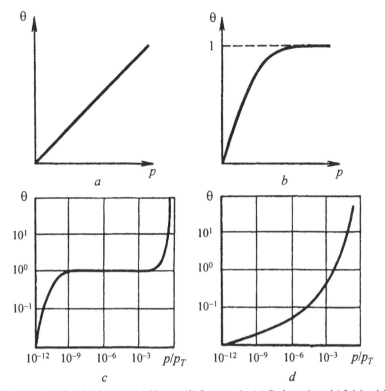

Figure 1.13 Adsorption isotherms: (*a*) Henry, (*b*) Langmuir, (*c*) S-shaped, and (*d*) island-like.

$$\theta = bp, \tag{1.53}$$

where

$$b = \frac{\gamma \tau_a}{a_m \sqrt{2\pi mkT}},$$

The equation of adsorption (1.53) is referred to as the Henry equation (Fig. 1.13, *a*). Low coverage $\theta \ll 1$ holds for low pressures or high temperatures.

At medium coverage ($\theta = 1$), when the adsorption heat is constant and greater than the condensation heat ($Q_a > E$), the equation of adsorption equilibrium can be written as

$$\frac{da}{dt} = \mu(1-\theta) - \theta\vartheta. \tag{1.54}$$

This formula implies that the condensation of molecules may only occur on a free surface. Under equilibrium conditions, i.e., $da/dt = 0$, Eq. (1.54) can be transformed to

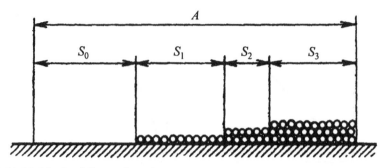

Figure 1.14 Model of multilayer adsorption.

$$\theta = \frac{bp}{1 + bp}. \tag{1.55}$$

Equation (1.55) is the equation of monomolecular adsorption, or the so-called Langmuir equation. It gives the most accurate result for chemisorption of gases on metallic surfaces and implies that only one layer of adsorbed molecules forms.

At high coverage, the adsorption is polymolecular. In the approximations that

1. the adsorption heat is constant in the first layer;
2. the adsorption heat in the second and next layers is equal to the condensation heat;
3. the probability of condensation on free surface is the same for all adsorbed layers,

we may use the model of multilayer adsorption as shown in Fig. 1.14.

Gas molecules adsorbed in several layers are shown shifted rightwards for the sake of simplicity in calculations for free surface. The equation of polymolecular adsorption has been obtained by Brunauer, Emmett, and Teller and is referred to as the BET equation:

$$\theta = \frac{Cp/p_T}{(1 - p/p_T)[1 + (C - 1)p/p_T]}, \tag{1.56}$$

where p is the equilibrium pressure in the gas phase, p_T is the saturation pressure of the adsorbate at temperature T, and C is the coefficient depending on the difference in the adsorption and condensation heats:

$$C = \exp\left(\frac{Q_a - E}{RT}\right).$$

Derivation of Eq. (1.56) is given in Appendix A.3. The adsorption and condensation heats of some gases are listed in Table C.6.

The adsorption isotherm plotted from Eq. (1.56) is S-shaped at $Q_a = E$ (Fig. 1.13, c) and island-like at $Q_a < E$ (Fig. 1.13, d). For an island-like adsorption isotherm, gas adsorption on covered surface areas is more probable than on free areas. Thus, areas with multilayered

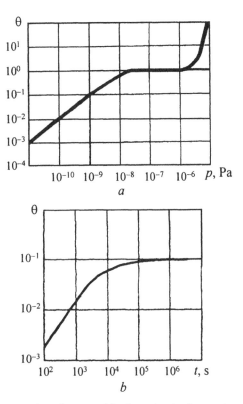

Figure 1.15 Surface coverage by oil vapor: (*a*) adsorption isotherm at room temperature and (*b*) temporal dependence at 1.3×10^{-9} Pa.

coatings form. At $p \to 0$, Eq. (1.56) at $b = C/p_T$ is similar to Eq. (1.53) and, with account of Eq. (1.54), we may correlate the saturation pressure and condensation heat.

Equation (1.56) at $a = G/A$ can be transformed to

$$\frac{p/p_T}{1 - p/p_T} = \frac{G}{a_m CA} + \frac{G(C - 1)p}{A a_m C p_T},$$

where G is the total amount of adsorbed gas and A is the total adsorbent surface area. Plotting experimental $G = f(p)$ curves in the $(p/p_T)/(1 - p/p_T)$- and p/p_T-coordinates, relative to which this equation is linear, we find the coefficients A and C, which allow surface area and adsorption heat to be calculated.

Figure 1.15 shows adsorption isotherm of vacuum oil vapor having adsorption heat $Q_a = 96$ MJ/kmol, condensation heat $E = 80$ MJ/kmol, and saturation pressure $p_T = 10^{-5}$ Pa. The isotherm was plotted from Eq. (1.56) for 293 K.

Real adsorbents usually have no uniform adsorption heat of surface adsorption centers, and the adsorption heat depends on the amount of sorbed gas. Experiments show that this dependence can often be written as

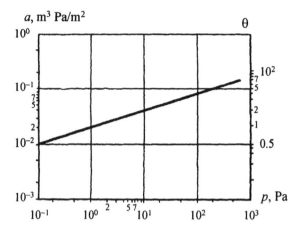

Figure 1.16 Isotherm of adsorption of water vapor on stainless steel at room temperature.

$$Q_a = Q_0 - K_Q \log\theta, \tag{1.57}$$

where Q_0 is the adsorption heat at $\theta \to 1$, K_Q is the coefficient, and θ is the coverage.

The solution of Eq. (1.54) with allowance for (1.57) leads to the Frendlich adsorption equation

$$\theta = dp^m. \tag{1.58}$$

Here

$$m = \frac{RT}{RT + 0.43 K_Q}; \text{ and } d = b^m(Q_0),$$

$b(Q_0)$ being as in Eq. (1.55) at τ_a corresponding to Q_0.

Figure 1.16 shows isotherm of water vapor adsorption on stainless steel. The coverage corresponds to a monomolecular coating 2×10^{-2} m^3 Pa/m^2. This isotherm can be described by Eq. (1.58) at $d = 1$ and $m = 0.35$ (p in Pa).

If adsorption equilibrium cannot be achieved during the experiment, the coverage will remain smaller or greater than the equilibrium one, depending on initial conditions. To determine θ as a function of time, one should solve the differential equation of monomolecular adsorption (Eq. (1.56)). This formula can be transformed to

$$\frac{d\theta}{dt} + A\theta - B = 0, \tag{1.59}$$

where

$$A = \frac{1}{\tau} + \frac{\gamma p}{a_m \sqrt{2\pi mkT}}; \quad B = \frac{\gamma p}{\sqrt{2\pi mkT}}.$$

The solution of Eq. (1.59) for the initial conditions $t = 0$ and $\theta = \theta_0$ is

$$\theta = \frac{B}{A}[1 - \exp(-At)] + \theta_0\exp(-At) \tag{1.60}$$

and can be rewritten as

$$\theta = \frac{bp}{1 + bp}[1 - \exp(-At)] + \theta_0\exp(-At). \tag{1.61}$$

Here b being as in Eq. (1.53).

At $t \to \infty$, the equilibrium coverage is

$$\theta_\infty = \frac{a_\infty}{a_m} = \frac{bp}{1 + bp}$$

which is similar to Eq. (1.55).

If $a = 0.99a_\infty$ is accepted to be the equilibrium state, the time required to achieve the equilibrium state at the initial condition $\theta_0 = 0$ is

$$t_p = \frac{4.6}{A} = 4.6 / \left[\frac{p}{a_m\sqrt{2\pi m k T}} + \frac{\exp(-Q_a/RT)}{\tau_0} \right]. \tag{1.62}$$

For nitrogen adsorption on graphite with the parameters $p = 1.33\times10^4$ Pa, $T = 293$ K, $a_m = 9.6\times10^{18}$ m^{-2}, $\gamma = 1$, and $\tau_0 = 10^{-13}$ s, we have $t_p = 3.10^{-11}$ s, which suggests that at high pressures and temperatures the adsorption equilibrium is achieved in almost no time.

At low pressures or low temperatures, the time required to achieve the equilibrium state is rather large. For oil vapors (M = 422 kg/kmole) at $p = 1.3\times10^{-9}$ Pa, $T = 298$ K, $Q_a = 96$ MJ/kmole, and $a_m = 6.67\times10^{17}$ m^{-2}, we have $t_p = 10^5$ s, which is comparable to the time of experiment (Fig. 1.15, b).

The rate of sorption q can be found by differentiating Eq. (1.60) over time:

$$q = \frac{a_m d\theta}{dt} = (B - \theta_0A)a_m\exp(-At). \tag{1.63}$$

If $B > \theta_0A$, the case is adsorption, and if $B < \theta_0A$, desorption occurs.

1.8 Gas Dissolution in Solids

The concentration of gas dissolved in a solid depends on temperature, pressure, and the type of lattice. For metals, in which there are homopolar metallic bonds of electrically positive atoms, the solubility depends on pressure and temperature as

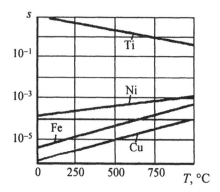

Figure 1.17 Solubility of hydrogen in metals at normal atmosphere pressure.

$$s = s_0 p^{1/n} \exp\left(\pm\frac{Q_s}{nRT}\right),$$ (1.64)

where n is the number of atoms in one gas molecule, Q_s is the activation energy of dissolution, and s_0 is constant.

The sign of the exponential term in Eq. (1.64) is '+' for gases that form chemical compounds with metals and '−' for gases that form true solutions in getters. The solubility of these latter gases increases with temperature (H_2 in Cu, Fe, and Ni), and the solubility of compound-forming gases decreases (H_2 in Ti). The solubility s as shown in Fig. 1.17 is calculated as the number of hydrogen atoms relative to the number of metal atoms. Note also that the solubility of hydrogen in Ti is much greater than in Ni, Fe, and Cu. At atmospheric pressure and ambient temperature, 1 cm³ of stainless steel may dissolve 1 cm³ of gas.

The pressure dependence of gas solubility in metals is a power function. The exponent of the function is $1/n$, because gases dissolve in metals in the atomic state, i.e., their molecules dissociate before dissolution. For oxygen, $n = 2$ and the reaction is $O_2 \leftrightarrow O + O$

According to the law of mass action, the equilibrium constant of this reaction $K_e = p_O^2/p_{O2}$, where p_O and p_{O2} are the atomic and molecular oxygen pressures. Evidently,

$$p_O = \sqrt{K_e p_{O2}}.$$ (1.65)

The solubility of gases in metals is proportional to the pressure of dissociated gases. For diatomic molecules, this gives a square-root proportion.

In nonmetals, the lattice atoms of which are bound by ionic and covalent bonds, gases dissolve in the molecular state. True solutions form, and the solubility depends on the temperature and pressure as

$$s = s_0 p \exp\left[-\frac{Q_s}{RT}\right].$$ (1.66)

The constants s_0 and Q_s, which characterize the solubility of the most important gases in metals and nonmetals are presented in Table C.11.

Absorption during the dissolution of gases in solids occurs by diffusion of gas molecules into the lattice or by grain boundary mechanism, the diffusion flow being proportional to the concentration gradient. For a steady gas flow through a wall with thickness $2h$, the concentration gradient is $ds/dx = (s_1 - s_2)/(2h)$, and hence

$$q = -D\frac{ds}{dx} = -D\frac{s_1 - s_2}{2h},\qquad(1.67)$$

where q is the number of gas molecules passing in unit time through unit cross-section area in the x-direction, D is the diffusion coefficient, and s_1 and s_2 are the gas concentrations at the wall boundaries.

The diffusion coefficient D is quite temperature-sensitive:

$$D = D_0\exp\left[-\frac{Q_D}{nRT}\right],$$

where Q_D is the diffusion activation energy, n is the number of atoms in one gas molecule (this parameter is effective only for metals; for nonmetals $n = 1$), and D_0 is constant.

D_0 and Q_D for diffusion of gases in structural materials (Table C.12) should be regarded approximate, because they are affected by internal stress in the materials. For diffusion over grain boundaries and structural defects, the type of mechanical treatment and grain sizes and orientations should be taken into account.

Substituting in Eq. (1.64) into Eq. (1.67), we obtain the equation of gas permeability of metals:

$$q = K_0\frac{p_2^{1/n} - p_1^{1/n}}{2h}\exp\left[-\frac{Q_D \pm Q_s}{nRT}\right],\qquad(1.68)$$

where $K_0 = D_0 s_0$ is the permeability constant.

The penetrability of hydrogen through a 1 mm thick stainless steel sheet at 298 K is 10^{-4} m^3 Pa/(s m^2).

If concentration gradient varies in time, then gas concentration in unit volume of a solid varies according to the differential equation of non-steady diffusion

$$\frac{\partial s}{\partial \tau} = D\frac{\partial^2 s}{\partial x^2}.$$

Insert $z = x/h$, where h is a characteristic dimension of the body, and $\tau = Dt/h^2$. The latter formulae can be written as:

$$\frac{\partial s}{\partial \tau} = \frac{\partial^2 s}{\partial z^2}.$$

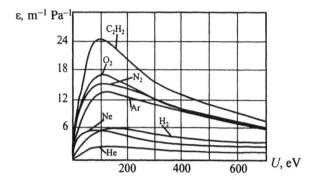

Figure 1.18 Efficiency of ionization by electron bombardment as a function of accelerating voltage.

For different forms of the body the equilibrium condition $\partial s / \partial \tau = 0$ begins after $\tau \geq 1$. This condition permits to define non-steady delay time.

$$t_D = \frac{h^2}{D}. \tag{1.69}$$

For hydrogen in stainless steel at room temperature, $D = 5 \times 10^{-12}$ m²/s. In a steel piece with characteristic dimension $h = 1$ mm, equilibrium is established in $50h$.

1.9   Electrical Phenomena

Passage of electric current through gases under difference in potentials originates from the transfer of electrons and positive ions. Without electric field, the energy distribution of the electrons, ions, and neutral molecules is uniform.

The mean free path length of electrons in vacuum can be calculated by analogy with molecules using Eq. (1.31). Taking into account that the mass and diameter of the electron are much smaller than those of a gas molecule, Eq. (1.31) can be simplified to

$$L_e = \frac{4}{\pi n_2 d_{T2}^2}, \tag{1.70}$$

where n_2 is the mean concentration of gas molecules, d_{T2} is the effective diameter of gas molecule, and L_e is the mean free path length of electrons.

Comparing Eq. (1.70) with the equation of free path length for gas (Eq. (1.24)), one can see that the mean free path length of electrons does not depend on their concentration, and, at the same concentration of gas molecules, the mean free path length of electrons is 5.6 times higher.

The ionization of residual gas molecules producing free electrons and positive ions is possible, when the molecules are exposed to α-, β-, or γ-radiation with an energy that is higher than the ionization energy of a given gas (Table C.13).

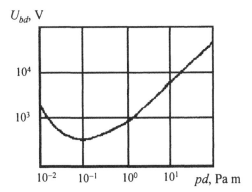

Figure 1.19 Breakdown voltage as a function of product of pressure and distance.

Electron bombardment is the most widely used method of ionization of residual gases. This process is quantitatively described by the ionization efficiency of molecules ε, i.e., the number of ion pairs formed by one electron over a path of 1 m at a pressure of 1 Pa. The dependence of ε on accelerating voltage (Fig. 1.18) has a typical maximum at 100–150 eV. The greater the atomic number, the greater the ionization efficiency.

Under difference of potentials U_e, ions and electrons acquire, in addition to $3kT/2$, energy qU_e, where q is the elementary charge. In order for the thermal energy of molecules to be equal to the energy of charged particles in an electric field, the temperature of the medium should be $T = 2qU_e/(3k)$. One may estimate that electrons accelerated by a voltage of 1 V have the same energy as at $T = 7800$ K without an electric field; 1 eV = 1.6×10^{-19} J.

The electrical conductivity of a gas gap under spontaneous discharge (without ionizing radiation) depends on pressure. Gases contain free electrons that are generated, for example, by space radiation.

In low vacuum, because of the small free path length, electrons affected by an electric field have not enough time to acquire the energy that is necessary to ionize gas molecules. Under these conditions, the electrical conductivity of a gas is low. In high vacuum, the number of charged particles is small, and the electrical conductivity is still low. In this situation, free electrons ionize the molecules of residual gases, and the thus formed secondary electrons provide for the spontaneous discharge.

The electrical conductivity of a gas gap is characterized by the breakdown voltage U_{bd}, which depends on gas, distance between the electrodes, and pressure. The breakdown voltage depends on the product of gas pressure and interelectrode distance, and not on any of these parameters alone. The dependence of U_{bd} on pd, where d is the distance, is plotted in Fig. 1.19 and has a typical minimum in the medium vacuum range (the so-called Paschen curve).

Passage of electric current through rarefied gases in medium vacuum causes gas to glow, depending on gas and pressure. This phenomenon has been used for identification of gas composition and pressure. At pressures of about 10^3 Pa, discharge appears between electrodes as a thin bright cord, which occupies the whole tube at about 10^2 Pa. Passing from the cathode to the anode, one may distinguish several characteristic regions of the discharge: the Aston dark space next to cathode, the bright cathode glow layer, the Crookes

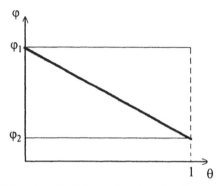

Figure 1.20 Electron work function of solids as a function of surface coverage: φ_1 for substrate and φ_2 for coating.

dark space, the negative glow zone, the Faraday dark zone, the positive glow zone, and the anode dark space.

The positive glow is quite intense, and its color testifies to the presence of a particular gas in the discharge space: air glows blue, oxygen yellow, nitrogen orange, helium pink, water vapor bluish-white, argon violet, and neon red.

The pressure of residual gases and the related surface coverage by adsorbed gas molecules greatly affect the electron emission of solids. The mean energy of electrons in solids is the Fermi energy (the Fermi level). Electron emission from solids is hindered by the potential barrier that is referred to as the electron work function. For most metals, the electron work function is within 2 through 6 eV. Gas or vapor adsorption on a surface changes its work function proportionally to coverage (Fig. 1.20). The work function may decrease or increase depending on the combination of substrate and coating materials. For example, cesium coating reduces the work function of tungsten by 2.7 eV, while nitrogen on tungsten increases it by 0.5 eV. Electrons with energies twice the sum of the Fermi energy and the work function may leave the solid. Electron emission depends on temperature as

$$j = AT^2 \exp\left(-\frac{\varphi}{kT}\right), \tag{1.71}$$

where j is the current density, A/cm^2, T is the absolute temperature, K; φ is the work function, J, $k = 1.23 \times 10^{-23}$ J/K, and $A \approx 120$ A/(cm^2 K^2).

In strong electric fields, electrons may overcome the surface potential barrier by tunneling. This electron emission is called field emission and obeys the equation

$$j = \frac{4\sqrt{\mu/\varphi}}{\mu+\varphi} \frac{e^3 E^2}{8\pi h\varphi} \exp\left(-\frac{8\pi\sqrt{2m}\varphi^{3/2}}{3heE}\right), \tag{1.72}$$

where E is the electric field intensity, e is the charge of the electron, h is the Planck constant, μ is the Fermi energy, and φ is the electron work function.

A current with a density of 10^2 A/cm^2 is generated by a field with an intensity of 10^7 V/cm. Field emission is used in autoelectronic (tunneling) electron microscopes with an emitter radius of less than 0.1 μm. The difference of potentials between the emitter and the luminescent screen being several thousands volts, the electrons will be emitted from the cathode to the screen. Polycrystalline surfaces have different electron work functions at edges with different crystallographic orientations. Adsorption of gas molecules will also change the electron work function on some parts of the cathode. The nonuniformity of electron current density caused by the nonuniformity of the electron work function over the surface produces image contrast. Thus, an image of the thin cathode end can be seen on the screen at a magnification of 10^6.

The autoelectronic (tunneling) microscope allows one to observe change in the polycrystalline structure of a surface due to migration of tungsten atoms at above 1400 K. Migration of barium atoms on tungsten can be observed at above 400 K. Oxygen, hydrogen, nitrogen, and many other gases adsorbed on the cathode may change its emission parameters.

Inverting the field polarity between the screen and the electrode needle, one changes the autoelectronic microscope to ion microscope, in which the screen acts as a cathode. Helium atoms bombarding the needle lose electrons due to surface ionization, and their positive ions move to the screen. Difference in the effective ionization of separate cathode areas allows atomic-scale resolution to be obtained on the screen.

10–200 MeV electron beams interact with solid surfaces and penetrate to the depth of only a few atomic layers. Diffraction of the reflected electrons produces images of superficial adsorbed layers on the screen. In slow electron diffraction instruments for surface study, the primary and secondary electron beams are separated by electric and magnetic fields. The secondary electron beam is additionally accelerated in order to obtain good image on the screen.

Surface bombardment with electron or ion beams produces secondary electron emission due to the reflection of the primary electron beam, imparting of energy to free electrons to the level enough for diffusing to the surface and overcoming the surface barrier, and interlevel electron transfers in the solid.

The energy of reflected electrons may be equal to the energy of the primary electrons. The energy of secondary electrons that diffuse from the bulk is from 1 to 5 eV. The energy of the electrons generated by the interlevel transfers (the Auger electrons) depends on emitter material and ranges from 10 to 20 eV. Analysis of Auger electron spectra allows quantitative elemental surface characterization of materials. The energy of Auger electrons is strongly affected by gas adsorption and is sensitive to the formation of even small fractions of a monomolecular layer. Therefore, Auger electron spectroscopy is a powerful tool in high-vacuum surface investigation.

There is yet no theory allowing the efficiency of secondary electron emission to be calculated on the basis of the fundamental properties of materials. At present, this efficiency is determined experimentally as the number of secondary electrons generated by one primary particle. This coefficient has a maximum at particular primary particle energies.

A vacuum-isolated target to which a floating potential is applied may acquire positive or negative charge, depending on the energy of primary electrons. At secondary electron emission coefficient greater that one, the target charge will be positive, and otherwise, it will be negative.

1.10   Test Questions

1.1   How will the equation of gas pressure change if the velocity of molecules evolved from a solid surface becomes half the velocity of incident molecules?

1.2   What is the maximal theoretical speed of vacuum pumps determined by?

1.3   What is the degree of surface coverage by nitrogen and oxygen molecules at room temperature and normal atmosphere pressure?

1.4   What is the difference between physical adsorption and chemisorption?

1.5   Why only one layer of gas molecules can be sorbed under chemisorption?

1.6   Indicate the main approximations of the mono- and polymolecular adsorption theories.

1.7   What is the difference between the regulariries of gas dissolution in structural and gettering materials?

1.8   What is the efficiency of ionization by electron bombardment?

1.9   Why is the meanfree path length of electrone greater than molecule's one?

1.10 What are the degrees of surface coverage with gas molecules?

Chapter 2

Theory

2.1 Degrees of Vacuum

The definition of vacuum as a state of gas at a pressure lower than the normal atmospheric pressure can be extended by classifying it into degrees depending on different physical processes. In vacuum, most of these depend strongly on the ratio of the numbers of molecular collisions with one another and with chamber walls.

The frequency of molecular collisions K_m is inversely proportional to the mean free path length: $K_m = v_{ar}/L$. From the total number of molecular collisions with chamber walls (Fig. 2.1) $nv_{ar}F/4$ for the nV molecules in the chamber, one may easily calculate the average number of collisions of one gas molecule with the chamber walls in unit time: $K_{av} = v_{ar}F/(4V) = v_{ar}/d_{ef}$, where F is the area of the wall contacting with the rarefied gas, V is the chamber volume, and $d_{ef} = 4V/F$ is the effective chamber size.

For gas molecules in a spherical volume with diameter D, the effective chamber size is $d_{ef} = 2/3D$, for a infinitely long pipe with diameter D, $d_{ef} = D$, and for two infinite parallel surfaces at distance D from each other $d_{ef} = 2D$.

The ratio K_{av}/K_m is referred to as the Knudsen criterion:

$$Kn = \frac{K_{av}}{K_m} = \frac{L}{d_{ef}}. \tag{2.1}$$

Depending on this criterion, vacuum is subdivided into low, medium, and high.

Low vacuum is such a state of a gas in which mutual molecular collisions are more frequent than molecular collisions with chamber walls. In this state, $Kn \ll 1$, Kn being usually taken to be less than 5×10^{-3}. The free path length of gas molecules is far smaller than chamber size. Gas flow in low vacuum occurs in viscous mode. The conductance of a pipe is proportional to pressure and depends on the fourth power of pipe diameter. The viscosity and heat conductivity of gases do not depend on pressure. The self-diffusion coefficient of gases is inversely proportional to pressure. For low-vacuum sputtering, the collisions of gas molecules with the sputtered molecules do not allow a screen placed across the molecular beam to be imaged on the chamber wall.

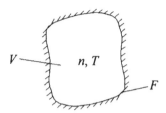

Figure 2.1 Parameters of an arbitrary shape vacuum chamber.

For medium vacuum, the frequency of mutual molecular collisions is equal to that of molecular collisions with chamber walls, with $L \approx d_{ef}$ and $Kn \approx 1$. The range of medium vacuum can be delimited as $5 \times 10^{-3} < Kn < 1.5$. The physical phenomena have transient character in this range. Medium vacuum provides the best conditions for electric breakdown, glow, and gas ionization.

High vacuum is the state in which gas molecules collide with chamber walls more frequently than with one another; $Kn > 1$. It is generally accepted that $Kn > 1.5$. The gas flow occurs in molecular mode. The conductance of a pipe depends on the third power of its diameter and not on pressure. The viscosity and heat conductivity coefficients are proportional to pressure, and self-diffusion does not depend on it. A screen placed across the molecular beam is clearly imaged.

For any degree of vacuum, changing temperature one may separate regions at which the surface coverage of solids in vacuum θ will be low ($\theta \ll 1$), medium ($\theta \approx 1$), and high ($\theta \gg 1$).

The range of high vacuum with low surface coverage has been specially referred to as ultrahigh vacuum: $Kn > 1.5$ and $\theta < 5 \times 10^{-3}$. The number of adsorbed molecules is small, and the properties of pure surfaces can be characterized, which is important for various scientific and technological applications.

In literature, a simplified classification of vacuum by pressure is often used. Despite the apparent simplicity, this classification is only applicable to vacuum chambers of definite shape, otherwise serious errors are involved in calculations.

2.2   Transport Phenomena

The viscosity, heat conductivity, and diffusion of gases are commonly referred to as transport phenomena. Viscosity, or internal friction of gases, is determined by the transport of impetus, heat conductivity by the transport of energy, and diffusion by molecular transport. The transport phenomena depend on the degree of vacuum. The steady-state transport rate of physical quantity A is

$$\frac{dA}{dt} = V_q \Delta A S, \tag{2.2}$$

where V_q is the volume of gas transported in unit time per unit area of a transport surface, ΔA is the difference in A for two transport surfaces, and S is the total area of the transport surface.

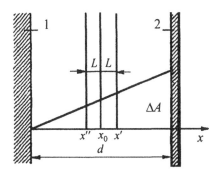

Figure 2.2 Schematic for calculation of transport phenomena in gases at low vacuum.

In low vacuum, the gas between the transport surfaces 1 and 2 (Fig. 2.2) can be subdivided into layers of similar thickness equal to the mean free path length. The difference of the transported physical quantity in two adjacent layers is

$$\Delta A = 2L\frac{dA}{dx}.$$

(2.3)

The steady-state transport rate (Eq. (2.2)) with account of Eq. (2.3) can be written as

$$\frac{dA}{dt} = -R\frac{dA}{dx}S,$$

(2.4)

where R is the transport coefficient. The negative signs appears because transport and it's gradient have opposite directions. Taking into account Eq. (1.35),

$$R = \frac{v_{ar}L}{2}.$$

(2.5)

If surface 2 (Fig. 2.2) moves at rate v_s relative to surface 2, impetus is transported between the gas layers. The internal friction force (viscosity) is determined by Eq. (2.4) at $A = nmv_s$.

$$F = -\eta\frac{dv_s}{dx}S,$$

(2.6)

where η is the dynamical viscosity coefficient. Taking into account Eq. (2.5), we obtain for low vacuum

$$\eta_l = \frac{v_{ar}Lnm}{2}.$$

(2.7)

Because mean free path length (Eq. (1.29)) is inversely proportional to pressure and molecular concentration (Eq. (1.10)) is proportional to pressure, one may conclude that the dynamical viscosity coefficient in low vacuum does not depend on pressure. Using Eqs. (1.10), (1.18), and (1.25), we transform Eq. (2.7) to

$$\eta_l = \frac{\sqrt{mkT}}{\pi^{3/2} d_m^2 (1 + C/T)}. \qquad (2.8)$$

This formula suggests that η_l depends on power of temperature T^x, where x ranges from 1/2 at high temperatures $T \gg C$ to 3/2 at low temperatures $T \ll C$. Dynamical viscosity coefficients of some gases at 273 K are presented in Table C.14.

The kinematic viscosity coefficient of a gas is the ratio of its dynamical viscosity coefficient to density, $\eta_k = v_{ar} L/2$.

The heat conductivity of gases depends on energy transfer. The energy in unit volume of a gas is $c_v nmT$, where the heat capacity of gas at constant volume is

$$c_v = \frac{k}{(\gamma - 1)m}. \qquad (2.9)$$

Here $\gamma = c_p/c_v$, which is 1.66 for monatomic gases, 1.4 for diatomic, and 1.3 for triatomic ones (Table C.15).

The equation of steady-state heat conductivity of gases can be derived from Eq. (2.4) at $A = c_v nmT$. Heat transport by gas molecules is

$$E_l = -\lambda \frac{dT}{dx} S, \qquad (2.10)$$

where λ is the heat conductivity coefficient. In low vacuum,

$$\lambda_l = c_v \eta_l. \qquad (2.11)$$

The heat conductivity coefficient of a gas at low vacuum, similarly to the dynamical viscosity coefficient, does not depend on pressure and has a similar temperature dependence. Substituting Eqs. (2.8) and (2.9) into Eq. (2.11), we obtain

$$\lambda_l = \frac{k^{3/2} T^{1/2}}{d_m^2 (1 + C/T)(\gamma - 1)\pi^{3/2} m^{1/2}}. \qquad (2.12)$$

Heat conductivity coefficients of some gases at low vacuum are listed in Table C.15. Heat transfer through a gas gap at low vacuum may also occur by convection and radiation.

Convection heat transfer from the surface of a hot wire at temperature T_w to chamber walls at temperature T is described as

$$E_k = \alpha(T_w - T)A, \tag{2.13}$$

where α is the heat exchange coefficient and A is the wire surface area.

Under free convection, due to the effect of gravity on the gas, which has nonuniform pressure because of temperature gradients, the heat exchange coefficient is

$$\alpha_F = ap^{3/2}, \tag{2.14}$$

where a is the experimental coefficient that depends on material, temperature and surface shape. The heat exchange coefficient of air under forced convection with lateral circumvention of a wire is

$$\alpha_F = \frac{Nu\lambda}{d},$$

where λ is the heat conductivity coefficient of air, d is the characteristic size (wire diameter), $Nu = k_1 Re^{k_2}$ is the Nusselt criterion, $Re = v_g d\rho/\eta$ is the Reynolds criterion, v_g is the gas flow speed, and k_1 and k_2 are constants that depend on Re as $k_1 = 0.45$ and $k_2 = 0.5$ at $Re < 10^3$ and $k_1 = 0.245$ and $k_2 = 0.6$ at $Re \geq 10^3$.

Radiation heat transfer in low vacuum can be determined using the Stefan–Boltzmann law:

$$E_r = 5.7R_e\left[\left(\frac{T_1}{100}\right)^4 - \left(\frac{T_2}{100}\right)^4\right]R_g, \tag{2.15}$$

where E_r is the heat flux, W/m^2, T_1 and T_2 are the temperatures of the outer and inner transfer surfaces, R_g is the geometrical factor ($R_g = 1$ for parallel planes and concentric cylinders), and R_e is the effective blackbody radiation.

$$R_e = \frac{1}{1/e_2 + (A_2/A_1)[(1/e_1) - 1]}. \tag{2.16}$$

where A_1 and A_2 are the areas of the outer and inner transfer surfaces and e_1 and e_2 are the radiation coefficients of these surfaces.

For smooth surface of stainless steel, $e = 0.1$ at 300 K and 0.06 at 77 K, and for copper these coefficients are 0.03 and 0.019, respectively. With screens, the effective blackbody radiation decreases proportionally to the number of the screens N. If $A_1 = A_2$ and $e_1 = e_2 = e$, the effective blackness is

$$R_N = \frac{R_e}{N+1} = \frac{e}{(2-e)(N+1)}. \tag{2.17}$$

The steady-state diffusion equation can be written, by accepting molecular concentration of gas as a transferred magnitude, as

$$P_n = -D_s \frac{dn}{dx} S,$$ (2.18)

where D_s is the self-diffusion coefficient in low vacuum. According to Eq. (2.5), we may write

$$D_s = \frac{v_{ar} L}{2}.$$ (2.19)

Using Eqs. (1.25), (1.18), and (1.10), we obtain

$$D_s = \frac{k^{3/2} T^{5/2}}{\pi^{3/2} m^{1/2} p d_m^2 (T + C)}.$$

Thus, the self-diffusion coefficient is inversely proportional to pressure at low vacuum. The temperature dependence of this coefficient is $T^{5/2}/(T + C)$, which can be written as T^x, x ranging from 1.5 for $T \gg C$ to 2.5 for $T \ll C$, i.e., self-diffusion enhances with increasing temperature.

The coefficient of interdiffusion of two gases D_i in low vacuum is calculated as

$$D_i = \frac{8\sqrt{2}}{3} \left(\frac{kT}{\pi} \right)^{3/2} \frac{1}{p(d_1 + d_2)^2} \sqrt{\frac{1}{m_1} + \frac{1}{m_2}},$$

where p is a total pressure, d_1 and d_2 are the effective diameters of the gas molecules with masses m_1 and m_2 (Eq. (1.26)). The interdiffusion coefficient D_i does not depend on mixture composition and is inversely proportional to the total pressure of the gases.

The equation of D_i is comfortable in the form $D_i = D_0 (T/T_0)^x (p_0/p)$, where $x = 1.5$–2.5, $T_0 = 273$ K, and $p_0 = 10^5$ Pa. D_0 for mixtures of air with various gases are shown in Table C.17.

In high vacuum, molecules move between surfaces 1 and 2 (Fig. 2.2) without mutual collisions. In this situation, $\Delta A = d \cdot dA/dx$, where d is the distance between the surfaces. For high vacuum, the transport coefficient described by Eq. (2.5) is as follows:

$$R = \frac{v_{ar} d}{4}.$$ (2.20)

Internal friction in high vacuum can be determined from Eq. (2.6), but the dynamical viscosity coefficient should be taken as

$$\eta_h = \frac{v_{ar} d n m}{4}.$$ (2.21)

Taking into account Eqs. (1.10) and (1.18), Eq. (2.21) can be transformed to

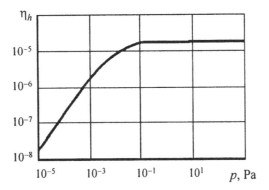

Figure 2.3 Coefficient of dynamical viscosity of nitrogen at room temperature.

$$\eta_h = dp \sqrt{\frac{m}{2\pi kT}}.$$ (2.22)

The dynamical viscosity coefficient in high vacuum is proportional to pressure and inversely proportional to the square root of absolute temperature. Heat transfer in high vacuum occurs by heat conductivity and radiation, the convection mechanism being negligible. The heat conduction of a gas can be calculated using Eq. (2.10). The heat conduction coefficient in high vacuum is

$$\lambda_h = R_\alpha c_v \eta_h,$$ (2.23)

where $R_\alpha = \alpha/(2 - \alpha)$, α being the accommodation coefficient (Table C.16). λ_h is the same function of temperature and pressure as η_h in Eq. (2.22). Transformation of Eq. (2.23) yields

$$\lambda_h = k_T p,$$ (2.24)

where

$$k_T = \frac{\alpha m^{1/2} d c_v}{(2 - \alpha)\sqrt{2\pi kT}}.$$ (2.25)

Diffusion of gases in high vacuum is determined by Eq. (2.18), the diffusion coefficient being independent of pressure and proportional to the square root of absolute temperature:

$$D_h = \frac{d v_{ar}}{4} = d \sqrt{\frac{kT}{2\pi m}}.$$ (2.26)

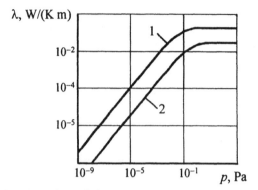

Figure 2.4 Heat flow through rarefied gases as a function of pressure: (*1*) hydrogen and (*2*) air.

Transient phenomena occur in medium vacuum due to a discontinuity of a transported physical quantity at the vacuum-solid interface. The discontinuity decreases with a decrease in pressure. Phenomenological description of these phenomena is quite complex. However, technical calculations may be satisfactory with approximate formulas accurate to ± 10%. Using Eq. (2.4) with the transport coefficient, we may write

$$R = \frac{v_{ar}Ld}{2(d+2L)}.$$
(2.27)

This formula is true for any degree of vacuum. At low vacuum ($2L \to 0$) it is similar to Eq. (2.5) and at high vacuum ($2L \to \infty$) it transforms to Eq. (2.20). The dynamical viscosity coefficient at any degree of vacuum (Fig. 2.3) is

$$\eta = \frac{\eta_l p}{Kn_1 + p},$$
(2.28)

where η_l is the dynamical viscosity coefficient for low vacuum (Eq. (2.8)) and $Kn_1 = 2L_1/d$, where L_1 from Eq. (1.29).

The heat conductivity coefficient for any vacuum (Fig. 2.4) is

$$\lambda = \frac{\lambda_l p}{Kn_1 + p},$$
(2.29)

where λ_l is the heat conductivity coefficient for low vacuum (Eq. (2.11)).

The self-diffusion coefficient for any vacuum (Fig. 2.5) is

$$D = \frac{D_l p}{Kn_1 + p},$$
(2.30)

where D_l is the self-diffusion coefficient for low vacuum (Eq. (2.19)).

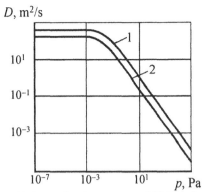

Figure 2.5 Self-diffusion coefficients of various gases at 293 K as functions of pressure: (*1*) hydrogen and (*2*) air.

2.3   Thermal Equilibrium of Pressures

The equilibrium ratio of the pressures and molecular concentrations in connected volumes having different temperatures depends on the degree of vacuum.

In low vacuum, the condition necessary for static equilibrium, i.e., the absence of gas flows in two connected volumes at different temperatures, is the equity of the pressures: $p_1 = p_2$. According to the equation of gas state, the ratio of the concentrations will then be $n_1/n_2 = T_2/T_1$.

In high vacuum, only dynamical equilibrium may be achieved, at which opposite gas flows between the volumes are equal. For vessels connected through a hole, the equations of the number of molecules colliding with the surface in unit time (Eq. (1.34)) and the arithmetical mean velocity of gas molecules (Eq. (1.18)) suggest the equilibrium condition be as follows:

$$A\frac{n_1}{4}\sqrt{\frac{8kT_1}{\pi m}} = A\frac{n_2}{4}\sqrt{\frac{8kT_2}{\pi m}},$$

where A is the hole area. Simplifying this formula, we obtain

$$\frac{p_1}{p_2} = \sqrt{\frac{T_1}{T_2}} \text{ and } \frac{n_1}{n_2} = \sqrt{\frac{T_2}{T_1}}. \tag{2.31}$$

This is also true for vessels connected through a tube. Thus, high vacuum equilibrium or thermal transpiration in connected volumes occurs at pressures proportional to the square root of the ratio of the absolute temperatures.

In medium vacuum, the ratio of pressures can be approximated as

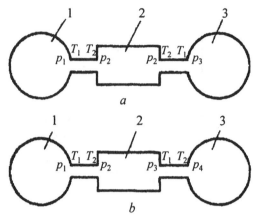

Figure 2.6 Schematic of connection of spaces having different temperatures.

$$\frac{p_1}{p_2} = \left(\frac{T_1}{T_2}\right)^{L/(d+2L)}, \tag{2.32}$$

where d is the effective size of the connecting hole or tube and L is the free path length at mean pressure. The relative error of this formula is max. 10%.

Equation (2.32) holds for low ($L \to 0$, $p_1 = p_2$) and high ($L \to \infty$, $p_1/p_2 = \sqrt{T_1/T_2}$) vacuum. The above relations are important for measurement of the pressures in connected vacuum chambers having different temperatures.

Let a manometric transducer be placed in the region at $T_1 = 293$ K of a vacuum system, and the pressure be measured in a volume cooled to the LN temperature ($T_2 = 77$ K). Such conditions are the case in vacuum systems with nitrogen traps and adsorption pumps. Then, according to Eq. (2.32), in high vacuum $p_2 = p_1/2$. If the pressure is measured by a manometer in the region of room temperature $T_1 = 293$ K, and $T_2 = 1273$ K, then $p_2 \approx 2p_1$.

Placing vessel 2 between vessels 1 and 3 (Fig. 2.6, a) that are at the same temperature and sequentially using Eq. (2.32), one may prove that the vessel 2 does not affect the pressures p_1 and p_3.

In high vacuum (Fig. 2.6, b) the ratio of pressures does not depend on the number of connected vessels $p_1/p_4 = \sqrt{T_1/T_2}$. If low vacuum is produced in vessel 2, e.g., by increasing its volume, p_2 becomes equal to p_3 and the ratio of pressures increases: $p_1/p_4 = T_1/T_2$. This effect is used in thermal vacuum pumps.

2.4   Calculation of Gas Flow by the Method of Continuum Mechanics

Calculation of a vacuum system may require solution of a direct task, i.e., determination of gas flow rate depending on geometrical size and pressure distribution. The inverse task is to determine the required geometrical parameters, given a gas flow rate. To solve these problems, one should know the laws of gas flow, which depend on the degree of vacuum.

Table 2.1 Gas flow modes in vacuum systems.

Flow modes	Boundaries	
	Lower	Upper
Viscous	Atmosphere pressure	$Kn \leq 5 \times 10^{-3}$
Molecular–viscous	$Kn > 5 \times 10^{-3}$	$Kn \leq 1.5$
Molecular	$Kn > 1.5$	$Kn \to \infty$

In low vacuum at high pressures, inertial gas flow may occur, which is analogous to turbulent flow dealt with in fluid dynamics. The inertia force of flowing gas produces eddies and leads to a complex distribution of gas velocity. The Reynolds criterion $Re = dv_g/\eta_k$ can be used to formulate the conditions of existence of inertial mode. Here d is the characteristic size of the element, v_g is the gas flow velocity, and η_k is the kinematic viscosity coefficient. The gas flow is inertial at $Re > 2200$. Alternatively, the inertial flow condition can be written by expressing v_g through gas flow Q: $v_g = 4Q/(\pi d^2 p)$. For air at room temperature the condition $Re > 2200$ can be written as $Q > 3 \times 10^3 d$, where Q is the gas flow, m^3 Pa/s and d is the pipe diameter, m. Such flows are quite a rare occurrence in vacuum systems; they may be obtained upon starting of some vacuum systems. We will therefore disregard this flow mode as atypical for vacuum equipment.

In low vacuum, viscous gas flow dominates for which the cross-section velocity distribution is controlled by internal friction. In high vacuum, internal friction tends to zero and the gas flow is such that separate motion of molecules is possible. This flow mode is called molecular.

In medium vacuum both internal friction and molecular transport are active, this transient flow mode being referred to as molecular–viscous.

The boundary conditions of the gas flow modes in vacuum systems are based on the Knudsen criterion, i.e., free path length L related to effective size d_{ef} (Table 2.1)

In any mode, each element of a vacuum system has its individual dependence of conductance on pressure, temperature, and element size. At low vacuum, for which the mean free path length of gas molecules is small, gas can be considered as a continuous medium whose movement is determined by inertia or internal friction forces. For high vacuum, these methods are still convenient for approximation of the basic laws of gas flow, although gas is no longer continuous.

Gas flow can be described by two differential equations of mass and force balance. The force balance equation leads to the equation of gas flow. The flow may be determined by the number of molecules passing through a pipe in unit time.

$$P = \frac{DA(p_1 - p_2)}{klT}, \qquad (2.33)$$

where A and l are the cross-section and length of the pipe and D is the diffusion coefficient.

In CI units (kg/s) the flow equation (2.33) transforms to mass flow equation

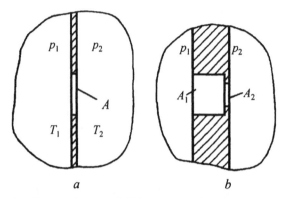

Figure 2.7 Shapes of orifices: (*a*) between infinite spaces and (*b*) between an infinite and a limited space.

$$G = Pm = \frac{DmA(p_1 - p_2)}{klT}. \tag{2.34}$$

According to the equation of gas state (Eq. (1.12)), the mass of gas Nm at a constant temperature is proportional to the product of gas pressure and its volume. Thus, one may introduce an arbitrary non-system unit of amount of gas, 1 m³ Pa, which is equal to the amount of gas in a volume of 1 m³ at a pressure of 1 Pa and a temperature of 273 K. For gas with molecular weight M this unit can be recalculated to kilograms using the relation 1 m³ Pa = 1.3×10^{-5} M/M_a (kg), where M_a is the molecular weight of air (29 kg/kmol). The arbitrary unit of gas flow 1 m³ Pa is often used in vacuum calculations.

If gas flow is expressed not in kg/s but in arbitrary gas flow units, then, according to Eqs. (1.12) and (2.34),

$$Q = G\frac{kT_0}{m} = \frac{DAT_0}{lT}(p_1 - p_2) = U(p_1 - p_2), \tag{2.35}$$

where Q is the gas flow, Pa m³/s and U is the conductance of the vacuum system, m³/s,

$$U = \frac{DAT_0}{lT}, \tag{2.36}$$

where T is the temperature, K and $T_0 = 273$ K.

Below we will dwell upon the calculation of the conductance of orifices and round-section pipes in different flow modes. A pipe whose length is much smaller than its diameter ($l \leq 0.01d$) will be considered a orifice. Let a orifice be placed in a wall separating two infinitely large spaces (Fig. 2.7, *a*). At low vacuum the viscous flow through the orifice in arbitrary units is

$$Q = \psi A p_1 \sqrt{\frac{RT_1}{M}}, \qquad (2.37)$$

where

$$\psi = r^{1/\gamma} \sqrt{\frac{2\gamma}{\gamma - 1}[1 - r^{(\gamma - 1)/\gamma}]}. \qquad (2.38)$$

A is the orifice area, m², and p_1 and T_1 are the pressure and temperature of space 1. The derivation of Eq. (2.37) is given in Appendix A.4.

A decrease in the pressure ratio $r = p_2/p_1 \leq 1$ causes the amount of gas passing through the orifice and the flow speed in the area p_2 to increase until the ratio p_2/p_1 reaches the critical value corresponding to the sound velocity. If the flow is adiabatic, the critical value is

$$r_c = \left(\frac{2}{\gamma + 1}\right)^{\frac{\gamma}{\gamma - 1}}. \qquad (2.39)$$

For diatomic gases and air ($\gamma = 1.4$), $r_c = 0.528$, for monatomic gases ($\gamma = 1.67$) $r_c = 0.487$, and for triatomic gases ($\gamma = 1.3$) $r_c = 0.546$.

At $p_2/p_1 < r_c$, the amount of gas is constant. At $p_2/p_1 > r_c$, the conductance of a orifice, according to Eqs. (2.35) and (2.37) is

$$U_{hv} = \frac{Q}{p_1 - p_2} = \psi \frac{A}{1 - r} \sqrt{\frac{RT_1}{M}}. \qquad (2.40)$$

For air and other diatomic gases at $\gamma = 1.4$, Eqs. (2.38) and (2.40) suggest

$$\psi = 2.65 r^{0.714}(1 - r^{0.236})^{1/2}, \qquad (2.41)$$

$$U_{hv} = \frac{91 A \psi}{1 - r} \sqrt{\frac{T_1}{M}}, \qquad (2.42)$$

where M is the molecular mass, kg/kmol, T_1 is the absolute temperature, K, A is the hole area, m², r is the pressure ratio p_2/p_1, and ψ is the function of r calculated according to Eq. (2.41) and shown in Fig. 2.8. For air ($M = 29$ kg/kmol) at room temperature ($T = 293$ K) Eq. (2.42) can be simplified to

$$U_{hv} = \frac{289 \psi A}{1 - r} \text{ for } 1 > r \geq 0.528. \qquad (2.43)$$

In hypercritical mode at $r < 0.528$, the value $\psi = \psi(0.528) = 0.69$. The conductance of a hole is then

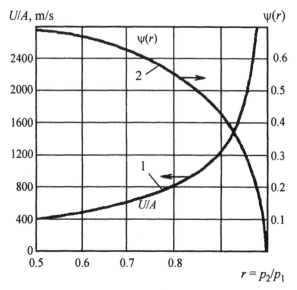

Figure 2.8 (*1*) specific air conductance of a orifice at 293 K and (*2*) the ψ-function of the orifice (Eq. (2.38)) for γ = 1.4 in viscous mode as functions of pressure.

$$U_{hv} = \frac{200A}{1-r} \text{ for } 0.528 > r \geq 0.1 .$$

(2.44)

For large pressure ratios ($r < 0.1$) we have

$$U_{hv} = 200A \text{ for } 0.1 > r \geq 0.$$

(2.45)

Usually, the pressure ratio r is not known *a priori* in conductance calculations, which are therefore performed using sequential iterations. To a first approximation it is safe to assume that the conductance is equal to the minimum value $U_{hv} = 200A$ m³/s and does not depend on r. For round holes this value is $U_{hv} = 160d^2$ m³/s.

In conventional vacuum systems designed for steady-state operation, the pressure ratio r is always greater than the coefficient of pump efficiency. At optimum pump efficiencies, the pressure ratio r is between 0.8 and 0.9, this allowing the conductance be accepted as U_v = 1000A m³/s to a first approximation. For round holes, $U_{hv} = 785d^2$ m³/s. The conductance of a orificee at high vacuum in molecular flow mode is calculated according to Eq. (2.35) as

$$U_{hm} = \frac{Q}{P_1 - P_2} = \frac{PkT}{m(p_1 - p_2)},$$

where $P = P_1 - P_2$ and P_1 and P_2 are the opposite mass flows through the hole. Taking into account that $P_1 = n_1 m v_1 A/4$ and $P_2 = n_2 m v_2 A/4$, the conductance can be expressed as

$$U_{hm} = \left(n_1 \sqrt{\frac{8kT_1}{\pi m}} - n_2 \sqrt{\frac{8kT_2}{\pi m}}\right) \frac{A}{4(n_1 - n_2)} . \tag{2.46}$$

At $T_1 = T_2 = T$ the latter equation simplifies to

$$U_{hm} = 36.4A \sqrt{\frac{T}{M}}, \tag{2.47}$$

where M is in kg/kmol, T in K, A in m^2, and U_{hm} in m^3/s.

Calculation of orifice conductance for air ($M = 29$ kg/kmol) at room temperature (298 K) according to Eq. (2.47) yields

$$U_{hm} = 116A. \tag{2.48}$$

Since for a round hole $A = \pi d^2/4$, then $U_{hm} = 91d^2$ m^3/s, where d is in m.

For low vacuum and molecular–viscous flow mode, one may use the approximate formula

$$U_{hmv} = U_{hm}b + U_{hv}, \tag{2.49}$$

where

$$b = \left(1 + 2.5\frac{d}{2L}\right) \Big/ \left(1 + 3.1\frac{d}{2L}\right),$$

which is also true in molecular and viscous flow modes.

Now we consider a orifice between an infinite and a limited spaces. In this case, a short tube-pipe with this hole (Fig. 2.7, b), according to the second thermodynamic law, should have the same conductance in opposite directions:

$$\frac{1}{U_{A_1}} + \frac{1}{U_p} + \frac{1}{U_{A_1, A_2}} = \frac{1}{U_{A_2}} + \frac{1}{U_p},$$

where U_{A_1} and U_{A_2} are the conductance of orifice with areas A_1 and A_2 from the side of the infinite spaces, U_{A_1, A_2} is the conductance of the orifice with area A_2 from the side of the pipe, and U_p is the conductance of the pipe inside the wall.

Solving the equation relative to U_{A_1, A_2} and taking into account that $U_{A_1}/U_{A_2} = A_1/A_2$, we obtain

$$U_{A_1, A_2} = \frac{U_{A_2}}{1 - A_2/A_1} . \tag{2.50}$$

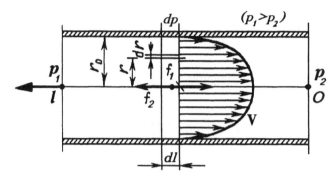

Figure 2.9 Schematic for calculation of gas flow in a pipe for viscous mode.

Using Eq. (2.50), one may transform conductances of holes in any flow modes for infinite spaces to conductances of holes connecting finite spaces. The shape of holes does not affect their conductance calculated using the present method.

We consider gas flow in pipes at low vacuum. In viscous flow mode the mean free path length of gas molecules is far smaller than pipe diameter. The gas layer adjacent to the pipe surface remains stationary, while other layers with thicknesses equal to the mean free path length move in steady-state mode at a rate determined by pipe radius.

In a round-section pipe with a steady-state flow, the gas cylinder formed by the pipe radius r and in increase dr (Fig. 2.9), the motive force due to pressure difference f_1 is in equilibrium with the internal friction force in gas f_2. The equilibrium condition can be written as $f_1 + f_2 = 0$ or

$$\pi r^2 dp + 2\pi r \eta \frac{dv}{dr} dl = 0.$$

Integrating over the pipe radius with initial conditions $r = r_0$ and $v = 0$, we obtain a parabolic distribution of gas velocity over the pipe section:

$$v = \frac{(r_0^2 - r^2)dp}{4\eta \, dl}.$$

The volumetric flow of the gas

$$V = \int_0^{r_0} v 2\pi r dr = \frac{\pi r_0^4 dp}{8\eta \, dl}. \tag{2.51}$$

Equation (2.51) allows the volumetric flow of a incompressible medium to be determined. For isothermally compressed rarefied gases, the condition $pV = \text{const}$ should also be taken into account. The gas flow Q in the pipe will be found as the product of gas

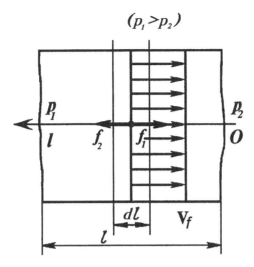

Figure 2.10 Schematic for calculation of gas flow in a pipe for molecular mode.

consumption V and gas pressure. We integrate Eq. (2.51) from $x = 0$ and $p = p_2$ to $x = l$ and $p = p_1$, disregarding the end effects for long pipes:

$$Q = pV = \frac{\pi r_0^4(p_1^2 - p_2^2)}{16\eta l}. \tag{2.52}$$

According to Eq. (2.35), we write the expression of conductance in viscous flow mode as

$$U_{pv} = \frac{Q}{p_1 - p_2} = \frac{\pi r_0^4 p_a}{8\eta l}. \tag{2.53}$$

Thus, the conductance of a round-section pipe in viscous flow mode is inversely proportional to the pipe length and dynamical viscosity coefficient and proportional to the mean pressure and the fourth power of the pipe radius. For air at 293 K and $\eta = 1.82 \times 10^{-5}$ N s/m², Eq. (2.53) can we transformed to

$$U_{pv} = 1.36 \times 10^3 \frac{d^4}{l} p_a; \tag{2.54}$$

where d and l are in meters, p in Pa, and U_v in m³/s.

In molecular flow mode the mean free path length is greater than the pipe diameter, and the molecules move independently and collide only with the pipe walls.

We assume that each of the molecules randomly moving in the pipe has constant component of flow velocity v_f directed along the pipe axis toward the lower pressure region (Fig. 2.10). The motive force $f_1 = dpA$, where A is the pipe cross-section.

The compensating force, which is equal to the change in the impetus of all the molecules due to their collision with the walls, is $f_2 = -BdlN_q mv_f$, where B is the pipe perimeter, and $N_q = nv_{ar}/4 = p/\sqrt{2\pi mkT}$ is the number of molecules colliding with the pipe wall in unit time. The equilibrium equation $f_1 + f_2 = 0$ takes the form

$$dpA - BdlN_q mv_f = 0$$

Introducing volume flow $V = v_f A$ and using Eq. (1.34, (1.10), and (1.18), we obtain

$$\frac{dp}{mpV}\sqrt{2\pi mkT} = \frac{B}{A^2}dl.$$

In steady state flow mode, the product pV in the denominator is constant. We integrate this formula from p_2 to p_1 and from 0 to l:

$$\frac{p_1 - p_2}{mQ}\sqrt{2\pi mkT} = \int_0^l \frac{B}{A^2}dl,$$

whence the gas flow is

$$Q = \frac{\sqrt{2\pi mkT}(p_1 - p_2)}{m\int_0^l \frac{B}{A^2}dl}.$$

using Eq. (1.18) for v_{ar}, we obtain

$$Q = \frac{\pi v_{ar}(p_1 - p_2)}{2\int_0^l \frac{B}{A^2}dl}.$$

The more exact equation of Q, with account of the velocity distribution function is

$$Q = \frac{4v_{ar}(p_1 - p_2)}{3\int_0^l \frac{B}{A^2}dl}.$$

Then the pipe conductance is

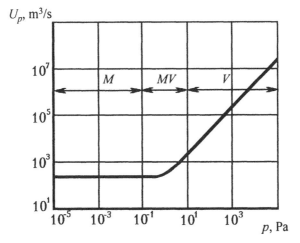

Figure 2.11 Air conductance of a pipeline with $d = 1$ m and $l = 1$ m at 293 K as a function of pressure for (M) molecular, (V) viscous, and (MV) molecular–viscous flow modes.

$$U_{pm} = \frac{Q}{p_1 - p_2} = \frac{4 v_{ar}}{3 \int_0^l \frac{B}{A^2} dl} \tag{2.55}$$

For a constant-section pipe,

$$U_{pm} = \frac{4 v_{ar} A^2}{3 B l}. \tag{2.56}$$

For a round-section pipe,

$$U_{pm} = \frac{\pi d^3 v_{ar}}{12 l} = 38.1 \frac{d^3}{l} \sqrt{\frac{T}{M}}, \tag{2.57}$$

where d and l are in meters, M in kg/kmol, T in K, and U in m³/s. Thus, the conductance of a pipe in molecular flow mode does not depend on pressure. For air at 293 K the conductance of a cylindrical pipe is

$$U_{pm} = 121 \frac{d^3}{l}, \tag{2.58}$$

using the same units.

At medium vacuum in molecular–viscous flow mode, the conductance of pipes can be calculated using the semiempirical Knudsen formula

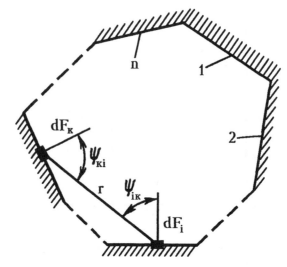

Figure 2.12 Schematic for calculation using the method of angular coefficients.

$$U_{pmv} = bU_{pm} + U_{pv},\tag{2.59}$$

here U_{pv} is the conductance in viscous mode, U_{pm} is the conductance in molecular mode, and b is the coefficient taking on 0.8 for transition to viscous mode and 1 for transition to molecular mode. The average value $b = 0.9$ can be accepted as universal for technical calculations. Figure 2.11 shows conductance of a round-section pipe as a function of gas pressure.

Conductance s of non-round-section pipes can be determined using the same methods that were used for round-section pipes. The resultant formulas for various pipe section shapes are listed in Table C.18.

2.5   Calculation of Gas Flow Using the Method of Integral Angular Coefficients

Molecular flows in high vacuum are analyzed using the method of integral angular coefficients. Most vacuum systems can be represented as combinations of homogeneous surfaces with constant adsorption and reflection coefficients β and ρ, with $\beta + \rho = 1$.

We accept that the cosine law is satisfied for molecular desorption from the walls. the angular distribution of molecules at the entrance of a vacuum system can also be approximated by the cosine law.

Let us consider a system of homogeneous surfaces that are visible from one another (Fig. 2.12). The molecular flow incident onto the ith surface is determined as

$$Q_i = \sum_{\substack{k=1 \\ k \neq i}}^{n} \varphi_{k,i} Q_{\rho k}, \tag{2.60}$$

where $Q_{\rho k}$ is the molecular flow desorbed from the kth surface; $\varphi_{k,i}$ is the angular coefficient determining the fraction of the molecular flow desorbed from the kth surface and incident onto the ith one.

The flow of molecules from the ith surface is

$$Q_{\rho,i} = Q_{di} + \rho_i Q_i, \tag{2.61}$$

where Q_{di} is the desorption of gas from the ith surface and ρ_i is the reflection coefficient of the ith surface.

The amount of gas adsorbed by the ith surface in unit time is

$$Q_{\beta i} = \beta_i Q_i, \tag{2.62}$$

where β_i is the adsorption coefficient of the ith surface.

To determine the angular coefficients, we write out the expression for the molecular flow incident from the unit area dF_k on the unit area dF_i. Taking into account the cosine law (Eq. (1.37)), we obtain

$$d^2 Q_{k,i} = \frac{dQ_k \cos \psi_{k,i}}{\pi} d\omega, \tag{2.63}$$

where dQ_k is the molecular flow from the surface dF_k, $d\omega$ is the solid angle from which the area dF_i is seen from the area dF_k; $d\omega = dF_i \cos \psi_{i,k}/r^2$. We accept that the unit areas are sufficiently small to satisfy the condition $r = \text{const}$. Substituting the expression of solid angle $d\omega$ into Eq. (2.63), we obtain

$$d^2 Q_{k,i} = \frac{dQ_k \cos \psi_{k,i} \cos \psi_{i,k}}{\pi r^2} dF_i, \tag{2.64}$$

whence the probability that a molecule desorbed from the area dF_k is incident onto the area dF_i, or the differential angular coefficient, is

$$d^2 \varphi_{k,i} = \frac{d^2 Q_{k,i}}{dQ_k} = \frac{\cos \psi_{k,i} \cos \psi_{i,k}}{\pi r^2} dF_i. \tag{2.65}$$

Integrating Eq. (2.65) over the area F_i, we obtain the local angular coefficient that determines the mass exchange between the unit area dF_k and the surface F_i:

$$d\varphi_{k,i} = \int_{F_i} \frac{\cos\psi_{k,i}\cos\psi_{i,k}}{\pi r^2} dF_i. \qquad (2.66)$$

The mean angular coefficient, or the probability that a molecule desorbed from the kth surface is incident onto the ith surface, is determined by the second integration over the area F_k:

$$\varphi_{k,i} = \frac{1}{F_k} \int_{F_k}\int_{F_i} \frac{\cos\psi_{k,i}\cos\psi_{i,k}}{\pi r^2} dF_k dF_i. \qquad (2.67)$$

This latter expression for the angular coefficient is only determined by the geometrical parameters of the system in question. These parameters are analogous to the angular coefficients of radiation heat exchange that are mutual, closed, and additive:

1. the angular coefficient of interacting surfaces are inversely proportional to their areas:

$$\varphi_{2,1}F_2 = \varphi_{1,2}F_1, \qquad (2.68)$$

which follows from Eq. (2.67);

2. for a closed surface subdivided into n interacting surfaces, the sum of angular coefficients of any surface relative to the other surfaces is equal to unity:

$$\sum_{\substack{k=1\\k\neq i}}^{n} \varphi_{k,i} = 1; \qquad (2.69)$$

3. the angular coefficient of two interacting surfaces, one of which is subdivided into n parts, is equal to the sum of the partial angular coefficients:

$$\varphi_{i,k} = \sum_{j=1}^{n} (\varphi_{k,i})_j. \qquad (2.70)$$

Table 2.2 shows angular coefficients for some surface shapes that are frequently dealt with in vacuum engineering.

Knowing the angular coefficients, one may compose a set of equations such as Eq. (2.61), whose solution allows, given ρ_i or β_i, all the mass exchange parameters to be determined: the total flow $Q_{\rho i}$ from the surface F_i, the flow of molecules Q_i incident onto the surface F_i, and the flow of molecules $Q_{\beta i}$ adsorbed by the surface F_i.

Table 2.2 Angular coefficients for calculation of vacuum systems.

Type of element	Coefficients	Notes
A body in another body	$\varphi_{1,1} = 0$ $\varphi_{1,2} = 1$ $\varphi_{2,1} = \dfrac{F_1}{F_2}$	;
Two coaxial disks	$\varphi_{1,2} = \dfrac{1}{2x^2}[1 + x^2 + z^2 - \sqrt{(1 + x^2 + z^2)^2 - 4x^2z^2}]$ $\varphi_{2,1} = \dfrac{x^2}{z^2}\varphi_{1,2}$ $\varphi_{1,2} = \varphi_{2,1} = \dfrac{1}{4x^2}(\sqrt{1 + 4x^2} - 1)^2$	$x = \dfrac{r}{a}$ $z = \dfrac{R}{a}$ $x = z$
Internal surface of a cylinder	$\varphi_{1,1} = 1 + y - \sqrt{y^2 + 1}$	$y = \dfrac{a}{d}$
Two cylinders	$\varphi_{1,2} = \dfrac{1}{2y}[2yt + (y + z)\sqrt{1 + (y + z)^2} - z\sqrt{1 + z^2}$ $\quad - (y + t + z)\sqrt{(y + t + z)^2 + 1} + (t + z)\sqrt{1 + (t + z)^2}]$ $\varphi_{2,1} = \dfrac{y}{t}\varphi_{1,2}$ $\varphi_{1,2} = \dfrac{1}{2y}[2yt + y\sqrt{1 + y^2} - (y + t)\sqrt{1 + (y + t)^2}$ $\quad + t\sqrt{1 + t^2}]$	$y = \dfrac{a}{d}$ $t = \dfrac{c}{d}$ $z = \dfrac{b}{d}$ $z = 0$
A cylinder and a disk	$\varphi_{1,2} = \dfrac{1}{2(1 - f^2)}[4t\sqrt{1 + t^2} - 4y\sqrt{1 + y^2}$ $\quad - \sqrt{(4t^2 + f^2 + 1)^2 - 4f^2} + \sqrt{(4y^2 + f^2 + 1)^2 - 4f^2}]$ $\varphi_{2,1} = \dfrac{1 - f^2}{4(t - y)}\varphi_{1,2}$	$f = \dfrac{d_2}{d_1}$ $t = \dfrac{a_2}{d_1}$ $y = \dfrac{a_1}{d_1}$

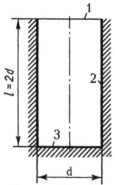

Figure 2.13 Schematic of a vacuum chamber with sorbing walls.

By way of example, we calculate the capture coefficient of a pump in the form of a cylinder with diameter d, height $l = 2d$, and sorbing walls 2 (Fig. 2.13). The molecules entering into the cylinder have a cosine-like angular distribution. Gas is not desorbed from the other surfaces. The reflection coefficient of the intake hole (surface 1) is $\rho_1 = 0$, and for the walls it is $\rho_{2,3} = 0.5$.

Set of equations (2.61) for the object in hand is

$$Q_{\rho 1} = Q_{d1},$$
$$Q_{\rho 2} = \rho_2(\varphi_{1,2}Q_{\rho 1} + \varphi_{2,2}Q_{\rho 2} + \varphi_{3,2}Q_{\rho 3}),$$
$$Q_{\rho 3} = \rho_3(\varphi_{1,3}Q_{\rho 1} + \varphi_{2,3}Q_{\rho 2} + \varphi_{3,3}Q_{\rho 3}). \tag{2.71}$$

To determine the angular coefficient $\varphi_{3,1}$ that determines the probability that molecules from surface 3 are incident onto surface 1, we use Table 2.2, which suggests

$$\varphi_{3,1} = \frac{l^2}{d^2}\left(\sqrt{1 + \frac{d^2}{l^2}} - 1\right)^2. \tag{2.72}$$

For $l = 2d$ in Eq. (2.72), $\varphi_{3,1} = 0.056$. To determine the other angular coefficients, we use the symmetry conditions $\varphi_{3,1} = \varphi_{1,3}$, $\varphi_{2,1} = \varphi_{2,3}$, and $\varphi_{1,2} = \varphi_{3,2}$, the logical condition $\varphi_{1,1} = \varphi_{3,3} = 0$, the mutuality condition $\varphi_{1,2} = \varphi_{2,1}(F_2/F_1)$, and the condition of closed state $\varphi_{1,2} = 1 - \varphi_{1,3}$ and $\varphi_{2,2} = 1 - \varphi_{2,1} - \varphi_{2,3}$:

$$\begin{bmatrix} \varphi_{1,1} & \varphi_{2,1} & \varphi_{3,1} \\ \varphi_{1,2} & \varphi_{2,2} & \varphi_{3,2} \\ \varphi_{1,3} & \varphi_{2,3} & \varphi_{3,3} \end{bmatrix} = \begin{bmatrix} 0 & 0.118 & 0.056 \\ 0.944 & 0.764 & 0.944 \\ 0.056 & 0.118 & 0 \end{bmatrix}.$$

Solving set of equations (2.71), we obtain $Q_{\rho 1} = Q_{d1}$, $Q_{\rho 2} = 0.82Q_{d1}$, and $Q_{\rho 3} = 0.077Q_{d1}$. The molecular flow leaving the object from surface 1 will be determined using the formula

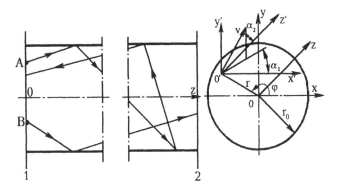

Figure 2.14 Vacuum system element of arbitrary shape.

$Q_1 = \varphi_{1,1}Q_{\rho 1} + \varphi_{2,1}Q_{\rho 2} + \varphi_{3,1}Q_{\rho 3} = 0.1Q_{d1}$. The capture coefficient of this pump (i.e., the probability that a molecule that entered into the pump through the intake hole is pumped) is $\eta = (Q_{d1} - Q_1)/Q_{d1} = 0.9$.

2.6 Modeling of Gas Flow

Modeling of gas flow in vacuum systems can be performed by statistical tests (using the Monte–Carlo method). This is a powerful numerical tool for solving mathematical problems by modeling of random quantities with a preset distribution law and it allows solution of any problems by reducing them to calculation of mathematical expectation. This method is widely used for the solution of steady-state and non-steady-state problems of gas flow in vacuum system elements in high or medium vacuum, for which each of the molecules moves randomly by sequential collisions with walls. The probability that molecules transfer from one element of a vacuum system to another is determined by the element geometry. The transition probabilities and gas concentrations allow determination of gas flows in vacuum systems.

We consider an application of the statistical test method for modeling of a steady-state gas flow through a vacuum system element that has one intake, one outlet, and arbitrary shape of the inner walls (Fig. 2.14). The source data for the calculation are the shapes of the elements, the distribution of molecules at the intake hole, and the angular distribution of molecules desorbed from the inner walls.

The shapes of the elements can be described, with appropriate restriction, by second-order equations such as

$$a_{11}x^2 + a_{22}y^2 + a_{33}z^2 + a_{12}y + a_{23}yz + a_{13}zx + a_{14}x + a_{24}y + a_{34}z + a_{44} = 0 . \quad (2.73)$$

In a properly chosen system of coordinates, Eq. (2.73) is noticeably simplified. The equation of a cylinder coaxial with the z-axis is $x^2 + y^2 = r_0$, the equation of a plane perpendicular to the z-axis is $z = 1$, etc.

The initial conditions for a system element are such that the element is connected with an infinite space. The molecular-kinetic theory suggests that molecules entering from an infinite space have a homogeneous distribution over the intake hole area and a cosine-like angular distribution.

We accept that the inner walls of an element desorb gas according to the cosine law. The random parameters to be modeled are the coordinates of molecule's entrance point and the two angles determining its movement in the system element from the entrance point or from the wall.

The entrance point coordinates for a round intake hole (Fig. 2.14) are the angle φ and the radius r. In order for the molecular distribution over the hole area to be homogeneous, it is required that the angle φ be homogeneously distributed in the range from 0 to 2π. Generating random value ξ in the range from 0 to 1, we obtain the random angle as

$$\varphi = 2\pi\xi[0, 1].$$

To determine the random radius of a molecule's entrance point, we write the number of molecules entering through ring dr at radius r as

$$dN = N\frac{2\pi r dr}{\pi r_0^2} = N\frac{2r}{r_0^2}dr,$$

where N is a total number of molecules having entered the element.

The probability of entrance through a circle with area πr^2 is

$$P(r) = \int_0^r \frac{2r}{r_0^2}dr = \frac{r^2}{r_0^2}. \tag{2.74}$$

Modeling the probability $P(r)$ by the new random value $\xi[0, 1]$ with a homogeneous distribution from 0 to 1, we obtain

$$r = r_0\sqrt{\xi[0, 1]}. \tag{2.75}$$

The direction of molecule's velocity v in the entrance point is determined by the angles α_1 and α_2 (Fig. 2.14). The angle α_1 is formed by the x'-axis and the projection of the velocity vector onto the $x'p'$-plane. The angle α_2 is formed by the z'-axis and the direction of the velocity v. According to the boundary conditions accepted, the random angle α_1 is homogeneously distributed from 0 to 2π:

$$\alpha_1 = 2\pi\xi[0, 1], \tag{2.76}$$

and the random angle α_2, which has a cosine-like distribution, can be found from Eq. (1.39):

$$\alpha_2 = \arcsin(\sqrt{\xi[0, 1]}) . \qquad (2.77)$$

Equations (2.74)–(2.77) allow φ, r, α_1, and α_2 that are necessary for solving such a problem to be modeled. If the surfaces are sorbing, additional parameters for modeling the sorption should be introduced.

The statistical model of steady-state gas flow described random molecular paths in an element of a vacuum system. For mathematical modeling of the path of gas molecules, we use the equation of a line in local systems of coordinates for each of the surfaces:

$$\frac{x'}{b'} = \frac{y'}{m'} = \frac{z'}{n'}, \qquad (2.78)$$

where b', m', and n' are the direction cosines of the lines in a local system of coordinates that are determined by the angles α_1 and α_2.

To find the point of molecule's collision with the surface of an element, Eq. (2.78) should be transformed to a global system of coordinates in which the equations describing the shapes of the elements should be written:

$$\frac{x - x_1}{b} = \frac{y - y_1}{m} = \frac{z - z_1}{n}, \qquad (2.79)$$

where x_1, y_1, and z_1 and the coordinates of molecule's desorption point in the global system of coordinates, and b, m, and n are the direction cosines in this system, which are determined from the transformation

$$\begin{bmatrix} b \\ m \\ n \end{bmatrix} = \begin{vmatrix} b_1 & m_1 & n_1 \\ b_2 & m_2 & n_2 \\ b_3 & m_3 & n_3 \end{vmatrix} \begin{bmatrix} b' \\ m' \\ n' \end{bmatrix},$$

where b_1, b_2, and b_3 are the direction cosines of the global axes relative to the x'-axis, m_1, m_2, and m_3 are these cosines for the y'-axis, and n_1, n_2, and n_3 for the z'-axis.

Combined solution of Eq. (2.79) with Eqs. (2.73) that determine the shape of a vacuum system element yields the coordinates of the intersection points. A line intersects a plane in one point, and a cylinder in two points. Logically, the intersection points beyond the inner surface can be rejected, in view of the restriction to Eq. (2.73) for the inner surface of an element. From the remaining points, the one at the minimum distance from the initial point in the molecular flow direction should be chosen.

For the intersection point determine in the above manner, the random direction of reflected flow should be determined. The angles are found by analogy with α_1 and α_2 in Eqs. (2.76) and (2.77) for the local system of coordinates in which the z'-axis is normal to the incidence surface. The motion of a molecule is traced until it leaves the vacuum system through the intake or the outlet hole.

The result of each test is described by the random value X that takes on 0 or 1. Let $X = 1$ be the case that a molecule passes through the system. The path of such a molecule begins at point B (Fig. 2.14).

If a molecule leaves the system through the intake hole, then $X = 0$ (path from the point A in Fig. 2.14). The probability that a molecule passes from the intake hole 1 to the outlet one 2 is the arithmetical mean of X over a large number of tests N:

$$P_{1 \to 2} = \frac{1}{N}(X_1 + X_2 + \dots + X_N), \qquad (2.80)$$

where $X_1, \dots, X_N$ are the values of X for the N tests.

The flow of gas through the element is

$$Q = Q_0 P_{1 \to 2}, \qquad (2.81)$$

where Q_0 is the inlet gas flow.

We statistically analyze the results. The parameters of the random value X are as follows. Let $M(X)$ and $M(X^2)$ are the mathematical expectations of the random values X and X^2:

$$M(X) = \sum_{i=1}^{2} P_i Y_i; \quad M(X^2) = \sum_{i=1}^{2} P_i Y_i^2,$$

where P_i are the probabilities that X takes on Y_i. The probability P_1 that corresponds to the case $Y_1 = 1$ is $P_{1 \to 2}$, and the probability P_2 corresponding to $Y_2 = 0$ is $1 - P_{1 \to 2}$. Hence,

$$M(X) = M(X^2) = P_{1 \to 2}. \qquad (2.82)$$

We will use the above mathematical expectations to find the variance of the random value X:

$$D(X) = M(X^2) - [M(X)]^2 = P_{1 \to 2}(1 - P_{1 \to 2}). \qquad (2.83)$$

The rms deviation of X is

$$\sigma = \sqrt{D(X)} = \sqrt{P_{1 \to 2}(1 - P_{1 \to 2})}. \qquad (2.84)$$

According to the Chebyshev equation, for any fixed $\varepsilon > 0$,

$$P(|X_{m,N} - M(X)| \le \varepsilon) \ge 1 - \gamma. \qquad (2.85)$$

At sufficiently large number of tests N, the arithmetical mean $X_{m,N}$ deviates from $M(X)$ by not more than ε with the probability not less than $1 - \gamma$, where

Table 2.3 Coefficient R.

δ	$1 - \gamma$			
	0.99	0.97	0.95	0.9
0.01	1.0×10^6	3.3×10^5	2.0×10^5	1.0×10^5
0.02	2.5×10^5	8.3×10^4	5.0×10^4	2.5×10^4
0.03	1.1×10^5	3.7×10^4	2.2×10^4	1.1×10^4
0.04	6.3×10^4	2.1×10^4	1.3×10^4	6.3×10^3
0.05	4.0×10^4	1.3×10^4	8.0×10^3	4.0×10^3
0.10	1.0×10^4	3.3×10^3	2.0×10^3	1.0×10^3

$$\gamma = \frac{D(X)}{N\varepsilon^2}. \qquad (2.86)$$

The relative error of the deviation of the arithmetical mean from the mathematical expectation is, with account of Eq. (2.86),

$$\delta = \frac{\varepsilon}{P_{1 \to 2}} = \frac{1}{P_{1 \to 2}}\sqrt{\frac{D(X)}{N\gamma}}. \qquad (2.87)$$

Using Eq. (2.83) for $D(X)$, we transform Eq. (2.87) so that

$$\delta = \sqrt{\frac{1 - P_{1 \to 2}}{P_{1 \to 2}N\gamma}}. \qquad (2.88)$$

At fixed γ, the calculation error decreases proportionally to the square root of the number of tests.

The number of tests is given by Eq. (2.88) as

$$N = R\frac{1 - P_{1 \to 2}}{P_{1 \to 2}}, \qquad (2.89)$$

where $R = 1/(\delta^2\gamma)$. At $P_{1 \to 2} = 0.5$, the number of tests is $N = R$. Table 2.3 shows numbers of tests R for different relative errors of δ and probabilities that $1 - \gamma$ criterion is true.

It can be seen from Table 2.3 that at $P_{1 \to 2} = 0.5$ an accuracy of 10% with a probability of 0.99 can be achieved after 10,000 tests and with a probability of 0.9 after 1000 tests.

The disadvantage of the statistical test method is the necessity of performing a large number of tests for obtaining acceptable accuracy. This requires the use of computers.

The advantage of the method is the universality of the calculation algorithm: calculation of a new element requires only its shape be analytically defined.

2.7   Gas Evolution

Gas evolution from solids to vacuum results from the sorption of gases during production or preliminary treatment of materials. Gas evolution may occur by two mechanisms, i.e., desorption and diffusion. The former mechanism is related to gas evolution from the surface, and the latter from the bulk of materials. Some gases are mainly evolved from the surface and others from the bulk. In atmosphere with a humidity of 50% the pressure of water vapor is 10^3 Pa. The surfaces of materials in air are covered with several monomolecular water layers. During chamber evacuation these molecules are desorbed, this process substantially increasing evacuation time. More complex phenomena occur due to chemical reactions. Adsorbed water may dissociate into hydrogen and oxygen during cathalytical reactions, with hydrogen being dissolved in solids.

Desorption gas evolution is unavoidable if pressure is reduced. The rate of gas desorption due to an abrupt change in pressure can be determined using Eq. (1.63):

$$q_d = a_m\frac{d\theta}{dt} = a_m(B - \theta_0 A)\exp(-At),   \tag{2.90}$$

where

$$A = \frac{1}{\tau} + \frac{\gamma p}{a_m\sqrt{2\pi mkT}}; \; B = \frac{\gamma p}{a_m\sqrt{2\pi mkT}}.$$

In Eq. (2.90), q_d is measured in the number of gas molecules evolved from unit surface area in unit time. In equilibrium ($q_d = 0$), the initial coverage degree is $\theta_0 = B/A$. Desorption requires that $\theta_0 > B/A$. For experimental data processing, it is convenient to transform Eq. (2.90) to

$$\log q_d = \log[a_m(B - \theta_0 A)] - 0.43At.   \tag{2.91}$$

Temporal dependences of q_d for different adsorption times are shown in Fig. 2.15.

At small adsorption heats, which correspond to small adsorption times and $A \gg 1$, adsorption equilibrium is achieved quite rapidly, the rate of desorption being determined by pressure change rate:

$$q_d = \frac{da}{dp}\cdot\frac{dp}{dt}.   \tag{2.92}$$

If adsorption equilibrium is achieved slowly, Eq. (2.92) is invalid, and q_d should be determined by differentiating Eq. (1.60) on the basis of a definite dp/dt law.

Diffusion gas evolution is due to gas transport through chamber walls. We consider that mass exchange occurs only in the superficial layer whose thickness is far smaller than that of the walls, the latter being treated as semiinfinite bodies. The length of a wall should be substantially larger than its thickness in order for the problem to be unidimensional.

q_d, m^3 Pa/(s m^2)

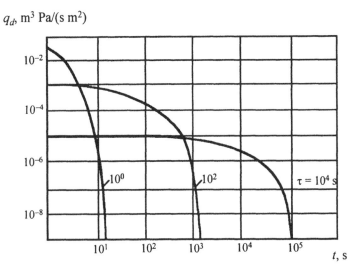

Figure 2.15 Specific desorption rate of water vapor as a function of adsorption time and heat at θ_0 = 5, T = 300 K.

Further assumptions are that the initial gas concentration in a wall is constant and equal to s_0. The pressure of gas over the wall will also be considered constant, and the equilibrium concentration of gas in the wall that corresponds to this pressure is s_m. The independence of s_0 and s_m on time corresponds to the first constant kind boundary conditions.

This case is a frequent occurrence in practice if a chamber is pumped at constant temperature and pressure, and the diffusion coefficient is sufficiently small: $D < 0.1h^2/t_{max}$, where h is the half-thickness of a chamber element and t_{max} is the maximum mass exchange time. The mathematical interpretation of this problem is as follows:

$$\frac{\partial s}{\partial t} = D\frac{\partial^2 s}{\partial x^2};$$

$$t = 0; \quad s(x,0) = s_0; \quad x = 0; \quad s(0,t) = s_m;$$

$$t = \infty; \quad s(x,\infty) = s_m; \quad x = \infty; \quad \frac{\partial s(\infty,t)}{\partial t} = 0. \tag{2.93}$$

The solution of Eq. (2.93) at the preset boundary conditions can be represented as a function of dimensionless time $\tau = Dt/h^2$:

$$s(x/h,\tau) = s_m + (s_0 - s_m)\text{erf}\left(\frac{x/h}{2\sqrt{\tau}}\right), \tag{2.94}$$

where

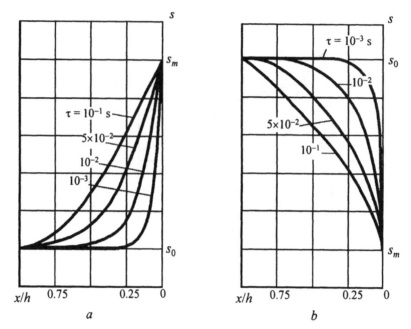

Figure 2.16 Distribution of concentration in a semi-finite body for various non-dimensional times: (*a*) absorption, (*b*) desorption.

$$\text{erf}(U) = \frac{2}{\sqrt{\pi}} \int_0^U e^{-U^2} du$$

is the Gauss error function.

Results of calculations using Eq. (2.94) for various ratios x/h for gas sorption ($s_m > s_0$) are shown in Fig. 2.16, *a*, and for gas evolution ($s_m < s_0$) in Fig. 2.16, *b*.

Since the gas flow through unit area toward vacuum is

$$q = -D\frac{ds}{dx}\bigg|_{x=0} = -\frac{(s_0 - s_m)D}{h}\frac{1}{\sqrt{\pi\tau}}, \qquad (2.95)$$

whence the specific amount of gas involved in mass exchange is

$$a = \int_0^t q\,dt = -2(s_0 - s_m)h\sqrt{\frac{\tau}{\pi}}, \qquad (2.96)$$

the mean gas concentration in a body being

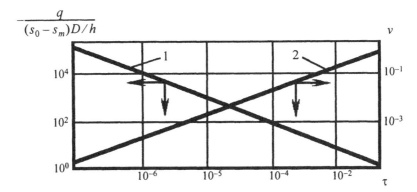

Figure 2.17 (*1*) dimensionless flow at gas-solid interface and (*2*) mass exchange efficiency as functions of dimensionless time.

$$s_a = s_0 - 2(s_0 - s_m)\sqrt{\frac{\tau}{\pi}}, \tag{2.97}$$

and the degree of mass exchange, i.e., the amount of gas involved in the exchange related to the total amount of gas in the body before the exchange, being

$$\sigma = \frac{|a|}{s_0 h} = 2\left(1 - \frac{s_m}{s_0}\right)\sqrt{\frac{\tau}{\pi}}. \tag{2.98}$$

The degree of mass exchange determines the absolute degree of gas depletion or gas saturation of a body and can be used as a criterion of a technological means for vacuum treatment of materials.

A good criterion for choosing the time of a technological process involving gas evolution or sorption is the efficiency of mass exchange

$$\nu = \frac{\sigma}{(s_0 - s_m)/s_0} = 2\sqrt{\frac{\tau}{\pi}}, \tag{2.99}$$

which is the degree of mass exchange achieved in a process related to the highest possible one.

Figure 2.17 shows specific dimensionless flow and mass exchange efficiency as functions of dimensionless time.

The above formulas are valid for solids of any shape at $\tau \leq 0.1$; hence the above limitation on diffusion coefficient. The body can be considered infinite in the *x*-direction. The maximum mass exchange efficiency at $\tau \leq 0.1$ is 36%.

If higher mass exchange degree should be obtained or $\tau > 0.1$, one should take into account the shape of bodies that evolve or sorb gas. Then new boundary conditions should be imposed onto Eq. (2.93). For typical shapes (sphere, plate, and cylinder) these conditions can be borrowed from literature.

Diffusion gas evolution greatly depends on the type of material and its preliminary treatment. One constructional means of avoiding diffusion gas evolution is usage of materials that were preliminarily degasing at high temperature (above 1000°C) in vacuum or hydrogen annealing ovens. Materials that do not withstand high-temperature treatment can be cooled during operation.

The above variant of gas evolution or sorption deals with inner parts of a vacuum chamber. However, chamber walls may not only desorb dissolved gases, but also allow gas to penetrate due to the difference of the pressures in the separated spaces.

We consider non-steady-state gas permeation through a vacuum chamber wall in the form of an infinite plate with asymmetrical boundary conditions. We also assume that the concentration of gas in the plate is s_0, the gas pressure from the vacuum is p_0, and the concentration and pressure of gas in the atmosphere are s_a and p_a, respectively. The initial concentration of gas in the plate is s_0. The mathematical interpretation of the problem is

$$\frac{\partial s}{\partial t} = D\frac{\partial^2 s}{\partial x^2} \tag{2.100}$$

with the boundary conditions

$$t = 0; \quad s(x, 0) = s_0; \qquad\qquad x = 0; \quad s(0, t) = s_0;$$

$$t = \infty; \quad s(x, \infty) = (s_a - s_0)\frac{x}{2h} + s_0; \quad x = 2h; \quad s(2h, t) = s_a.$$

The solution can be written as a function of dimensionless time $\tau = Dt/h^2$:

$$s(x, t) = s_0 + \frac{s_a - s_0}{2h} + \frac{2(s_a - s_0)}{\pi}\sum_{k=1}^{\infty}\frac{(-1)^k}{k}\sin\left(\frac{k\pi x}{2h}\right)\exp\left(-\frac{\pi^2 k^2\tau}{4}\right). \tag{2.101}$$

The rate of gas permeation at time τ is then

$$q = -D\frac{\partial s(0, t)}{\partial x} = -\frac{D(s_a - s_0)}{2h} - \frac{2(s_a - s_0)D}{\pi}\sum_{k=1}^{\infty}(-1)^k\exp\left(-\frac{\pi^2 k^2\tau}{4}\right). \tag{2.102}$$

At $t \to \infty$, Eq. (2.102) transforms to a steady-state dependence such as Eq. (1.67) The total amount of gas penetrating into the chamber in time t is

$$G = -\int_0^t q\,dt = -\frac{(s_a - s_0)\tau h}{2} + \frac{(s_a - s_0)h}{3} + \frac{4(s_a - s_0)h}{\pi^2}\sum_{k=1}^{\infty}\frac{(-1)^k}{k^2}\exp\left(-\frac{\pi^2 k^2\tau}{4}\right). \tag{2.103}$$

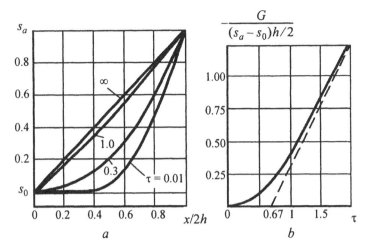

Figure 2.18 Gas permeation of a vacuum chamber wall: (*a*) gas concentration distribution and (*b*) amount of evolved gas.

Figure 2.18, *a* shows the distribution of gas concentration in a vacuum chamber wall, and Fig. 2.18, *b* shows amount of gas evolved from the wall into the vacuum in time τ.

At $\tau > 0.67$, Eq. (2.102) can be simplified to

$$q = \frac{(s_a - s_0)D}{2ht}\left[t - \frac{(2h)^2}{6D}\right].\tag{2.104}$$

This latter simplified formula allows estimation of the amount of gas evolved into the vacuum:

$$G = qt = \frac{(s_a - s_0)D}{2h}\left[t - \frac{(2h)^2}{6D}\right].\tag{2.105}$$

results of calculations using Eq. (2.105) are shown by dashes in Fig. 2.18, *b*. At $\tau \leq 0.67$, the penetrability of a gas is assumed to be zero, the delay time being

$$t_d = \frac{\sqrt{0.67}h}{\sqrt{D}} = 0.82\frac{h}{\sqrt{D}}.\tag{2.106}$$

2.8   Basic Equation

Figure 2.19 shows schematic of the simplest vacuum chamber consisting of pumped space 1, the manometers 2 and 3, pump 4, and connecting pipe 5. Gas flow from the pumped space to the pump is due to the difference of pressures $p_0 - p_1$ ($p_0 > p_1$).

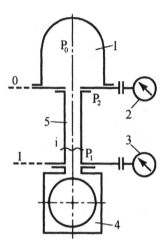

Figure 2.19 Schematic of the simplest vacuum system.

The pumping speed of pump S_i for an arbitrary section of the connecting pipe can be determined as the volume of gas passing through this section in unit time:

$$S_i = \frac{dV_i}{dt}.$$

The volume of gas passing in unit time from the pumped space to the pipe through the section 0 at pressure p_0 is referred to as the pumping speed for the chamber, or the effective pumping speed of the pump:

$$S_e = \frac{dV_2}{dt}.$$

The volume of gas removed by the pump in unit time through the intake pipe section (section I) at pressure p_1 is referred to as the proper pumping speed of the pump:

$$S_p = \frac{dV_1}{dt}.$$

The ratio of the pumping speed for the chamber to the proper pumping speed of the pump is the pump utilization coefficient:

$$K_e = \frac{S_e}{S_p}. \tag{2.107}$$

The flow of gas through a pipe section or throughput for this section:

$$Q = p_i S_i.$$

If the gas flow remains the same for any pipe section, the mass balance equation $dQ/dx = 0$ is the flow continuity equation. The flow is the same in any section of the vacuum system, the source of gas being in the pumped space (this is typical of vacuum systems with concentrated parameters)

$$Q = p_1 S_p = p_0 S_e. \qquad (2.108)$$

For a particular portion of a vacuum system consisting of a set of pipes, ventages, trap, etc. one may write the formula (Eq. (2.35))

$$Q = U_0(p_0 - p_1), \qquad (2.109)$$

where U_0 is the conductance of the portion considered and p_0 and p_1 are the pressures at the ends of the portion.

The conductance of a system is the coefficient relating the flow to pressure difference. Numerically, it is equal to the amount of gas passing through a portion of a vacuum system at unit pressure difference at the portion ends. The resistance of a vacuum system is the quantity inverse to its conductance:

$$Z = \frac{1}{U_0}.$$

Let us find the relation between the three basic parameters of a vacuum system, i.e., the proper pumping speed of the pump S_p, the effective pumping speed S_e, and the conductance of the system between the pump and the chamber U_0. Using Eqs. (2.108) and (2.109), one may write

$$S_p = \frac{Q}{p_1} = U_0 \frac{p_0 - p_1}{p_1};$$

$$S_e = \frac{Q}{p_0} = U_0 \frac{p_0 - p_1}{p_0}.$$

Rewriting these formulas as

$$\frac{1}{S_e} = \frac{p_0}{(p_0 - p_1)U_0}, \quad \frac{1}{S_p} = \frac{p_1}{(p_0 - p_1)U_0}$$

and subtracting the second equation from the first one, we obtain

$$\frac{1}{S_e} - \frac{1}{S_p} = \frac{1}{U_0}. \qquad (2.110)$$

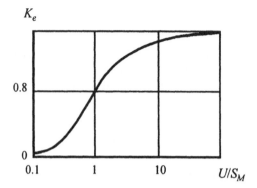

Figure 2.20 Pump utilization coefficient as a function of ratio of conductance to proper pump speed.

Equation (2.110) is referred to as the basic equation of vacuum engineering. It may also be written as

$$S_e = \frac{S_p U_0}{S_p + U_0}.$$

Given $S_p = U_0$, then $S_e = 0.5S_p$. If $U_0 \to \infty$, then $S_e \to S_p$; at $U_0 \to 0$, $S_e \to 0$.

Introducing the pump utilization coefficient K_e as in Eq. (2.107) into the main equation, we obtain two useful relations:

$$K_e = \frac{U_0}{S_p + U_0};\tag{2.111}$$

$$U = \frac{S_p K_e}{1 - K_e}.\tag{2.112}$$

Graphical interpretation of Eq. (2.111) is given in Fig. 2.20. The maximum value of the pump utilization coefficient is unity.

2.9   Test Questions

2.1   Why are the dynamical viscosity and heat conductivity coefficients of gases do not depend on pressure at low vacuum?

2.2   How does the conductance of pipes depend on pressure in viscous flow mode? What conductance of a pipe should be accepted as a first approximation in vacuum calculations?

2.3   Which heat transfer mechanisms are the most effective in high vacuum?

2.4   How does transport phenomena depend on degrees of vacuum?

2.5   How thermal equilibrium of pressures depend on degrees of vacuum?

2.6   What is the difference between degrees of vacuum?

2.7   Explain the physical sence of Knudsen criterium.

2.8   How to determine an effective demension of vacuum chamber?

2.9   Give the determination of ultrahigh vacuum.

2.10 How does conductance of pipes depend on diameter, length and pressure in the molecular flow mode?

Chapter 3

Measurement of Vacuum

3.1   Classification of Measurement Methods

The range of pressures used in modern vacuum engineering is very wide (10^5 to 10^{-12}) Pa. The measurement of pressure in this range cannot be provided by one device. In practice, the pressure of rarefied gases is usually measured by various manometric converters that are distinguished by the principle of action and the class of accuracy.

Devices for the measurement of pressure are referred to in vacuum engineering as vacuum gauges or vacuummeters. They usually consist of two parts: a manometric transducer and a measuring device. By the method of measurement vacuum gauges can be subdivided into absolute and relative. The indication of absolute devices does not depend on the type of gas and can be determined in advance. Devices for relative measurements use the dependence of the parameters of some physical processes that occur in vacuum on pressure. They require reference devices. Vacuum gauges measure total pressure of gases that are present in a vacuum system. Figure 3.1 shows the ranges of working pressure for various types of vacuum gauges.

Measurers of partial pressure, as well as those of total pressure, are characterized by the lower and upper measured pressures, sensitivity, as well as their unique parameter, resolution which determines working mass range. The resolution is the ratio of molecular weight of gas M to its least detected change ΔM,

$$\rho_M = \frac{M}{\Delta M}.$$

Depending on the type of device, the values of ρ_M, M/ρ_M, $M\rho_M$ can remain constant in the entire range of measurement. Experimentally, resolution is determined from a mass spectrum. The width of a peak is measured at 10 or 50 % of peak height. Measurement of partial pressure in vacuum systems is performed using two methods: ionisation and sorption.

Ionization is the method based on ionization and separation of positive ions according to the ratio of ion mass to its charge. One may simultaneously or sequentially measure components of ion current corresponding to partial pressures of various gases present in a vacuum system.

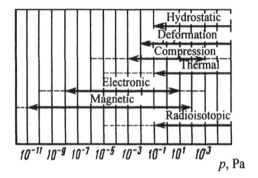

Figure 3.1 Working pressure ranges of vacuum gauges.

Ion current is separated into components using difference in the velocities of ions of various gases accelerated by difference of potentials and hence having similar energy $mv^2/2 = Uq$, whence the expression for the velocity of an ion is

$$v = \sqrt{\frac{2Uq}{m}}. \tag{3.1}$$

The velocity of an ion is determined by the m/q ratio. In most cases, during ionization of gases, slow electrons form positive single-charge ions, and one may therefore consider, with an accuracy that is acceptable for most measurements, that the velocity of ions in an electrical field is inversely proportional to the square root of molecular weight of the gas.

The $M_e = M/n_q$ ratio is referred to as the mass number of an ion. Here M is the molecular mass, expressed in atomic units (a.u.); n_q is the number of elementary charges per ion; 1 a.u. is 1/16 of the mass of the main ^{16}O oxygen isotope. For single-charged ions the mass number coincides with the molecular mass. For example, for single-charged ion $CO_2 + M_e$ = 44 a.u.. As a result of measurements of ion currents that correspond to various mass numbers, a mass spectrum (Fig. 3.2) is obtained.

The sensitivity of ionization gas analyzers is determined as the ratio of change in ion current in the collector circuit to the change in partial pressure of a gas and is expressed in A/Pa and depends on the type of gas.

The upper measured pressure is determined by the deviation from a linear dependence of ion current and a corresponding partial pressure due to the scattering of ions in the analyzer. The largest allowable deviation is 10%. The maximum working pressures do not usually exceed 10^{-3} to 10^{-2} Pa. At higher pressures analysis of gases requires expanding systems or an auxiliary vacuum system to reduce the density of the analyzed gas mixture without changing the percentage of its components.

The lower level of measured partial pressures is the minimum absolute pressure measured by a device. The relative partial pressure of a gas that can be measured at the given signal-to-noise ratio is referred to as the sensitivity threshold. A signal twice as high as the noise level can be accepted as a true one.

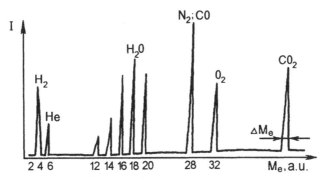

Figure 3.2 Mass-spectrum of residual gases in a vacuum system.

Depending on the character of electrical and magnetic fields used, the ionization methods of partial pressure measurement can be subdivided into static and dynamic. Static methods use permanent electromagnetic fields, and dynamic methods use variable fields.

Static gas analyzers include magnetic and panoramic ones, and dynamic analyzers include omegatron, time-of-flight, quadruple, and monopole mass filters. Most common are magnetic and quadruple spectrometers. The resolution of industrial devices is within 40–200 a.u.

The sorption method of partial pressure measurement uses the analysis of adsorbed gases. In one of its variants, the desorption method, because of the various adsorption heats of residual gases, heating of a surface following a certain temporal law is accompanied by consecutive desorption of the components of the gas mixture. A disadvantage of the method is the impossibility of detecting poorly adsorbed gases He, Ne, and H_2.

Another variant of the sorption method is Auger-spectroscopy, which is a method of analysis of solid surfaces from the characteristic energy of electrons emitted due to electronic transitions between energy levels in atoms. The surface under test is subjected to ionizing radiation, usually primary electron beams with energies 3–5 times the ionization potential of an appropriate level. The yield depth of Auger-electrons is, on the average, about 1 nm, i.e., the information obtain using this method refers to the surface of the test specimen. The composition of the surface allows determination of the composition of well-adsorbed substances in the gaseous phase. This method has found wide application for the test of the elemental structure of surfaces, evaluation of the efficiency of various methods of substrate cleaning, the processes of diffusion and migration, epitaxy, adsorption, and desorption.

Gas flow is the mass of gas passing in unit time through a given cross-section of a vacuum system element. In the International System the unit of gas flow is kg/s. Flows of individual substances can also be measured from the number of gas molecules passing through a given section of a vacuum system element in unit time.

At a constant temperature, a practical unit of gas flow m^3Pa/s is frequently used. For air at 273 K, 1 $m^3 Pa/s = 1.3 \times 10^{-5}$ kg/s. The steady-state throughput of a gas expressed in m^3 Pa/s, can be written in the form

$$Q = U(p_1 - p_2),$$

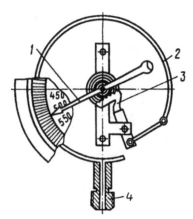

Figure 3.3 Deformation pipe converter (Bourdon tube).

where U is the conductance of a vacuum system. This equation can be used for the measurement of steady-state throughput using the method of two manometers on the basis of difference in pressures at a vacuum element of known conductance.

The expression for the determination of both steady-state and non-steady-state gas flows can be written in differential form:

$$Q = \frac{d(pV)}{dt} = p\frac{dV}{dt} + V\frac{dp}{dt}. \qquad (3.2)$$

According to Eq. (3.2), gas flows can be measured using two methods: the method of constant pressure and the method of constant volume. For the method of constant pressure, flow is measured from the rate of change in gas volume. For the method of constant volume, a flow is measured from the rate of change in pressure.

The above methods of gas flow measurement are considered as absolute. Indirect methods, i.e., thermal, radioisotopic, and ionization, are calibrated using absolute methods.

3.2   Mechanical Methods

The mechanical methods include the deformation, hydrostatic, and viscosity methods. The deformation method is based on the measurement of deformation of membrane elements due to a difference in pressure. This deformation can be measured using lever or tooth gears, change in the capacity of a capacitor, or the inductance of a coil. At small deformations the convenient for measurements linearity between pressure and deformation is retained. For the zero method of measurement, the deformation of the working element is equal to zero due to the compensation of the measured pressure by the pressure of other gases or by the electrostatic force. The possibility of measuring deformation with the desired accuracy makes these manometers absolute. The sensitivity of the method is increased with a decrease in the rigidity of the working element. It is limited by the mechanical strength and the presence of plastic deformations.

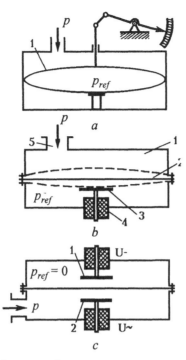

Figure 3.4 Deformation membrane transducer.

The upper measurement limit can be equal to atmospheric pressure. The lower limit is determined by temperature deformations and the elastic aftereffect of the element. The elastic aftereffect is the delay in the restoration of a deformed element to the initial condition after removal of the load.

The Bourdon tube (Fig. 3.3) is a deformation manometer in the form of a spiral tube 2, rolling under the action of atmospheric air during pumping of an inner volume, due to the difference in the curvature radii, and, hence, the areas of the outer and inner tube surfaces.

We calculate the forces F_1 and F_2 that act onto the surfaces with the larger and the smaller radii, respectively:

$$F_1 = (p_a - p)A_1 \; ; \quad F_2 = (p_a - p)A_2 \; ,$$

where p_a is the atmospheric pressure, A_1 and A_2 are the areas of the outer and the inner surfaces of a part of the spiral tube. The measuring equation connects the movement of the tube end x and the difference in pressure $p_a - p$

$$\Delta F = F_1 - F_2 = (p_a - p)(A_1 - A_2) = cx \; ,$$

where c is the rigidity of the tube.

Manometer measures pressure within 10^5–10^3 Pa. The measurement of pressure below 10^3 Pa is hindered by the fact that a tube with a small rigidity should be strong enough to

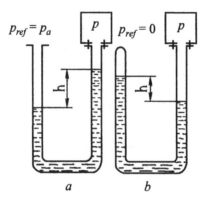

Figure 3.5 Liquid manometers: (*a*) opened-arm and (*b*) closed-arm.

withstand atmospheric pressure. The error of the measurement is 5% and is limited by the elastic aftereffect of the tube. The manometer is connected to a vacuum system through union 4. The movement of the spiral tube is usually registered by a lever-and-arrow mechanism, in which the end of the spiral tube is connected through gear sector 3 to arrow 1.

Deformation membrane transducers are distinguished by the method of membrane movement registration and pressure measurement. In barometers, the circuit shown in Fig. 3.4, *a* is used. Inside the hermetic volume formed by membranes 1, reference pressure p_{ref} is produced. If the measured pressure is not equal to p_{ref}, the membrane is deformed, and the arrow moves proportionally to the difference in pressure such an instrument measures difference in pressure and is therefore referred to as differential.

The same principle is used in the membrane converter (Fig. 3.4, *b*), but there the capacitor method is applied. The membrane 2 in this device tightly divides the casing 1 into two chambers, in one of which the reference pressure p_{ref} is produced, and the other is connected through the branch pipe 5 to a vacuum system. Electrode 3 is connected to the bottom of the chamber through insulator 4 and forms, in combination with the membrane, a pressure-controlled capacitor. This converter can measure absolute pressure in a vacuum system, provided $p_{ref} = 0$.

Membrane converter with two electrodes (Fig. 3.4, *c*) works at zero membrane deformation. Electrode 2 is fed with alternating voltage, allowing one to determine the capacity and position of the membrane. To electrode 1, direct voltage is applied, which, at the expense of electrostatic force, brings the membrane to the initial position, and thus compensates the effect of the difference in pressure. The difference in pressure in this device is proportional to the direct voltage squared applied to electrode 1.

The measurement range of membrane transducers is 10^5–10^{-1} Pa. However, because the linearity of the indication is retained only at moderate membrane deformations, one device can measure pressure within the limits of 2–3 orders of magnitude.

For the hydrostatic method the measured difference in pressure is compensated by the weight of a column of liquid with height h:

$$(p_{ref} - p) = g\rho h.$$

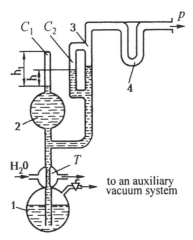

Figure 3.6 Compression manometer.

where g is the gravity acceleration and ρ is the density of a liquid.

As a working liquid for filling considered manometers, mercury and oil can be used. Oil manometers have high sensitivity, because the density of oil is about 15 times lower than that of mercury. However, oil dissolves gases well, and its careful degasing is required before work.

The measurement limits of mercury manometers is 10^5–10^3 Pa, and for oil ones, it is 10^4–10^0 Pa. The error of counting the level h can be to 0,1 mm. More accurate measurement of this level is useless because of the inconstancy of surface tension, density fluctuations, temperature gradients, etc.. The sensitivity of a manometer to difference in pressure is basically limited to the viscosity of the working liquid.

Opened-arm manometers (Fig. 3.5, a) are convenient for the measurement of pressure close to the atmospheric one. In this case, $p_{ref} = p_a$ and the height h is minimal. The indication of these manometers depends on atmospheric pressure. In closed-arm manometers (Fig. 3.5, b), pressure $p_{ref} = 0$ is produced before filling with a working liquid, this allowing to directly measure the absolute pressure of a gas in a vacuum system. In this case the indication of a device does not depend on atmospheric pressure. Closed-arm manometers for pressures lower than 2×10^4 Pa have smaller size than opened-arm manometers for the same working pressures.

Hydrostatic manometers with preliminary gas compression are referred to as compression manometers. Compression manometer, known as a McLeod gauge, (Fig. 3.6) consists of a measuring cylinder 2 with capillary C_1, a vessel 1 with mercury, and a connecting pipeline 3 with capillary C_2. Through a nitrogen trap 4 the manometer is connected to a vacuum system. Before the beginning of measurements the cylinder 2 is connected with a vacuum system through pipe 3. From the cylinder 1, under the pressure of atmospheric air, mercury goes upwards in pipe T, disconnects the cylinder 2 from the vacuum system, and compresses the gas in the cylinder to pressure that can be directly measured from the difference of the mercury levels in the closed and the reference capillaries C_1 and C_2. After compression, the pressure is measured in exactly the same manner as and in usual closed-arm mercury manometers.

The equation of compression manometers on the basis of the Boyle–Mariotte law has the following form:

$$pV_0 = (p + \rho gh)V, \tag{3.3}$$

where V_0 is the initial volume of the compressed gas, h is the difference of levels in the refrence and closed capillaries, $V = \pi d_c^2 h_1/4$ is the final volume of the gas after compression, and d_c is the capillary diameter. Solving Eq. (3.3) relative to pressure p, we have

$$p = \frac{\pi \rho g d_c^2 h h_1}{4(V_0 - \pi d_c^2 h_1/4)}.$$

Provided that $\pi h_1 d_c^2/4$ is far smaller than V_0,

$$p = \rho g \frac{\pi d_c^2}{4V_0} h h_1. \tag{3.4}$$

If the mercury in the closed capillary of the manometer is always raised to the same level, then h_1 is constant and Eq. (3.4) can be written as $p = C_1 h$; $C_1 = \rho g \pi d_c^2/(4V_0)$. This method of pressure measurement is referred to as the method of linear scale.

To expand the measurement limits, one may use the method of square-law scale, for which the compression in the manometer is made in such a way that the mercury in the reference capillary C_2 be always established at the same level with the sealed end of the closed capillary C_1. Then $h = h_1$ and Eq. (3.4) can be written in the form $p = C_1 h^2$; $C_2 = \rho g \pi d_c^2/(4V_0)$

The measurement range of compression manometers is 10^1–10^{-3} Pa. The difficulties in measurement of lower pressures are connected with the inconstancy of capillary depression of mercury (decrease of the level of mercury in the capillary in comparison with the mercury level in a broad vessel connected with the capillary), the pumping action of the mercury vapor flow from the manometer to the trap, and by the difference in the shapes of the sealed capillary end and the mercury meniscus, which fact limits the minimum h_1 to 5–10 mm. Moreover, for technological reasons, the capillary diameter should not be smaller than 1 mm, and the volume of the measuring volume is determined by the strength of glass and is usually not greater than 1 l, allowing a maximum compression of 2.5×10^5.

For measurement of higher pressures one should use manometers with very long (or variable section) capillaries. At pressures of higher than 10 Pa, one may use usual hydrostatic manometers without preliminary gas compression. To reduce the pumping action of the mercury vapor flow, the pipe T is made in the form of a water-cooled capillary.

Compression manometers are absolute devices and are used as a reference for the calibration of other devices because their indication does not depend on the sort of gas. However, compression manometers cannot measure pressures of those substances whose saturation pressure at the measurement temperature is less than the pressure in the measuring capillary after compression; it is also impossible to measure pressure continuously. Another disadvantage of these manometers is that they should be connected to vacuum systems through nitrogen traps.

3.3   Thermal Methods

The operation principle of thermal transducers is based on the pressure dependence of heat conductivity through rarefied gases. Heat is transferred from a thin metal filament to a volume located at room temperature. The metal filament is heated in vacuum by an electrical current. The thermal balance equation of the device can be presented in the following form:

$$I_w^2 R = E_c + E_{hg} + E_r + E_{hm},$$ (3.5)

where I_w is the current through the filament, R is the resistance of the filament, and E_c, E_{hg}, E_r, and E_{hm} are the losses of heat due to convection, heat conductivity of the gas, radiation from the filament, and heat conductivity of the filament material, respectively. Convective heat exchange in medium and high vacuum can be neglected, i.e., $E_c \approx 0$, and the loss of heat by radiation

$$E_r = K_r(T_w^4 - T_v^4)A,$$ (3.6)

where A is the surface area of the filament, K_r is the radiation coefficient of the filament material; and T_w and T_v are the temperatures of the filament and the gas envelope.

Thermal losses on the materials of the filament and the electrodes connecting the filament with the casing of the transducer are

$$E_{hm} = b(T_w - T_v)f,$$

where b is the heat conductivuty of the filament material and f is the cross-section of the filament. Heat losses through the gas gap are

$$E_{hg} = \alpha(T_w - T_v)A,$$

$\alpha = \lambda/d$, where λ is the heat conductivity of the gas, d is the gas cap.

At low vacuum, the heat conductivity of gases does not depend on pressure. The pressure corresponding to the transition to low vacuum is the upper measurement limit of thermal manometers, because at high vacuum heat conductivity decreases proportionally to pressure ($\alpha = K_t p$). The equation of a thermal transducer with account of Eqs. (3.5), (3.6), and (2.24) can be written as

$$p = \frac{I_w^2 R - (E_c + E_r)}{K_t(T_w - T_v)A}.$$ (3.7)

Accurate measurement of pressure requires that E_{hg} be significant compared to $E_r + E_{hw}$, i.e., that the sum of these losses be substantially smaller than the heat loss of the filament. Therefore the condition

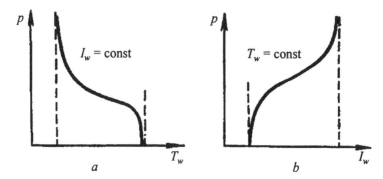

Figure 3.7 Calibration curves of a thermal transducer: (*a*) at constant heating current and (*b*) at constant filament temperature.

$$I_w^2 R - (E_{hm} + E_r) \geq 0.01 I_w^2 R$$

determines the lower measurement level of these manometers. From Eq. (3.7) it can be seen that pressure is a function of two variables: the heating current and the filament temperature.

There are two operation methods of thermal manometers, i.e., constant filament temperature and constant heating current. The calibration curves of thermal manometers that is shown on Fig. 3.7 *a, b* for both operation are parabolic and hyperbolic in the middle part, respectively.

The ends of the calibration curves at the upper and lower measurement limits are not described by Eq. 3.7 and transform to lines parallel to the axis of pressure.

To expand the measurement range of a thermal transducer one should reduce its size, thus increasing L/d and shifting the boundary of low vacuum to higher pressures. The pressure dependence of convective heat exchange coefficient is used in the measurements at low vacuum. Disadvantage of this method is the dependence of the device indication on its position.

The lower measurement limit of thermal transducers can be improved by reducing the role of $(E_r + E_{hw})$ in the sum of heat losses. This can be achieved by reducing filament temperature and the diameter of the contacts connecting the filament with the cylinder.

The indication of thermal converters is determined by the product $K_t p$. The transducer will indicate the same value with the following conditions:

$$K_{t1} p = K_{t2} p = \ldots = K_{ti} p = \ldots = K_{tn} p. \tag{3.8}$$

Commercial devices are calibrated to dry air. If it is necessary to measure the pressure of other gases, one should take into account the relative sensitivity of a device to a particular gas

$$p_i = p_a \frac{K_{ta}}{K_{ti}} = p_a q_i, \tag{3.9}$$

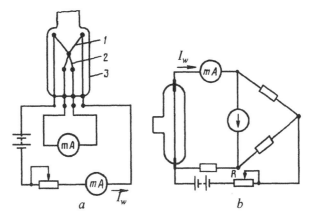

Figure 3.8 Schematics of thermal transducers: (*a*) thermocouple and (*b*) resistance.

where p_a and K_{ta} are the pressure and heat conductivity of air and $q_i = K_{ta}/K_{ti}$ is the relative sensitivity of a thermal transducer to a particular gas. q_i of various gases in relation to air can differ by several times (Table C.15).

If the transducer measures the pressure of a gas mixture, its indication will be expressed in the air equivalent p_a:

$$K_{t1}p_1 + K_{t2}p_2 + \ldots + K_{ti}p_i + \ldots + K_{tn}p_n = K_{ta}p_a. \qquad (3.10)$$

Because the definition of relative sensitivity (3.9) suggests that $p_a = p_{mix}/q_{mix}$, Eq. (3.10) can be written as

$$\frac{p_{mix}}{q_{mix}} = \frac{p_1}{q_1} + \frac{p_2}{q_2} + \ldots + \frac{p_n}{q_n}.$$

Dividing both sides of this equation by p_{mix}, we obtain

$$\frac{1}{q_{mix}} = \frac{V_1}{q_1} + \frac{V_2}{q_2} + \ldots + \frac{V_n}{q_n},$$

where $V_1, V_2, \ldots, V_n$ are the volume concentrations of appropriate gases, and $\sum_{i=1}^{n} V_i = 1$. Thus, the relative sensitivity for a gas mixture is determined by the formula

$$\frac{1}{q_{mix}} = \sum_{i=1}^{n} \frac{V_i}{q_i}. \qquad (3.11)$$

The thermal transducers are subdivided by the method of temperature measurement into thermocouple and resistance. In thermocouple transducers (Fig. 3.8, *a*) the temperature of filament 1 is measured by a thermocouple 2. Electrodes are located in a glass or metal cylinder 3 that has a branch pipe for connection to a vacuum system. Thermal EMF of the thermocouple is measured by a millivoltmeter, and the heating current of the filament is adjusted by a resister and is measured by a milliampermeter.

In a resistance transducer (Pirany), the temperature dependence of filament resistance is used. The transducer is directly connected to a circuit (Fig. 3.8, b). The filament heating current is measured by a milliampermeter connected to the same leg of the bridge as the transducer, and the filament temperature is measured from the current of the galvanometer in the measuring diagonal of the bridge. The heating current is adjusted by a resistor.

The transducers can work both in constant heating current mode and in constant filament temperature mode. An advantage of thermal transducers is that they measure the total pressure of the gases and vapors that are present in a vacuum system, and provide continuous pressure measurement. The lag of the indication associatedted with the thermal lag of the filament ranges from several seconds at low pressures to several milliseconds at high pressures.

As devices for relative pressure measurements, thermal converters are usually calibrated by compression manometers. The working pressure range of conventional transducer is $5 \times 10^3 - 10^{-1}$ Pa. Transducer made with the use of micromechanics and tensoresistors located on a thin silicon membrane have the best lower level of measured pressure (10^{-3} Pa). Parameters of thermal vacuum gauges are shown in Table A.1.

3.4   Electrical Methods of Total Pressure Measurement

Electrical methods for the measurement of low total pressures are based on the laws of electrical phenomena in vacuum. The most widely used is the ionization method that uses the proportionality between the pressure and discharge current between electrodes, to which a difference of potentials is applied. The lower measurement level of these transducers is usually restricted by the background current that has the same direction with the measured current. The main origin of background currents is secondary electronic emission from the ion collector. The upper measurement limit is determined by the allowable deviation from the linearity of the calibration characteristic, i.e., the dependence of discharge current on pressure.

Several types of ionization transducers exist: electronic, magnetic, and radioisotopic. In electronic transducers, gas is ionized by thermal electrons emitted by a heated cathode. In magnetic transducers ionization is effected by electrons obtaind by autoelectronic emission from a cold cathode; in radioisotopic transducers α or β radiation is used for the same purposes. The sensitivity of ionization transducers, as well as the efficiency of ionization, depend on the type of gas.

If a transducer is calibrated to air and is used for the measurement of pressure of other gases, it is necessary to take into account the relative sensitivity R. From the condition of the equality of ion currents at measurement of pressure of various gases, we obtain

$$K_1 p_1 = K_2 p_2 = \ldots = K_i p_i = \ldots = K_a p_a ,$$

Table 3.1 Relative sensitivity of ionization transducer.

	Type		
	Electronic	Magnetic	Radioisotopic
N_2	1.00	1.00	1.00
H_2	0.43	0.43	0.23
He	0.16	0.15	0.21
Ar	1.30	1.40	1.20
CO_2	1.60	1.30	1.50
O_2	0.85	0.85	–
Acetone	–	–	2.70
Hg	2.50	–	–
CH_4	1.50	–	–
Xe	2.8	3.50	–
Ne	0.27	0.26	–
CO	1.04	–	–
H_2O	–	–	0.88

whence $p_i = p_d/R_i$, where $R_i = K_i/K_a$ is the relative sensitivity. Relative   sensitivities of ionization transducers are given in Table 3.1.

For the measurement of the pressure of a gas mixture from the condition of the equality of ion currents, we have

$$K_{mix}p_{mix} = K_1p_1 + K_2p_2 + \dots + K_ip_i + \dots + K_np_n .$$

Dividing this latter equation by K_a, we   obtain

$$\frac{K_{mix}p_{mix}}{K_a} = R_1p_1 + R_2p_2 + \dots + R_ip_i + \dots + R_np_n ,$$

whence

$$R_{mix} = \sum_{i=1}^{n} R_i V_i , \text{ where } V_i = \frac{p_i}{p_{mix}} .$$

The action principle of electronic transducers is based on the proportionality between the pressure and ion current that is produced as a result of the thermal electrons ionization of residual gases. Two basic circuits of electronic transducers exist, i.e., with internal and outer collector.

Internal collector circuit (Fig. 3.9, *a*) is similar to a usual triode. The ion collector is a grid to which a negative voltage of several tens of volts relative to the cathode is applied, and to the anode, a positive voltage of 100–200 V. Electrons moving from the cathode to

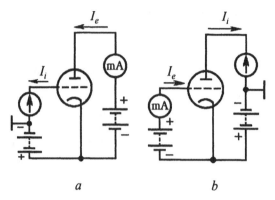

a   *b*

Figure 3.9 Schematics of electronic transducers: (*a*) with internal collector and (*b*) with external collector.

the anode (current I_e) collide with molecules of residual gases, and the resultant positive ions fall onto the grid to produce the ion current I_i that is measured by a galvanometer.

In the external collector circuit (Fig. 3.9, *b*) the potentials of the grid and the anode are in the inverse relation, and the collector becomes an anode. Electrons moving from the cathode to the grid orbit around its wires, thus increasing the length of their trajectory and hence the probability of ionization of residual gas molecules. This makes the external collector circuit more sensitive, notwithstanding that part of positive ions formed between the grid and the cathode do not participate in pressure measurement.

We now consider the equation of electronic transducer

$$dN = np\varepsilon dr \tag{3.12}$$

where dN is the number of positive ions, n is the number of electrons, dr is the elementary length of an electron trajectory, ε is the ionization efficiency, which is equal to the number of positive ions formed by one electron in unit path at unit pressure (Fig. 1. 18).

Introducing into Eq. (3.12) the electronic current $I_e = n/t$, then

$$\frac{dN}{t} = I_e p\varepsilon dr. \tag{3.13}$$

Integrating (3.13) over the length of the trajectory of an electron with energy greater than the ionization potential, we obtain the expression of ion current

$$I_i = I_e p \int_{r_1}^{r_2} \varepsilon dr,$$

which will be rewritten as

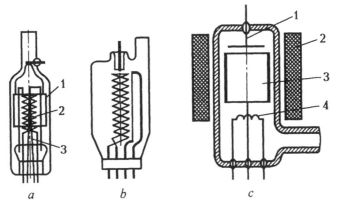

Figure 3.10 Constructive schematics of electronic transducers: (*a*) with external collector, (*b*) with axial collector, and (*c*) with magnetic field.

$$I_i = K_i I_e p, \qquad (3.14)$$

where K_i is the specific sensitivity of the electronic transducer and

$$K_i = \int_{r_1}^{r_2} \varepsilon \, dr,$$

Eq. (3.14), which is referred to as the equation of electronic transducer, suggests that, in order for the measured ion current be proportional to pressure, a constant electronic current should be maintained during the measurements. Then the sensetivity of electronic transducer

$$K_e = I_e K_i.$$

is the slope tangent of the calibration curve.

Schematically, the design of electronic transducer with external collector is shown in Fig. 3.10, a. The ion collector 1 is in the form of a cylinder with electric contact on the top and grid 2 is a double spiral with two contacts for degasing by passing electrical current. The cathode 3 is tungsten. Sensitivity at an emission current of 5 mA for a typical transducer makes $\sim 10^{-3}$ A/Pa. The limit pressures that can be measured using this instrument are 1 to 10^{-5} Pa.

The upper measurement limit of typical electronic transducers is ~ 1 Pa and corresponds to violation of the linearity of the calibration characteristic, when the mean free path length of an electron in the volume of the device becomes smaller than the interelectrode distance. To expand the upper limit one may reduce the distance between the electrodes. In some devices, the upper measured pressure was increased in such a way to 10^2 Pa. To avoid cathode burning at such high pressures, it should be made from rare-earth metal oxides.

The lower measurement level is determined by background currents in the collector circuit that are due to the photoelectrons emitted by soft X-ray radiation of the anode grid and the UV radiation of the heated cathode. The X-ray radiation of the anode grid is a result of its electronic bombardment. Autoelectronic emission of the collector occurs at a potential difference of 200–300 V between the collector and the anode grid and gives additional contribution to the background current.

Background electronic currents have identical direction with the ion current and hence show the same effect on pressure measuring devices. The maximum background current is the current of X-ray radiation proportional to the emission current:

$$I_b = K_s I_e,$$

where K_s is the proportionality coefficient.

With account of the existence of background currents the equation of electronic transducer can be written in the following form:

$$I = I_i + I_b = I_e(K_s + K_i p),$$

and the lower level of measurement can be determined by the ratio of ion and background currents:

$$\frac{I_i}{I_b} = \frac{K_i p}{K_s}$$

Thus, to shift the lower measurement level, it is necessary to increase K_i or to reduce K_s.

For reduction of background currents and, hence, the factor K_s, the axial collector transducer was suggested (Fig. 3.10, *b*), in which the cathode and collector exchanged places, which solution has considerably reduced the solid angle at which the X-ray radiation of the grid is incident on the collector. This has resulted in a reduction of K_s approximately by 10^3 time in comparison with the design shown in Fig. 3.10, *a*, and shifted the lower level of measurement to 10^{-8} Pa.

The specific sensitivity K_i can be increased by placing the transducer in a magnetic field (Fig. 3.10, *c*). Electrons from the cathode in this case move in spirals. In an electronic transducer with a magnetic field produced by coil 2 and directed parallel to the axis of the anode 3, the cathode 4 thermionic, and the collector 1 is located in the top of the volume. Due to an increase in sensitivity, this transducer has the measurement level by 2–3 orders of magnitude lower than that shown in Fig. 3.10, *a*.

The pumping speed of electronic converters is 10^{-3}–10^{-1} l/s, which, in closed designs of transducers connected by pipes with small conductance, may result in appreciable measurement errors. The additional sources of error are connected with the chemical interaction of gases with the heated cathode and ion desorption due to electronic bombardment of the gases chemically absorbed by the anode. Characteristics of vacuum gauges with electronic transducers are given in Table A.1.

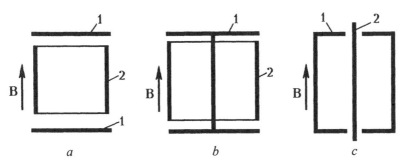

Figure 3.11 Electrode systems of magnetic transducers: (*a*) Penning cell, (*b*) magnetron, and (*c*) inverse-magnetron.

The action principle of magnetic transducers is based on the pressure dependence of the current of a spontaneous gas discharge in intersecting magnetic and electric fields. Electrode systems that ensure the maintenance of spontaneous gas discharge at high and super-high vacuum may have several designs.

The Penning cell (Fig. 3.11, *a*) consists of two disc-shaped cathodes 1 and a cylindrical anode 2; in the magnetron transducer (Fig. 3.11, *b*), unlike the Penning cell, the cathodes are connected by a central rod; the inverse-magnetic (Fig. 3.11, *c*) the central rod acts as an anode, and the outer cylinder becomes a cathode. All the electrodes are in a permanent magnetic field. A positive voltage 2–5 kV relative to the cathode is applied to the anode, the cathode is grounded and connected with the inlet of a direct current amplifier.

The electrons emitted from the cathode as a result of autoelectronic emission move in the intersecting electric and magnetic fields of magnetron or inverse-magnetron transducers by cycloids formed by a circle with the diameter

$$D = \frac{2me}{qB^2},$$

that rolls over a circle of radius *r* with the angular rotation frequency

$$\omega = \frac{qB}{m}$$

and the tangential speed

$$V_t = \frac{E}{B},$$

where E is the intensity of the electric field, B is the magnetic induction, and m and q are the mass and charge of the electron. In the Penning cell electrons move by spiral trajectories between the cathode plates.

The magnetic induction is chosen greater than the critical value that corresponds to the equality of electrode diameter and the diameter of circle over which electrons move, and is in modern devices ~ 0,1 Tl.

Colliding with a molecule of a residual gas, electron loses part of energy on its ionization and moves in the radial direction relative to the anode. Because the radial speed of electrons is considerably smaller than the tangential one, a negative volume charge may occur in the discharge gap at low pressures.

Positive ions formed as a result of collision molecules with electrons, move to the cathode. Since their mass is much greater than that of the electron, the magnetic field shows only a little effect on their trajectories. The collision of positive ions with the cathode produces secondary electrons, the current of which is proportional to the ion current.

Thus, the discharge current of the magnetic transducer is

$$I_d = I_b + I_i + I_s,$$

where I_b is the background current of autoelectronic emission, I_i is the ion current, and I_s is the current of secondary electronic emission.

The current of autoelectronic emission does not depend on pressure and can therefore be considered as the background current; the ion and secondary electron emission currents depend on pressure as follows:

$$I_i + I_s = ap^n,$$

where constant $a = 10^{-2}$–10^{-1} A/Pa and $n = 1$–1.4. Taking into account this dependence and neglecting the background current, we obtain the measuring equation of the magnetic transducer

$$I_i = K_i p,$$

where K_i is the sensitivity of the device; $K_i = ap^{n-1}$. The discharge current of a magnetic transducer depends on pressure in a nonlinear manner.

The upper measurement limit is connected with the restriction of the maximum discharge current by the ballast resistance that protects the device from arc discharge. To expand the upper measurement limit one should reduce the anode voltage and the size of the discharge gap. Usually, the upper measurement limit is 10–100 Pa.

The lower measurement limit is determined by the ignition time of the discharge and the background current. In modern devices it is 10^{-11} Pa. To reduce the background current, special screens 3 (Fig. 3.12) are located in the gap between the cathode 2 and the anode 1, where intensity of the electric field is the maximum. Large part of the background current in this case passes to the casing, and not to the device that measures the discharge current.

To ensure discharge ignition at low pressure, it is necessary to increase the anode voltage and the size of the discharge gap. To facilitate the ignition in ultrahigh vacuum, the screen plates are pinned with sharp needles that enhance autoelectronic emission. The most reliable way of obtaining fast ignition is to use heated elements, the switching of which results in a sharp increase of pressure or thermoelectronic emission.

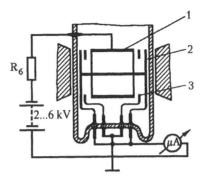

Figure 3.12 Schematic of magnetron transducer with reduced background currents.

In strong magnetic fields ($B > 0.1$ Tl) the constant n in the equation of device sensitivity tends to unity. The working range of a device both at low and high pressures is therefore expanded.

Magnetic transducers, as well as electronic ones, have unequal sensitivity to various gases. The relative sensitivity coefficients of magnetic transducers

$$R_g = \frac{K_g}{K_a}$$

for a number of gases are shown in Table 3.1.

Pumping speed changes for various transducers depending on the type of gas and operation modes from 10^{-2} to 1 l/s, which is considerably greater than for electronic transducers. This results in an increase of the error of measurements if between the transducer and the vacuum chamber has a resistance.

An advantage of magnetic transducers over electronic ones is the high safety in operation due to the replacement of heated cathode by cold one, and a disadvantage is the instability of work connected with fluctuations electron work function due to cathode pollution.

These instabilities are especially appreciable for operation in vacuum systems with oil vapors whose products of decomposition during ion bombardment, as well as and the dielectric film that cover electrode surfaces, may change the constant of the transducer by several times. Self-cleaning magnetic transducers that use alternating current are largely free of this defect. In such transducers the cathode and anode interchange positions pursuant to feeding voltage half-cycles, and their surfaces are cleaned by ion bombardment. Characteristics of some magnetic manometers are given in Table A.1.

In radioisotopic transducers gas is ionized by α- or β-radiation of radioactive isotopes, such as: ^{226}Ra, ^{239}Pu, ^{238}Pu, ^{3}H etc. It is especially effective to use α-radiation. The energy of α-particles (double-charged positive helium ions) that are produced during radioactive fission, is $(4.5–5.5) \times 10^6$ eV. Therefore, radioisotopic transducers do not require heated cathode and high voltage, unlike electronic and magnetic transducers.

The stability of device operation is provided by the independence of radioactive fission on ambient temperature and the physical and chemical effect of the gases in the transducer.

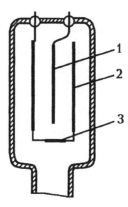

Figure 3.13 Radioisotopic transducer.

Due to these properties and unlimited service life, radioisotopic transducers can be considered to be the best devices for the measurement of vacuum from the metrological viewpoint.

At interaction with gas molecules α-particles cause their ionization, and the number of formed positive ions is proportional to pressure. The equation of the radioisotopic transducer is

$$I_i = K_g p,$$

we K_g is the sensitivity of the transducer, which is dependent on the type of gas. Transducer usually calibrated to air or nitrogen (for air $K_a = 10^{-10}$–10^{-11} A/Pa). For determination of the sensitivity to other gases it is possible to use the values of $R_g = K_g/K_a$ listed in Table 3.1.

The radioisotopic transducer (Fig. 3.13) consists of a rod-shaped collector 1 and cylindrical anode 2 and the isotope source 3. Emitted from the radioisotope source, α-particles collide with molecules of residual gases to form positive ions, which, under a difference of potentials between the anode and the collector (50 V) are directed to the collector and produce ion current that is proportional to pressure.

The lower working pressure of these devices is determined by the background current due to α-particle bombardment of the collector. This has two components, one of which is connected with the positive charge of the α-particles, and the other with the secondary electron emission current from the collector. The lower measurement limit of radioisotopic manometers is 10^{-1}–10^{-2} Pa. To extend the range of working pressures, in some transducers the collector is protected by screens or is located outside of the zone of direct flight of the α-particles. Thereby the lower level of measured pressures can be additionally reduced by two orders of magnitude. This can also be achieved by increasing transducer size, which is accompanied by an increase in the sensitivity without changing the background current.

The upper measured pressure is determined by the independence of ion collector current on gas pressure if the path of an α-particle with the initial energy E_α until complete loss of energy is smaller than the transducer size:

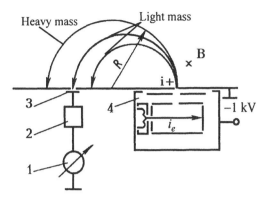

Figure 3.14 Static magnetic static gas analyzer.

$$L_\alpha = \frac{E_\alpha L_1}{E_i p}, \tag{3.15}$$

where E_i is the ionization energy of a gas molecule and L_1 is the mean free path length of the α-particle at unit pressure. At $E_\alpha \approx 4.79 \times 10^6$ eV (for radium), $E_i = 15$ eV, $L_1 = 5 \times 10^{-1}$ cm/Pa, and $p = 10^5$ Pa, we obtain from Eq. (3.15) $L \approx 1.6$ cm. Thus, the maximum path length of α-particles in a transducer with the maximum pressure equal to the atmospheric one should not exceed 1.6 cm.

The range of working pressures of radioisotopic transducers is 10^5–10^{-2} Pa, but this range can rarely be obtained in one device. Wide-pressure-range transducers have two chambers with large and small sizes. Characteristics of radioisotopic transducers are presented in Table A.1.

3.5 Electrical Methods of Measuring Partial Pressure

Static magnetic gas analyzers (mass-spectrometers, Fig. 3.14) are based on spatial separation of a monoenergetic ion beam in a perpendicular uniform magnetic field. The process of gas analysis includes ion formation, acceleration of the ions and the formation of an ion beam, separation of the ions by mass numbers, and the measurement of ion current intensity.

Ions are formed by electronic bombardment of neutral gas molecules in an ion source 4 at the expense of the emission current i_e. The ion source is fed by a negative potential relative to the ground (1 kV), which moves the ion beam i^+ into the drift space. The positive ions that passed the same accelerating voltage enter the drift space according to Eq. (3.1) with velocities $v = \sqrt{2Uq/m}$.

In the drift space, where a cross magnetic field with the induction B acts, positive ions move, according to the Lorentz force

$$F_i = qvB,$$

in the direction determined by the rule of the left hand, over circles of constant radii R. Thus, the Lorentz force is compensated by the centrifugal force

$$F_2 = \frac{mv^2}{R}.$$

From the condition of the equality of forces F_1 and F_2 we find the expression for the radius of ion trajectory

$$R = \frac{mv}{qB},$$

which, with account of Eq. (3.1) can be written in the form

$$R = \frac{1}{B}\sqrt{\frac{2Um}{q}}.$$

Thus, radius of an ion trajectory is directly proportional to the root square of the ion mass-to-charge ratio. Solving last Eq. relativel to m/q, we obtain the equation of static mass-spectrometer with a magnetic deviation:

$$\frac{m}{q} = \frac{R^2 B^2}{2U}. \tag{3.16}$$

As a result of the interaction with the magnetic field only those ions fall on the collector 3, the trajectory radius of which fits the diaphragm slot before the collector. Changing the trajectory radius of an ion R by changing the accelerating voltage U or the magnetic induction B, one may produce conditions for ions with various mass numbers to fall on the collector. The ion current of the collector after the amplifier 2 is measured by the output device 1. In the analyzer circuit considered, the ions deviate through 180 arc deg. In practice, deviation angles can be greater or smaller than 180 deg.

As follows from Eq. (3.16), magnetic separation of a mass-spectrum by magnetic induction B is the most effective, but, because of the higher speed, simplicity of the feeding circuit, and smaller sizes of the magnetic system, the preference in commercial devices is given to electrostatic separation by the accelerating voltage U.

The resolution of a mass-spectrometer depends on the size of the device and the expansion of the ion beam:

$$\rho_M = \frac{M_e}{\Delta M_e} = \frac{R}{s_1 + s_2 + \sigma(R)},$$

Table 3.2 The relative sensetivity coefficients.

	Analizer	
	Magnetic	Omegatron
N_2	1.00	1.00
H_2	0.46	0.45
He	0.17	0.21
Ar	1.30	1.20
CO_2	1.10	1.37
O_2	0.83	0.70
C_2H_2	–	1.40
CH_4	0.84	0.67
Ne	0.25	0.30
CO	1.08	1.10
H_2O	1.05	0.65
C_2H_4	0.59	1.04
C_2H_6	1.16	1.30

where s_1 and s_2 are the width of slots on the output from the ion and before the ion collector, respectively, $\sigma(R)$ is the total expansion of ion beam in the plane of the slot s_2 as a result of aberrations. The aberrations originate from the scatter of ions of similar mass by speeds, inhomogeneities in the distribution of magnetic field by the height of the ion beam, the effect of space charge, etc.. For mass numbers from 2 to 150, resolution at the half of peak height $\rho_M \geq 50$.

The sensitivity of a mass-spectrometer increases with slot width, which is usually chosen from 0,1 to 1 mm. For conventional devices the sensitivity K_i for measuring of nitrogen is 10 A/Pa. The relative sensitivity of magnetic analyzers

$$C_i = \frac{K_i}{K_{N_2}}$$

depends on the type of gas (Table 3.2). The maximum working pressure 10^{-2} Pa is determined by the constancy of the sensitivity coefficient, and the minimum pressure 10^{-9} Pa, by the background currents. The sensitivity threshold is 10^{-3} % and it depends on noise level of the electric amplifier or multiplier and the mass-spectrum background of the device.

The simplicity and reliability of magnetic analyzers permits one to apply them in systems for monitoring and management vacuum technological processes. Technological characteristics of the MSD-1 device, which was specially developed for the computer-assisted management of technological processes are listed in Table A.2.

Resonance gas analyzer (Farvitron) refers to static devices for the determination of partial pressures. The separation of ions by mass numbers is effected from the frequency of ion fluctuations in an electrostatic field with a parabolic distribution of the potential. The dependence of frequency of ion fluctuations on mass number in such a field can be written in the following form:

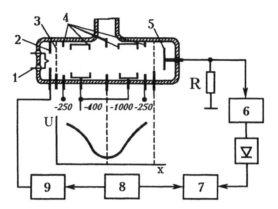

Figure 3.15 Resonance gas analyzer (Farvitron).

$$f = K \sqrt{\frac{qU}{m}}, \qquad\qquad (3.17)$$

where U is the voltage at the central electrode of the electrostatic field with a parabolic distribution of the potential, K is the proportionality coefficient determined by the geometry of the analyzer, m is the ion mass, and q is the ion charge.

The block diagram of a resonance gas analyzer is shown in Fig. 3.15. Electronic current between the cathode 1 and the grid anode 3 passes through the modulating diaphragm 2. The positive ions that are formed as a result of the electronic bombardment of neutral molecules of residual gases are accelerated by the anode in the direction of the electrostatic field with a parabolic distribution of the potential, which is formed by electrodes 4.

Modulating the electronic current by the fluctuation frequency of the ions with the given mass numbers according to (3.17), one may form an ion package, the number of ions in which increases continuously after each cycle of the ionizing electronic current.

The package of ions, the fluctuation frequency of which coincides with the frequency of the modulating voltage at the electrodes is referred to as the resonance one. This package produces voltage at the electrode 5, which is amplified by a high-frequency amplifier 6 and passes through a demodulator to the vertically deviating system a cathode oscillograph 7. The horizontal deviation in the oscillograph is provided by the time-base generator 8 at a frequency of 50 Hz, which synchronizes the horizontal sweep of the oscillograph with the time-base development of the high-frequency voltage at the generator 9, which is connected to the modulating electrode 2. The whole spectrum of gases can be simultaneously formed on an oscillograph.

To maintain a constant number of modulation cycles during the registration of a signal given by ions of any mass, the sweep of frequency follows the law

$$f = \frac{a}{t^b},$$

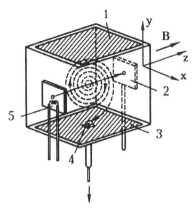

Figure 3.16 Cyclotron gas analyzer (Omegatron).

where a and b are constants and t is time. Selecting the exponential parameter b one may ensure definitive correlation between the output signals and the partial pressures of various gases.

The resolution of resonance analyzers (ρ_M = 15 –20, the range of mass numbers is 2–200, and range of working pressures is 10^{-3}–10^{-7} Pa. Characteristics of the APDP-2 gas analyzer are presented in Table A.2.

Cyclotrone gas analyzers (omegatrones) are dynamic analyzers of partial pressure. Their action principle is the movement of positive ions in mutually perpendicular constant magnetic and high-frequency electric fields. The emission current of thermal cathode 5 (Fig. 3.16) forms an electron beam on the z-axis of the device, which is directed parallel to the power lines of the magnetic field B to the anode 2. The positive ions formed in the electron beam, move over spiral trajectories in the intersecting constant magnetic and high-frequency electric fields between the plates 1 and 3. The equation of movement of ions in the perfect omegatron can be written in the following form:

$$m\ddot{x} = -q\dot{y}B;$$
$$m\ddot{y} = qE\sin(\omega t) + q\dot{x}B;$$
$$m\ddot{z} = 0,$$

where E and ω are the intensity and the angular frequency of the high-frequency electric field and t is time.

The collector 4, which located in a hollow in the bottom plate, detects only those ions whose rotation period coincides with the cycle of the electric field. Such ions are called resonance. The trajectory of resonance ions is an unswinging spiral with the radius

$$R = \frac{Et}{2B}.$$

In every rotation, resonance ions acquire additional energy from the electric field. Other ions move over restricted trajectories and do not reach the collector.

Using Eq. (3.14) for radius of an ion trajectory, we determine the rotation period and frequency of a resonance ion:

$$T = \frac{2\pi R}{v} = \frac{2\pi m}{2Bq};$$
$$f = \frac{1}{T} = \frac{\Omega}{2\pi}$$

($\Omega = Bq/m$ is the cyclotrone resonance frequency). In a constant magnetic field ($\Omega m/q$ = const, and, therefore, the scale of mass numbers for frequency time-base is hyperbolic. The ion current passing through the collector is measured by an electronic amplifier.

The resolution of Omegatron is inversely proportional to ion mass:

$$\rho_m = \frac{qB^2 R_c}{2Em},$$

where R_c is the distance from the electron beam to the collector. With an increase in the mass of ions, the resolution decreases. In practice, at $\Delta Me = 1$ a.u. one may achieve $\rho_M = 35-45$.

The sensitivity to nitrogen at an electronic current of 10 μA is within $(4-10)\times10^{-7}$ A/Pa. At large electronic currents the sensitivity decreases because of an increase in the space charge of non-resonance ions. The relative sensitivity

$$C_i = \frac{K_i}{K_{iN_2}}$$

depends on the type of gas (Table 3.2). The maximum working pressure of omegatron is $(1-4)\times10^{-3}$ Pa, and the minimum one is 10^{-8} Pa. Technical characteristics of omegatron partial pressure measuring devices (OPPM) are listed in Table A.2.

In time-of-flight dynamic gas analyzers ions are separated by mass using the difference in the time of passing a gap (l_0) by ions of various gases (Fig. 3.17, a)

$$t = \frac{l_0}{v} = l_0\sqrt{\frac{m}{2qU}}.$$

Ions are formed by the electronic current passing between the thermocathode 6 and the anode 7. The grid 5 is fed with potential U_5 that accelerates ions in the drift space between the grounded grids 4 and 3. Passing the grid 5, ions acquire equal additional energy and various velocities according to their mass numbers. In the gap between the grounded grids 4 and 3, the ions are subdivided into groups with similar mass numbers.

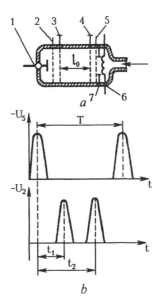

Figure 3.17 Time-of-flight gas analizer (Chronotron).

After the space of drift, groups of ions, with the grid 2 being grounded, sequentially reach the ion receptor 1, which serves as an electronic multiplier. The duration of the pulse that forms the mass-spectrum, is very small (0.1 μs), and the frequency of the pulses may be up to 10 kHz. The multiplier is connected to a broadband amplifier whose output signal passes to the vertically deviating plates of the oscilllograph. The time between the pulse and the moment of ion current peak detection characterizes the mass number.

For continuous registration of only one spectrum component, the grid 2 is fed with a positive potential that alternates with the negative pulses of voltage U_2 (Fig. 3.17, b), during which the group of ions with a certain mass number reaches the receptor 1. Other ions are decelerated by the positive potential of the grid. Changing the periods t_1, t_2 etc. between the pulses fed to the grids 5 and 2, one may separate components of a mass-spectrum.

The resolution of pass-time gas analyzers is

$$\rho_M = \frac{M_e}{\Delta M_e} = \frac{t}{2\Delta t},$$

where (Δt is the time necessary for a group of ions with the same mass number to pass the drift space and t is the duration of ionizing or forcing pulses. The scatter of initial ion velocities results in the necessity of reducing Δt.

The gas analyzer considered is refered to as Chronotron. This analyzer have half peak height resolutions of $\rho_M = 100$, the range of mass numbers 1 to 600 a.u., and the minimum measured pressure 10^{-8} Pa. A disadvantage of chronotron is the large length of the drift space $l_0 \geq 0.5$ m.

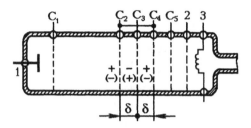

Figure 3.18 Radiofrequency mass-spectrometer (Topatron).

The design of pass-time gas analyzer shown in Fig. 3.18 is known as the RF mass-spectrometer, or topatron. The positive ions that are formed in the ionization zone (cathode 3 and anode 2), are accelerated by the grid C_5 in the direction of the three-grid cascade C_2–C_4. Between the grid C_3 and the grids connected together C_2 and C_4, a high-frequency voltage is fed. The distances between the grids C_2, C_3 and C_3, C_4 are identical and equal to δ.

Between the grids C_4 and C_3, ions are affected by the high-frequency electric field

$$F_1 = qE\sin(\omega t + \theta),$$

where E is the maximum intensity of the electric field between the grids, and θ is the initial phase of the high-frequency field, which corresponds to the moment of ion crossing the plane of the grid C_4 at $t = 0$. Between the grids C_3 and C_2 the force acting on the ions is the same, but with the inverse sign

$$F_2 = -qE(\omega t + \theta).$$

The maximum augment of energy is acquired by those ions that cross the grid C_3 at the moment of change in polarity, and for which the condition $\omega\delta/v + \theta = \pi$ is satisfied.

For such ions, the time of passing the three-grid cascade and the cycle of the high-frequency voltage coincide. Such ions are called resonance ones. Other ions acquire smaller augment of energy or are decelerated. The grid C_1 has a negative potential that can be overcome on the path to the collector 1 only by the resonance ions, which have acquired an augment of energy.

The mass-spectrum is displayed using an oscillograph connected to the output of the multiplier. The sweep by ion masses is effected by changing high-frequency voltage.

In this device, the ion current of grid C_5, which is proportional to the total pressure of a gas mixture, is frequently measured, and, therefore, the device is convenient for measuring either total or partial pressure. Because of the small resolution of the three-grid cascade, several cascades can be installed sequentially. For four cascades, $\rho_M = 45$. The range of registered mass numbers is 2–200 a.u., and the minimum measured pressure is 10^{-6} Pa.

In the electric mass filter, ions are separated by masses using a high-frequency electrical field of the quadruple that is formed by four parallel electrodes with round sections (Fig. 3.19, a). Opposite electrodes are connected in pairs, and voltage $\pm(U + V\cos(\omega t))$ is applied between them.

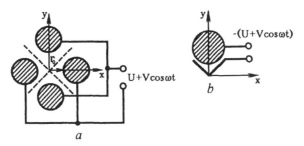

Figure 3.19 Mass filters: (*a*) quadruple and (*b*) monopole.

The equations of ion movement in a quadruple mass filter have the following form:

$$m\ddot{x} + 2q(U + V\cos(\omega t))\frac{x}{r_0^2} = 0\ ;$$

$$m\ddot{y} - 2q(U + V\cos(\omega t))\frac{y}{r_0^2} = 0\ ;$$

$$m\ddot{z} = 0\ . \tag{3.18}$$

The last equation of set (3.18) suggest that the velocity of ions along the analyzer *z*-axis remains constant. The two former differential equations are referred to as the Matthew equations. Substituting the new variables

$$\varepsilon = \frac{\omega t}{2}\ ;\quad a = \frac{8qU}{mr_0^2\omega^2}\ ;\ g = \frac{4qV}{mr_0^2\omega^2}$$

set of equations (3.18) can be rewritten as

$$\ddot{x} = (a + 2g\cos(2\varepsilon))x\ ;$$

$$\ddot{y} = -(a + 2g\cos(2\varepsilon))y\ ;$$

$$\ddot{z} = 0. \tag{3.19}$$

For *a* and *g*, lying within the crosshatched area in Fig. 3.20, Eq. (3.19) have stable solutions $x(\varepsilon)$ and $y(\varepsilon)$. Then the ions in the analyzer fluctuate with amplitudes smaller than the radius r_0 and reach the ion collector. The ions that move over unstable trajectories may infinitely increase the amplitude of their fluctuations and are neutralizes at the analyzer electrodes.

The resolution of the analyzer depends on the ratio of the voltages $\gamma = U/V$:

$$\rho_M = \frac{M_e}{\Delta M_e} = \frac{0.75}{1 - \gamma/\gamma_{max}}\ . \tag{3.20}$$

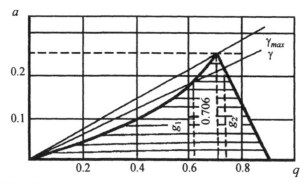

Figure 3.20 Stability diagram.

The maximum resolution is achieved at $\gamma \sim \gamma_{max} = 0.168$. At $\gamma \geq \gamma_{max}$ ions of all masses have unstable trajectories. Equation (3.20) suggests that to obtain high resolution one should maintain the ratio $\gamma = 0.999\,\gamma_{max}$. Thus, the voltages U and V change during the sweep of a mass-spectrum in a common manner, and their ratio should be maintained quite accurately.

The mass number of the ions corresponding to γ_{max} and having stable trajectories, is related to field parameters as follows:

$$M_e = \frac{1.39 \times 10^{-5} V}{f^2 r_0^2},$$

where f is the frequency, MHz, V is the amplitude of the high-frequency voltage, V, and r is the field radius, m. The sweep of a mass-spectrum is effected by changing the amplitude of alternating voltage on the electrodes. The accelerating voltage in the ion source providing the necessary initial speed of ions along the z-axis should not exceed the following value:

$$U_a = 4.2 \times 10^2 L^2 f^2 M_e \rho_M^{-1}, \tag{3.21}$$

where L is the length of the analyzer, m.

If the accelerating voltage exceeds this value, the ions moving over unstable trajectories will not have enough time to reach the electrodes of the analyzer. From Eq. (3.21) it can be seen that the resolution is proportional to the mass number of ions.

A modification of the electric mass filter is the monopolar analyzer (Fig. 3.19, b), which uses one quadrant of the quadruple. This analyzer consists of cylindrical electrodes with a negative voltage and a corner-like grounded electrode, whose working surface is aligned with the zero equipotential lines of the quadruple, as shown by dashes in Fig. 3.19, a. The monopolar analyzer allows passage of only those ions that enter the analyzer during the negative half-cycle of the high-frequency voltage. During the positive half-cycle, the ions fall onto the corner-like electrode and are neutralized.

Modern quadruple and monopolar gas analyzers (Table A.2) work at mass numbers from 1 to 500, their resolution is at a level of 10% of peak height $\rho_M > 500$, the sensitivity is 10^{-2} A/ Pa, and maximum and minimum working pressures are 10^{-2} and 10^{-12} Pa, respectively.

3.6   Sorption Methods

Analysis of the residual gases that are present in vacuum chambers on the basis of the difference in their adsorption heats is referred to as desorption mass spectrometry. For this analysis, in a chamber with the test gas and a conventional manometer converter, an electric current heated tungsten filament should be placed, which, before the work, should be degased by heating to 2500 K. After cooling, molecules of residual gases are adsorbed on the filament surface. The surface coverage by molecules of residual gases, according to Eq. (1.61), can be calculated at $\theta_0 = 0$ using the formula

$$\theta_i = \frac{b_i p_i}{1 + b_i p_i}(1 - \exp(-A_i t_a)),\qquad(3.22)$$

where p_i is the partial pressure the ith of gas, t_a is time of establishment of adsorption equilibrium,

$$b_i = \frac{f_i \tau_i}{a_{mi}\sqrt{2\pi mkT}}\,;\ A_i = \frac{1}{\tau_i} + \frac{f_i \tau_i}{a_{mi}\sqrt{2\pi mkT}}\,;\ \tau_i = \tau_0 \exp\!\left(\frac{Q_{ai}}{RT}\right),$$

τ_i is the mean time of the presence of a molecule in adsorbed state, and Q_{ai} is the adsorption heat the ith gas. The time t_a is chosen on a level sufficient for the exponent in Eq. (3.22) be negligible. The minimum time required is determined experimentally. With an increase in t_a the output signal first grows linearly, and then at $t_a > 4.6/A_i$ it becomes constant. In this case Eq. (3.22) can be simplified to

$$\theta_i = \frac{b_i p_i}{1 + b_i p_i}\,.$$

At heating of the filament desorption of gases occurs, and in the chamber in which the filament and the manometric transducer are placed the pressure increases by

$$\Delta p_i = \frac{a_{mi}\theta_i AkT}{V}\,,$$

where A is the surface of the filament; a_{mi} is the number of molecules necessary for the formation of a monomolecular layer of the ith gas, V is the volume of the chamber.

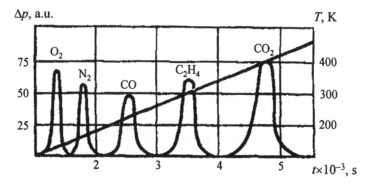

Figure 3.21 Mass-spectrum obtained using desorption mass-spectrometry.

An increase in pressure Δp_i for gases with various adsorption heats occurs at various filament temperatures, this fact being used for determining the composition and concentrations of the residual gases.

If the adsorbing surface was cooled before to the LN temperature, then one may analyze gases with low adsorption heats. Figure 3.21 a shows a mass-spectrum obtained using a desorption mass-spectrometer. The change in the filament temperature is also shown.

The susbtantial difference of this mass-spectrum from those obtained using ionization gas analyzers is that the peaks are arranged not in the order of increasing mass numbers, but in the order of increasing adsorption heats. This device resolves well the peaks of N_2 and CO which have identical mass numbers.

Analyzing the mass-spectra of gas mixtures one should take into account the presence of surface adsorption centers with various desorption heats. A gas may then have several thermal desorption peaks. To increase the sensitivity of a device one should increase the adsorbing surface, for example, by addition of active adsorbents or by etching.

If an atom is affected by radiation with an energy that is sufficient for the ionization of one of its inner electronic shells, the thus formed vacant place can be filled by an electron with a lower binding energy from another shell. The energy released due to a single transition may exceed the binding energy of an electron in one of the outer shells, and the electron will then be emitted. Thus, if a vacancy formed as a result of an external effect at the Kth level is filled by electrons from the Mth level, and the energy released is imparted to an electron of the Nth level, this process is called a *KMN* Auger transition. The kinetic energy of the electron emitted as a result of this transition, is determined by the expression

$$E_{KMN} = E_K - E_M - E_N,$$

where E_K and E_M are the binding energies of the electrons on the Kth and the Mth shells of the atom and E_N is the binding energy of the Nth level in the atom with the single-ionized Mth level.

The absolute energy of emitted electrons may range, depending on the atomic number of element and the type of transition, from 20 to 3000 eV.

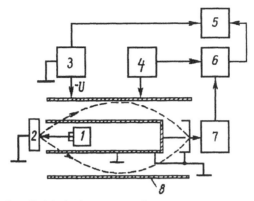

Figure 3.22 Electrostatic cylindrical mirror-type analyzer.

The block diagram of a cylindrical mirror type Auger spectrometer is shown in Fig. 3.22. The primary electron beam 1 is incident onto the target 2. Secondary electrons through the aperture in the inner cylinder fall on the inlet of the electronic multiplier 7. The energy of the electrons passing through the analyzer aperture depends on the voltage U supplied by the source 3 to the outer cylinder 8.

This voltage is used for the sweep of the electronic energy spectrum, which is recorded by two-coordinate self-recorder 5. The sine-like voltage generator 4 and the amplifier with the synchronous detector 6 allow differentiation and noise filtering of the output signal. Figure 3.23 shows a stainless steel spectrum obtained on an analyzer of this type. The abscissa axis is the energy of emitted electrons, and the axis of ordinates is their relative intensity.

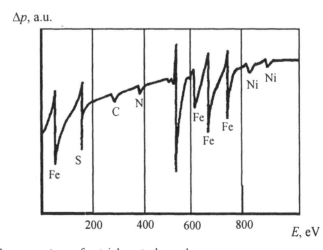

Figure 3.23 Auger spectrum of a stainless steel sample.

3.7 Calibration of Transducers

Transducers for indirect pressure measurements, the indication of which cannot be calculated with practically required accuracy are calibrated by comparing their indication with that of reference devices, i.e., mechanical transducers and compression manometers. These transducers are thermal, electronic, magnetic discharge and radioisotopic.

Devices for pressure measurement that are used as primary for the calibration of other devices are referred to as reference measurement means of the first category. Their constants or calibration curves are calculate from their size and other parameters involved into the measurement equation. Reference devices previously calibrated using other reference devices are called reference measurement means of secondary category. The reference measurement means of the first category have a calibration error of not more than 1–5 %, and those of the secondary category, to 10–15 %.

To expand the pressure range in which reference devices are calibrated, one may use the methods of isothermal expansion and flow division. For the isothermal expansion method, a chamber of small volume, the pressure in which is previously measured by a reference device, is connected to a chamber of greater volume, the initial pressure in which can be accepted as equal to zero. The Boyle–Mariotte law allows one to easily calculate the final pressure.

Further decrease in pressure can be obtained by increasing the number of expansion steps. It is thus important to take into account the errors associated with leakage, sorption by chamber walls, and the pumping action of the calibrated transducer. These errors determine the lower level of calibration pressure. The additional error of pressure measurement, which is connected with the accuracy of measuring the volume, does not usually exceed 0.5%. This method is frequently used in the pressure range 10^{-2}–10^{-3} Pa, for which leakage and gas evolution in the calibrated volume can be neglected.

At low pressure ($< 10^{-7}$ Pa), for which it is impossible to directly compare the indication of the calibrated device and a compression manometer one may use the method of flow division. This method implies that the gas flow in a dynamic vacuum system passes through sequentially connected cells (the first of which is shown in Fig. 3.24) and decreases gradually at the expense of work auxiliary vacuum pumps. The equation of flows for this cell is

$$Q = Q_1 + Q_2,$$

where $Q = U_1(p_0 - p_1)$ is the flow of gas, passing to the cell through pipeline with the conductance U_1, at the pressure difference on its ends p_0 and p_1, $Q_1 = S_1 p_1$ is the flow of gas, pumped by the auxiliary pump connected to the cell and having the effective pumping speed S_1, and $Q_2 = U_2 (p_1 - p_2)$ is the flow of gas, removed from the cell through the pipeline with the conductance U_2 and the pressures on its ends p_1 and p_2.

With of the expressions for Q, Q_1, and Q_2, Eq. (3.22) can be presented in the form

$$U_1(p_0 - p_1) = S_1 p_1 + U_2(p_2 - p_1),$$

whence

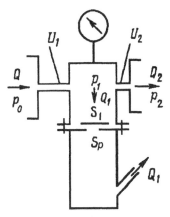

Figure 3.24 Vacuum system cell for calibration of manometric transducers using the flow division method.

$$p_1 = \frac{U_1 p_0 + U_2 p_2}{S_1 + U_1 + U_2}. \tag{3.23}$$

If the construction ensures that $S_1 \gg U_1 + U_2$, and $U_1 p_0 \gg U_2 p_2$, then $p_0 \gg p_1 \gg p_2$ and Eq. (3.23) can be simplified to

$$p_1 = \frac{U_1}{S_1} p_0.$$

In a number of consistently connected cells the pressure is gradually reduced from cell to cell, and for the nth cell it can be determined using the formula

$$p_n = \frac{U_1 U_2 \ldots U_n}{S_1 S_2 \ldots S_n} p_0.$$

The pressure p_0 can be measured using a compression manometer.

A disadvantage of this method is that its implementation requires pumps with very low limiting pressure and high pumping speed, because for obtaining stable value S_i, where $i = 1, 2, .., n$ the pump is connected to a chamber through a diaphragm and the condition $S_p \gg S_i$ is satisfied.

Calibration using the method of flow division allows the errors connected with leakage and sorption to be neglected while the gas flow in the calibration chamber considerably exceed the total leakage and gas evolution. The use of pumps with high pumping speed allows one calibrate manometer transducers with an error of 4–5% at the lowest pressures.

Because of the difference in ionization efficiency, possible separation of mixture components at input, occurrence of molecular particles, repeated ionization, and the presence

Table 3.3 Relative intensities of spectral lines of pure gases β_{ji}.

Mass number, a.u.	Gas											
	H_2	He	CH_4	H_2O	Ne	N_2	CO	C_2H_4	C_2H_6	O_2	Ar	CO_2
2	1.00		0.01	0.01								
3	0.01											
4		1.00										
12			0.20				0.02	0.02				0.03
13			0.60					0.04	0.01			
14			0.14			0.08	0.01	0.06	0.03			0.03
15			0.80						0.05			
16			1.00	0.02			0.01			0.10		0.08
17				0.25								
18				1.00								
20					1.00						0.15	
22					0.10							0.22
25								0.12	0.04			
26								0.62	0.23			
27								0.65	0.33			
28						1.00	1.00	1.00	1.00			0.14
29						0.01	0.02	0.02	0.22			
32										1.00		
40											1.00	
44												1.00

of false peaks that are inherent to some devices, line intensities in a spectrum may not correspond to the partial pressures of the gases in the mixture.

For quantitative analysis, devices are previously calibrated to each component of the mixture. A sample of gas is introduced into a standard volume, from which it passes to an ion source. The pressure in the standard volume is measured by a usual manometer transducer. The main line of the spectrum of this component is chosen and from this line the sensitivity of the device is determined. After that the intensity ratios of all the lines in the spectrum to the intensity of the main line of the gas spectrum β_{ji} are determined, where j is the main mass number. Results of these tests for an omegatron are given in Table 3.3.

Spectra of pure gases allow the structure and partial pressure of mixture components to be determined from the intensities of the mixture mass-spectrum. Assuming that the mass-spectrum of a mixture is formed by additive superimposition of mass-spectra of the separate components, one may write the set of linear equations

$$I_j = \sum_{i=1}^{n} I_{ji} = \sum_{i=1}^{n} K_{ji} p_i, \qquad (3.24)$$

where I_j is the intensity of the line in the gas mixture mass-spectrum corresponding to the mass mass number j, I_{ji} is the intensity of the line in the mass-spectrum of the ith gas

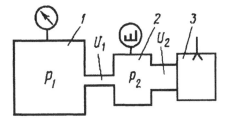

Figure 3.25 Pressure reduction schematic.

corresponding to the mass number j, p_i is the partial pressure of the ith component of the mixture, n is the number of components in the mixture, and K_{ji} is the sensitivity to the ith gas for the mass number j:

$$K_{ji} = \beta_{ji} K_i, \tag{3.25}$$

where K_i is the sensitivity of the main line of the spectrum and i is the serial number of the component in the mixture.

In calculations relative sensitivities (Table 3.2) for the main lines are used:

$$C_i = \frac{K_i}{K_r}, \tag{3.26}$$

where K_r is the main line sensitivity for the reference gas (usually, nitrogen or argon), C_i depends only on the type of gas analyzer, and K_r is determined individually for each device.

Substituting Eqs. (3.25) and (3.26) into Eq. (3.24), we obtain

$$I_j = K_r \sum_{i=1}^{n} \beta_{ji} C_i p_i. \tag{3.27}$$

Set of equations (3.27) has not solution if $j < i$, one solution at $j = i$, and several solutions at $j > i$. In this latter case, which is a usual practical occurrence, those partial pressures p_i are considered for which the squared residual of Eqs. (3.27) are the smallest.

For gas analysis in the pressure range from 10^5 to 10^{-1} Pa use the circuit of pressure reduction (Fig. 3.25) is used. From cylinder 1, filled with the test gas mixture, through capullary U_1, the gas mixture passes to chamber 2, from which it is pumped by pump 3 through pipeline U_2. The chamber 2 is connected to a gas analyzer, to which the gas mixture passes at pressure p_2, which differs from pressure p_1. If $p_1 \gg p_2$ and the conductance of pipeline U_2 is considerably smaller than the pumping speed of the pump for all the gases of the mixture, the condition of the equality of the inlet and the output flows of the chamber 2 suggests that $U_1 p_1 = U_2 p_2$, whence $p_2 = p_1 U_1 / U_2$. This ratio does not take into account the effect of adsorption processes and pumping action of the analyzer to the pressure in the chamber 2.

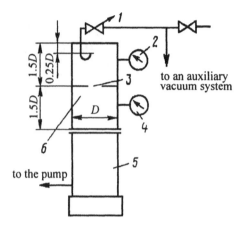

Figure 3.26 Schematic of gas flow measurement using the method of two manometers.

The conductance ratio U_2/U_1 and, hence, the degree of decrease in pressure p_1/p_2 may be as high as 10^3–10^6. In molecular mode of gas flow in pipelines the U_2/U_1 ratio does not depend on pressure and the molecular mass of gas, i.e., the partial pressure of each component of the mixture in the chamber 2 is proportional to its partial pressure in the chamber 1. The percentage of the gas mixture in the analyzer will be retained the same as in the test volume.

3.8   Measurement of Gas Flows

The method of two manometers is based on the measurement of difference in the pressures on an element with known conductance according to Eq. (2.35) and is used in the measurement of steady-state or gas flows that slow vary in time.

Figure 3.26 shows circuit of the measurement of the throughput and speed of vacuum pumps by the method of two manometers. The gas flow that is pumped by pump 5 is measured from difference in the pressures on the diaphragm 3 with known conductance U. The diaphragm is installed in the colibration chamber 6, the recommended size of which is shown in the schematic. The pressures p_1 and p_2 are measured by manometers 2 and 4, respectively. The gas flow is adjusted executed using controlled leak 1 that is connected to an auxiliary vacuum system. The working pressure of the system auxiliary vacuum is higher than that of the main one. The throughput of the pump, or the gas flow passing through its inlet branch pipe, is calculated using Eq. (2.35), and the pumping speed is

$$S = \frac{Q}{p_2} = \frac{U(p_1 - p_2)}{p_2}.$$ (3.28)

To expand the measurement limits the conductance of the diaphragm can be changed either continuously, for example, with the help of an iris diaphragm used in still cameras, or

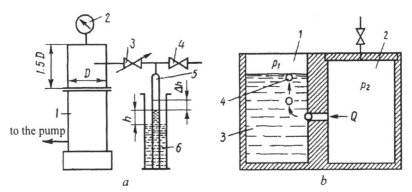

Figure 3.27 Schematic of gas flow measurement using the method of constant pressure: (*a*) oil burette method and (*b*) gas bubbles method.

in steps, using rotary disks with apertures of various diameters. For an element with a small conductance, the apertures the manufacture of which is accompanied by technical difficulties of shape maintenance, are replaced with long pipelines (capillaries).

The method of constant pressure for gas flow measurement uses a liquid burette, such as the one shown in Fig. 3.27, *a*. The pressure in the measuring volume 5 of the liquid burette 6 is

$$p_{meas} = p_a - \rho gh,$$ (3.29)

where p_a is the pressure of the external environment; ρ is the density of the liquid in the burette, and h is the difference of the levels of the liquid in the measuring volume and the external cylinder of the burette. If the condition that $p_a \gg \rho gh$ is satisfied, one may consider that p_{meas} = const and the requirement of constant pressure is fulfilled. Then, according to Eq. (3.2), the gas flow is determined using the equation

$$Q = p_{meas}\frac{\Delta V}{\Delta t} = K_b p_{meas}\frac{\Delta h}{\Delta t},$$ (3.30)

where Δh and ΔV are the changes of the level of the liquid and the volume of the measuring cylinder of the burette for the time Δt, K_b is the constant of the burette, $K_b = \pi R_{meas}^2$, and R_{meas} is the radius of the measuring volume.

The valve 4 is necessary for lowering the liquid level in the burette and recurrence of experiments. The leak 3 is intended for adjusting the gas flow. The burette is usually filled with vacuum oil that has low saturation pressure at room temperature.

The scheme of gas flow measurement by the method of constant pressure that uses the formation of gas bubbles is shown in Fig. 3.27, *b*. During leaking of a gas from the volume 2 to volume 1 at $p_2 > p_1$ the gas flow can be calculated from the speed dN/dt of the occurrence and volume V_b of gas bubbles 4 that form in the liquid 3:

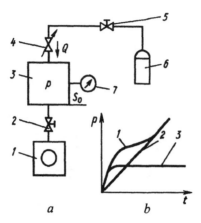

Figure 3.28 Schematic of gas flow measurement using the method of constant volume: (*a*) vacuum schematic and (*b*) pumping curves.

$$Q = V_b p_1 \frac{dN}{dt}. \tag{3.31}$$

Evacuation of the space over the liquid increases the sensitivity of measurement, because it is accompanied by an increase in the volume of bubbles.

Example of using the method of constant pressure is determination of gas flow from known pumping speed of a vacuum chamber S_0. Vacuum pump 1 (Fig. 3.27, *a*) is connected through a valve or a diaphragm to the pumped object, in which the pressure of gas is adjusted by the valve 4 and the leak 3.

The pumping speed of a chamber, according to the main equation of vacuum engineering (2.110) is

$$S_0 = \frac{U_0 S_p}{U_0 + S_p},$$

where U_0 is the conductance of the vacuum system from the pump to the pumped object and S_p is the pumping speed of the pump.

The measurement should be conducted in molecular mode, for which the conductance U_0 depends only on the type of gas and temperature, and not on pressure. The instability of pumping speed of the pump can be neglected if $S_p \gg U_0$; then Eq. (2.110) suggests that $S_0 = U_0$. As $S_0 = dV/dt$, then, according to Eq. (2.108), the gas flow is

$$Q = p S_0.$$

To determine the gas flow by the method of constant volume one may use schematic shown in Fig. 3.28, *a*. In this case the pump 1 and the valve 2 are used as an auxiliary vacuum system for producing vacuum in the volume 3. During the measurement the valve 2 is

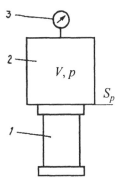

Figure 3.29 Determination of pumping efficiency using the method of constant volume.

closed. The gas passes to the volume 3 from the cylinder 6 through the leaker 4 and increases the pressure. If gas flow is constant, pressure increases linearly (curve 2 in Fig. 3.28, *b*). The gas flow is determined from the rate of this increase in pressure:

$$Q = V\frac{dp}{dt}. \qquad (3.32)$$

At the moment when pumping of the vacuum chamber is stopped ($t = 0$) adsorbed gases are evolved and the pressure increases (curve 3 in Fig. 3.28, *b*). The total change in pressure is shown by curve 1. Reliable measurement of a gas flow by the method of constant volume requires gas evolution be small in comparison with the measured gas flow. This can be achieved by careful preliminary pumping of the vacuum chamber.

The method of constant volume can be used for the determination of the throughput and pumping speed of vacuum pumps. Pump 1 (Fig. 3.29) is directly connected to the pumped object 2. The pumping curve (Fig. 3.30, *a*) allows p_i to be determined in every moment of

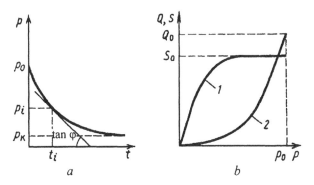

Figure 3.30 Experimental determination of vacuum pump efficiency: (*a*) pumping curve and (*b*) (*1*) pumping speed S and (*2*) pump throughput Q as functions of inlet pressure *p*.

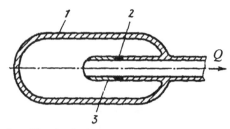

Figure 3.31 Schematic of a calibrating leak.

time t_i, and $(dp/dt)_i$ be determined from $\tan \varphi$. If the condition $V \gg S_p$ is satisfied, the gas flow of a pump can be calculated, in every moment of time, using the Eq. (3.32). The speed of the pump $S_{pi} = Q_i/p_i$ (Fig. 3.30, b).

Indirect methods of gas flow determination require preliminary calibration to obtain the required accuracy. The sensitivity of devices to gas flow is, for indirect measurements,

$$K_Q = \frac{\alpha}{Q}, \tag{3.33}$$

where α is the indication of the device in the most sensitive scale and Q is the gas flow determined by and absolute method.

During calibration, the range of gas flows is determined in which the linearity of the calibration characteristic is retained. Checking of calibration while in service is made using calibrated leaks, a possible design of which is shown in Fig. 3.31. The leak is a glass cylinder 1, filled with helium at a pressure of 10^5 Pa. Quartz tube 3 is soldered into the cylinder through connector 2. Helium diffuses through fused quartz. For flows of 10^{-7}–10^{-10} m^3 Pa/s leak checking may be performed once a year.

Example of an indirect gas flow measurement method is the thermal (khatarometric) method. Heat transfer in the range of low vacuum with forced convection depends on the speed of gas flow, and, hence, the temperature of the heated filament depends on gas flow. Because exact calculation of the dependence between temperature and gas flow is a difficult task, gas flow gauges are calibrated to absolute devices.

3.9 Leak Detection Methods

During the manufacture of vacuum systems, because of pores or cracks in materials, leaks may occur. Because of the small size of leak-causing defects, they cannot be detected visually.

To locate leaks, several methods have been developed: the trace gas method, the high-frequency discharge method, and the luminescent, radioisotopic, and bubble methods.

The trace gas method has found the widest use. After obtaining vacuum in the test object, it is blown around by a gas that starts to fill the vacuum system instead of air if a leak is

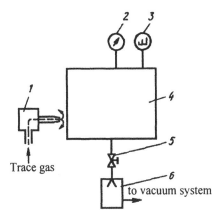

Figure 3.32 Schematic of vacuum tests using the trace gas method: (*1*) probe, (*2*) manometer, (*3*) gas analyzer, (*4*) test volume, (*5*) valve, and (*6*) pump.

present. Change in the composition of residual gases can be detected using of a vacuum gauge, the indication of which depends on the type of gas, or a mass spectrometer set to the trace gas.

To avoid an increase in pressure, the test object should be continuously pumped. The test scheme shown in Fig. 3.32. The source of trace gas 1 locally blows around the external surface of the test object 4. In the evacuated volume with nonhermetic casing, the difference in the indications of the vacuum gauge due to a change in the pressure of air and the trace gas is

$$\Delta\alpha = \frac{K_t Q_t}{S_t} - \frac{K_a Q_a}{S_a} = \frac{K_a Q_a}{S_a}\left(R\frac{Q_t S_a}{Q_a S_t} - 1\right), \tag{3.34}$$

where Q_a and Q_t are the flows of air and the trace gas, respectively, S_a and S_t are the effective speeds of the pump in the evacuated object for air at and the test gas, respectively, K_a and K_t are the sensitivities of pressure measurement for air and the trace gas, respectively, and R is the relative sensitivity of pressure measurement for the trace gas. To maximize the signal it is necessary to choose such trace gas that the expression $RQ_t S_a/Q_a S_t$ be maximally different from unity.

The flow ratio of trace gas Q_t to air Q_a can be written as

$$\frac{Q_t}{Q_a} = \frac{\Delta p_t U_t}{\Delta p_a U_a}, \tag{3.35}$$

where U_a and U_t are the conductances of the leak for air and the test gas, respectively, and Δp_a and Δp_t are the test difference in pressures for air and the trace gas, respectively. In molecular gas flow mode ratio (3.35) can be transformed to

$$\frac{Q_t}{Q_a} = \frac{\Delta p_t}{\Delta p_a}\sqrt{\frac{T_t M_a}{T_a M_t}},\tag{3.36}$$

where T_a and T_t are the temperatures of air and trace gas, respectively, and M_a and M_t are the molecular masses of air and trace gas, respectively. If an ionization manometer is used for the determination of gas tightness of a vacuum system pumped by a vapor jet pump, replacement of air to argon or helium increases the manometer indication by 50%, and replacement of air to carbon dioxide reduces the indication by 50%. The main trial gas used for leak detection is helium. Due to its chemical inactivity it is safe in work, its small molecular weight provides for good penetration through leaks, and the low adsorptivity reduces the required time of tests.

Electrical compensation of the indication of a vacuum gauge during the measurement of pressure of air before its replacement by a test allows the sensitivity of gas flow measurement to be increased. In this case the background signal is only determined by the stability of power supply.

A substantial decrease of background signal is provided by a manometer that is separated from the vacuum system by a selective membrane through which only the test gas can penetrate. For example, a palladium membrane heated to 700–800°C transmits well hydrogen, though remaining unpenetrable for the other gases. The sensitivity to gas flow is determined in this case by the residual pressure of the gases in volume of the manometer transducer.

For the S_a/S_t ratio in Eq. (3.34) be maximally different from unity, it is necessary to use pumps with gas-dependent speed. For example, an adsorption pump cooled by liquid nitrogen pumps helium, neon, and hydrogen considerably poorer than air.

If a gas analyzer is used for the detection of the trace gas, the background signal is determined by the partial pressure of the test gas in air, and the difference in the indication is

$$\Delta\alpha = \frac{K_t(Q_t - Q_b)}{S_t},$$

where Q_b is the background flow of the trace gas. Choosing such trace gas the content of which in air is low, one may consider that $Q_b = 0$. The parameter $\Delta\alpha$ is then much higher than for a vacuum gauge.

The trace gas is detected in the vacuum system by special gauges sensitive to the gas chosen. For example, in a diode with a platinum anode heated to 800–900°C, the presence of halogens produces electric current emitted by the anode of positive ions.

High-frequency discharge in medium vacuum changes colour depending on the type of gas. If a vacuum system is configured with a glass discharge tube, the replacement of air for petrol or acetone vapors changes the colour of the discharge in the tube from pink to grey.

Trace substances capable of forming negative ions by capturing electrons from gas discharge are used in electron-trapping vacuum gauges. Such a substance is for example freon. Electrons and negative ions are separated by deviation in a permanent magnetic field.

When using the trace gas method one should take into account the time response. The time during which a stable signal is achieved is 5–6 times the time of pumping $\tau = V/S_a$, where V is the volume of the vacuum system. The application of this method is limited to small leakages, because normal operation of vacuum gauges and gas analyzer requires high vacuum in test objects.

The accuracy of leak localization using trace gas probing is low. For more accurate location of a leak, the test surface is covered with an easily removable vacuum sealant, which locates a leak when penetration of the trace gas is stopped. A leak can thereby be located accurate to several millimeters.

Leaking of vacuum chambers by the trace gas method can also be detected at higher than atmospheric pressures. In this case, a probe with a device for suction of the mixture that contains the trace gas is supplied with electron-capturing, gas analyzing, or halogene detector of the trace gas. In electron-capturing atmospheric detectors, the current of the discharge diode that operates in the carrier gas, for example, argon, is reduced in the presence of the trace gas due to the recombination of positive ions of the carrier gas and negative ions of the trace gas. Katharometer analyzers use the difference of heat transfer coefficients of air and trace gases. The sensitivity of tests at high pressure is usually lower than for vacuum tests.

Instead of trace gases, trace liquids can be used, such as alcohol, ether, petrol, acetone, etc. The large lag of such tests that is determined by the time of penetration of the test liquid through thin capillaries is a noticeable drawback of trace liquids. At capillary radius of 10^{-4} cm the time of penetration of a test liquid reaches several hours and grows proportionally to decrease in capillary radius. The use of trace liquids and tests at high pressures is only expedient for the detection of rough leakages.

In the method of high-frequency discharge, approaching of the electrode of a high-frequency transformer to the place of a leakage, a directed discharge is formed due to a decrease in the pressure of air at the leak and more favorable conditions for electric breakdown of the gas gap. This method is convenient for the detection of leakage in glass vacuum systems.

The luminescent method uses the penetration of a luminophoric solution in capillary leaks. The test object is exposed to the luminophoric solution for a long time. After removal of the solution from the surface, filled capillaries can be easily observed as points or bands in mercury-quartz lamp light. Lumogenic solution gives yellow or red glow that can be easily distinguished from spurious signals of greenish glow due to air bubbles in glass or light-blue glow of surface pollution by grease.

For the radioisotopic method of leak detection, tested objects are exposed, for some time, to a radioactive gas. After removal of the radioactive gas and careful cleaning of the surface from radioactive pollution, only leaking devices continue radiating. This method is applied for automatic tightness checking of small-sized semiconductor devices.

The bubble method refers to the simplest ones. Excess pressure of a gas is produced in the test object, which should be immersed in a liquid. Leaks then show themselves by the formation of bubbles. The diameter of a bubble in the place of its formation is equal to the diameter of an appropriate leak cappilary. Immersion of test objects in a heated liquid is accompanied by an increase in the pressure accroding to the equation of gas state

$$\frac{p_2 - p_1}{p_1} = \frac{T_2 - T_1}{T_1},$$

where p_1, T_1 and p_2, T_2 are the pressure and temperature of a gas before and after heating, respectively.

Water can be heated without the formation of bubbles to 80°C, which corresponds to an excess pressure of 2×10^4 Pa. For oil, the temperature can be increased to 200°C, allowing one to obtain an excess pressure of 6×10^4 Pa.

The requirements to tightness of vacuum systems are formulated in view of the conditions of their operation. Technical documents for the development of a vacuum system operating under the conditions of continuous pumping specifies the working pressure p_w, at which the technological process will be performed. For the effective use of pumps it is necessary to provide the limiting pressure

$$p_{lim} \geq 0.1 p_w. \tag{3.37}$$

We assume that the vacuum system is well-degased and gas evolution can be neglected in comparison with leaking; then the maximum permissible leak rate in the vacuum system is

$$Q_m \leq p_w S_p = 0.1 p_w S_p, \tag{3.38}$$

where S_p is the pumping speed.

If the technical requirement concerns the maintenance of not general, but partial pressure p_p of any component of a mixture, the volume concentration of which is γ, then

$$Q_m \leq \frac{0.1 p_w S_0}{\gamma}. \tag{3.39}$$

In vacuum systems with volume V that operate during time Δt without continuous pumping, for the allowable increase in pressure Δp

$$Q_m \leq V \frac{\Delta p}{\Delta t}. \tag{3.40}$$

If the requirements concern partial pressures, then, similarly to Eq. (3.39) we obtain

$$Q_m \leq V \frac{\Delta p}{(\Delta t \gamma)}. \tag{3.41}$$

If the conditions of a tightness test are the same as operating conditions, then Eqs. (3.38)–(3.41) determine the requirements to the threshold sensitivity of the test. If the control of tightness is performed using a test gas or at other temperature and pressure difference, the requirement to the threshold sensitivity should be specified using Eq. (3.35).

The requirements to tightness of assembly units and parts depends on the requirements to tightness of the installation as a whole. For individual tightness check of elements during manufacture, considering simultaneous occurrence of more than two similar leaks in assembled installation to be improbable, one may accept the allowable leakage of an element

$$Q_{le} = 0.5Q_l. \tag{3.42}$$

Now we consider examples of determining tightness requirements.

Example 1. A working volume continuously pumped with the effective speed $S_0 = 0.1$ m³/s. The technical conditions set the working pressure p_w in the installation to 10^{-4} Pa. What are the requirements to the threshold sensitivity of the leak detector?

The limiting pressure of the installation, is, according to Eq. (3.37), $p_{lim} = 0.1p_w = 0.1 \times 10^{-4} = 10^{-5}$ Pa. The flow pumping at the limit pressure is $Q = p_{lim}S_0 = 10^{-5} \times 0.1 = 10^{-6}$ m³ Pa/s. Considering that the installation is well degasified and gas evolution can therefore be neglected, leakage into the installation should not exceed $Q_l = 10^{-6}$ m³ Pa/s.

In view that assembly units pass independent tightness tests, then according to Eq. (3.42), the allowable leakage is $Q_{le} = 5 \times 10^{-7}$ m³ Pa/s.

Tightness tests are performed by vacuum testing, with trace gases. In this case the trace gas will be helium. Then, according to Eq. (3.35), at similar pressure difference and temperature, the threshold sensitivity of the leak detector should not be less than

$$Q_{He} = Q_{le} \sqrt{\frac{M_a}{M_{He}}} = 5 \times 10^{-7} \sqrt{\frac{29}{4}} \approx 1 \times 10^{-6} \text{ m}^3 \text{ Pa/s}.$$

Example 2. A vacuum device with the volume $V = 0.01$ m³ without continuous pumping should be stored for one year ($\Delta t = 3.5 \times 10^7$ s) in a hydrogen atmosphere at the pressure difference $\Delta p_{H_2} = 2 \times 10^5$ Pa. The allowable increase in pressure because of leaking during the storage time is 10^{-2} Pa. What are the requirements to the threshold sensitivity of the leak detector?

The flow that can cause the above increase in pressure is, according to Eq. (3.40),

$$Q_l = \frac{V\Delta p}{\Delta t} = \frac{0.01 \times 10^{-2}}{3.5 \times 10^7} = 2.9 \times 10^{-12} \text{ m}^3 \text{ Pa/s}.$$

If the vacuum test is performed with helium as a trace gas, the trace gas method should provide, according to Eq. (3.35), the threshold sensitivity

$$Q_{He} = \frac{Q_l \Delta p_{He}}{\Delta p_{H_2}} \sqrt{\frac{M_{H_2}}{M_{He}}} = \frac{2.9 \times 10^{-12} \cdot 10^5}{2 \times 10^5} \sqrt{\frac{2}{4}} = 10^{-12} \text{ m}^3 \text{ Pa/s}.$$

At present, mass-spectrometer leak detectors are widely used that have the highest sensitivity. The minimum leaks that can be detected using such devices, are 10^{-13} m³ Pa/s. Figure 3.33 shows a mass-spectrometer chamber of a leak detector intended for work with helium as a trace gas. The electrons emitted by the cathode 9 pass to the ionization chamber 8. The power supply of the cathode 11 is connected to the analyzer through flange 10. If a vacuum system leaks when exposed to the trace gas, helium molecules penetrate through flange 5 into the ionization chamber. Positive ions of helium by accelerating voltage are directed to the magnetic analyzer chamber 6.

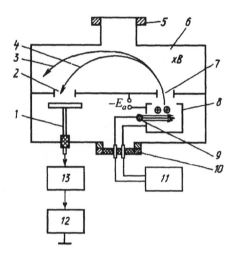

Figure 3.33 Mass-spectrometer chamber of a leak detector.

The accelerating voltage E_a and the magnetic induction B are selected such that helium ions passing through the inlet slot 7 and moving over the trajectory 4 reached the target aperture 2. Residual gases discharge on the analyzer walls over the trajectory 3. Unlike partial pressure analyzers, which should have high resolution and be specially set to different mass numbers, a leak detector is set only to one gas. The source and target slots can be broadened to increase the sensitivity of the leak detector. This method of increasing sensitivity can be used for helium, which has not substances with close mass number in air.

The ion collector 1 is connected to the electrometer amplifier 13 that increases the voltage drop at the high-ohmic resistance. The ion flow measuring unit 12, after additional amplification of the output signal of the electrometer amplifier shows results on an indicator or an automatic recorder.

The leak detector is connected to the test object by flange 1 (Fig. 3.34). For preliminary calibration helium leak 12 can be used that is connected through valve 13. For throttling of

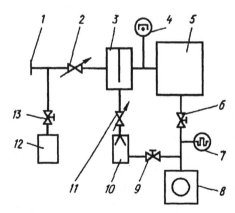

Figure 3.34 Vacuum schematic of a mass-spectrometer leak detector.

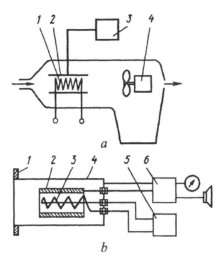

Figure 3.35 Halogen leak detector transducers: (*a*) atmospheric and (*b*) vacuum.

large flows counter flow mode is used when leaker valve 2 is applied. The trap 3 with the pump 10 connected through valve 11 is used for producing a working pressure of 10^{-2} Pa that is measured by transducer 4 and the mass-spectrometer chamber 5. Preliminary vacuum pump 8 provides, through valve 9, the operation of the high-vacuum vapor-jet pump 10 and through valve 6 the by-pass pumping chamber 5. The serviceability of pump 8 can be determined using manometer 7.

The leak detector can be connected to vacuum systems either from the part of high vacuum or from the part of low vacuum. The vacuum system of the leak detector is intended for pumping the mass-spectrometer chamber and, as a rule, it cannot be used for pumping of the test object, which should have its own pumping system.

The atmospic transducer of a halogen leak detector is a direct incandescence diode (Fig. 3.35, *a*), whose ion emitting electrode is the platinum spiral 1 that is heated to 800–900°C, and the collector of ions forming in the presence of halogens is the cylindrical electrode 2 that wraps the filament. The ion current is measured by amplifier 3. Air with halogens leaving the chamber is sucked into the gauge by fan 4.

The vacuum transducer (Fig. 3.35, *b*) is connected to the vacuum system through flange 1 and contains, like the atmospheric one, spiral 3 and ion collector 2 that are connected through electrical contacts to vacuum in case 4 with the measuring block 6, which indicates a leak by a output device or a sound signal, and the power supply 5. Technical characteristics of the leak detectors are listed in Table A.3.

3.10   Test Questions

3.1   What is the difference between absolute and indirect methods of measuring low pressures?

3.2   What is the origin of the occurrence of background currents in ionization transducers?

3.3   What are the methods of increasing the sensitivity of ionization transducers?

3.4   What is the resolution of partial pressure analyzers?

3.5   Which properties of gase molecules allow their division in the sorption method of partial pressure analysis?

3.6   Which methods are used for the determining the speed of vacuum pumps?

3.7   How can one reduce the discharge glowing time in magnetic transducers?

3.8   How can one reduce the effect of pumping on the accuracy of manometric transducers?

3.9   Which ways are used to sweep mass-spectra in magnetic analyzers?

3.10 How can one divide the partial pressure of ions with similar mass numbers in ionization analyzers?

Chapter 4

Mechanical Methods of Vacuum Production

4.1 General Characteristic of Vacuum Pumps

Vacuum pumps are subdivided by purpose into ultrahigh-vacuum, high-vacuum, medium-vacuum, and low-vacuum, and by the principle of action, into mechanical and physico-chemical. Working pressure ranges of mechanical types of vacuum pumps are given at Fig. 4.1. The main parameters of any vacuum pump are the pumping speed, the limiting pressure, the lowest working pressure, the highest working pressure, the highest pressure of start, and the highest output pressure.

The limiting (ultimate) pressure of a pump p_{lim} (Fig. 4.2) is the minimum pressure that the pump can produce when working without a pumped volume. The pumping speed of a pump tends to zero as the working pressure approaches the limiting one. The limiting pressure of the majority of vacuum pumps is determined by the gas evolution of the materials from which the pump is made, flow of gases through leaks, and other phenomena occurring during pumping.

The lowest working pressure of a vacuum pump p_{wl} is the minimum pressure at which pump retains the nominal working mode. The lowest working pressure is by about one order of magnitude higher than the limiting one. The use of a pump at pressures between the limiting and the lowest ones is economically disadvantageous because of the deterioration of the specific characteristics.

The highest working pressure of a vacuum pump p_{wh} is the maximum pressure at which the pump retains the nominal working mode. In the working ranges from the lowest to the highest pressure the use of the pump is effective. The working pressure ranges are basically determined by the action principle of a pump, and dashes lines show data on laboratory experience.

The starting pressure of a vacuum pump p_s is the maximum pressure at the inlet section of the pump at which it can be started. The starting pressure often appreciably exceeds the highest working pressure. For some types of pumps, for example, magnetic discharge ones, this difference may be as high as 2–3 orders of magnitude.

The parameters of vacuum pumps are shown by the main characteristic of a vacuum pump, i.e., the dependence of pumping speed on inlet pressure (Fig. 4.2). The main characteristic of a pump can be approximated, using its parameters, by the equation

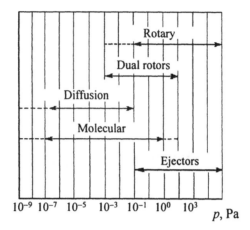

Figure 4.1 Working pressure ranges of mechanical vacuum pumps: (solid line) commercial and (dashed) laboratory.

$$S_p = S_m\left(1 - \frac{p_{lim}}{p} - \frac{p}{p_s}\right).\tag{4.1}$$

Mechanical vacuum pumps can be subdivided into displacement and molecular ones. Displacement pumps operate by periodically changing the volume of the working chamber. There are some construction variants of such pumps: piston, liquid-ring, and rotory ones.

Molecular pumps operate by imparting gas molecules an impulse from a rapidly moving solid, liquid, or vapor surfaces. They, in turn, can be subdivided into water-jet, ejectors, diffusion, molecular drag, and turbomolecular. Characteristics of such pumps can be calculated on the basis of the laws of internal friction in gases.

Additional important parameter of mechanical vacuum pumps – breaking pressure is the highest output (backing) pressure p_h, which is the maximum pressure at the output section of the pump at which it can operate. At output pressures lower than p_h, the pumping speed

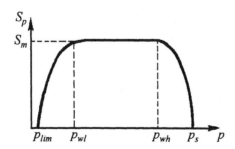

Figure 4.2 The main characteristic of a vacuum pump: the dependence of pumping speed on inlet pressure.

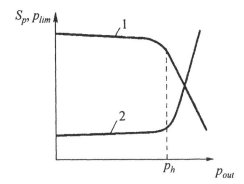

Figure 4.3 Pumping speed and limiting pressure as functions of output pressure: (*1*) pumping speed and (*2*) limiting pressure.

and the limiting pressure of pumps is only slightly depend on output pressure. If the output pressure exceeds significantly p_h the pump fails, i.e., its limiting pressure and pumping speed are sharply impaired (Fig. 4.3).

4.2 Volume Pumping

Displacement pumping includes the following basic operations: (1) admittance of the gas by expanding the working chamber, (2) reduction of the volume of the working chamber and compression of the gas in it, and (3) removal of the compressed gas from the working chamber to the atmosphere or a roughing pump.

The working diagram of volume pumps is the dependence of volume V_c and pressure p_c in the working chamber on time (Fig. 4.4). During the time t_1, the gas is ingested (area I), in the period from t_1 to t_3 the gas is compressed (area II), and from t_3 to t_4 the gas is let out (area III); after that the cycle is repeated. Curves 1 and 3 are the time dependences of volume and pressure in the working chamber, respectively. Curve 2 is the change in pressure for operation by letting in a ballast gas at the moment of the time t_2.

The ballast gas is let in during pumping of water or organic solvent vapors whose saturation pressure p_T at the working temperature in the pump is in the range from the limiting

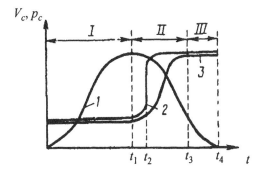

Figure 4.4 Working diagram of volume pumps.

pressure p_{lim} to the maximum output pressure. Letting in a ballast gas reduces the degree of compression and prevents condensation of pumped vapors in the pump.

The geometrical speed of volume pumping is the product of working chamber volume and the cycle frequency of pumping n, $S_g = V_c n$.

The resistance of the inlet pipe reduces pumping speed. According to the main equation of vacuum engineering (2.110), the highest speed of a pump

$$S_m = \frac{S_g U}{S_g + U},$$

where U is the conductance of the inlet pipe.

The real pumping speed is even lower because of the back flow of gases in the pump. The back flows occur due to gas leakages and the presence of detrimental space in the pump, and in high vacuum also because of the diffusion of a working liquid vapor from the pump to the pumped volume.

The throughput of pumping in molecular mode of gas flow is equal to the difference of the forward and back flows:

$$Q = Q_f - Q_b, \tag{4.2}$$

where $Q = S_p p$ and $Q_f = S_m p$.

At $p = p_{lim}$ throughput $Q = 0$, in which case, according to Eq. (4.1), one may write

$$Q_b = Q_f = S_m p_{lim}.$$

Substituting this expression into Eq. (4.2), we obtain Eq. (4.1) at $p \ll p_s$

$$S_p = S_m \left(1 - \frac{p_{lim}}{p}\right).$$

Considering Eq. (2.110) this expression can be written in the following form:

$$S_p = \frac{S_g U}{S_g + U}\left(1 - \frac{p_{lim}}{p}\right) = K_\lambda S_g,$$

where $K_\lambda = U\,(1 - p_{lim}/p)/(S_g + U)$ is the efficience coefficient.

The real pumping speed is K_λ times lower than the geometrical pumping speed. To increase this speed one should increase the volume of the pumping chamber V_c and the conductance of the inlet pipe U. There is the optimum cycle frequency n above which a pump may be overheated or the pumping chamber will not be filled with the pumped gas. Modern pumps work at $n = 1400$ rpm, with 50 cycles electrical supply and 1700 with 60 cycles.

To determine the limiting pressure, we write the equation of material balance

$$G_{ini} + G_{ad} = G_1,$$

where G_{ini} is the amount of gas in the pumped volume before the beginning of pumping cycle, G_{ad} is the amount of gas, adding to the pumped volume during pumping, and G_1 is the amount of gas in the pumped volume and the pump after one pumping cycle. This equation can be rewritten as

$$p_0 V + \alpha_1 V_d p_{out} = p_1 (V + V_c),$$ (4.3)

where V and V_c is the volume of the pumping chamber and chamber in the pump, V_d is the volume of the 'dead' space, p_{out} is the output pressure, α_1 is the coefficient taking into account gas evolution in the pump and the degree of gas backflow, and p_0 and p_1 are the initial and the final pressure in the pumped volume, respectively.

We solve Eq. (4.3) relative to pressure p_1:

$$p_1 = p_0 \frac{V}{V + V_c} + \alpha_1 \frac{V_d}{V} p_{out} \frac{V}{V + V_c}.$$

With one more pumping cycle, we obtain

$$p_2 = p_1 \frac{V}{V + V_c} + \alpha_1 \frac{V_d}{V} p_{out} \frac{V}{V + V_c}$$

$$= p_0 \left(\frac{V}{V + V_c} \right)^2 + \alpha_1 \frac{V_d}{V} p_{out} \left[\frac{V}{V + V_c} + \left(\frac{V}{V + V_c} \right)^2 \right].$$

After n pumping cycles, we obtain the following expression:

$$p_n = p_0 \frac{1}{[1 + (V_c / V)]^n} + \sum_{m=1}^{n} \alpha_1 \frac{V_d}{V} p_{out} \frac{1}{[1 + (V_c / V)]^m}.$$ (4.4)

Here the first term at $n \to \infty$ tends to zero, and the second term can be calculated using the formula of the sum of infinitely declining geometrical proggression:

$$\sum_{m=1}^{\infty} \frac{1}{[1 + (V_c / V)]^m} = \frac{V}{V_c}.$$

At $n \to \infty$, Eq. (4.4) can be simplified to

$$p_{lim} = \alpha_1 \frac{V_d}{V_c} p_{out}.$$

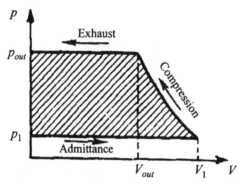

Figure 4.5 pV-diagram of volume pumps.

This latter formula suggests that the limiting pressure of a pump is determined by the ratio of the volume of the dead space to the volume of the working chamber, the output pressure, and the coefficient α_1 that takes into account gas evolution in the pump and gas backflow from the compression chamber to the suction chamber. The overall gas flow of a pump is composed of gas evolution from the main parts, evolution of gases dissolved in the working liquid, and leaking through the connections in the pump casing. The limiting pressure of volume pumping can be improves by reducing the total back flow from a pump. This can be achieved by using a trap that prevents the penetration of working liquid vapors to the pumped volume, or by sequentially connecting two steps of pumping for reducing the backflow of gases from the compression chamber to the suction chamber.

The power consumed by a rotory pump, W, is spent for compression of the pumped gas to the output pressure (useful power W_u) and for overcoming the friction forces (power losses W_{fr}): $W = W_u + W_{fr}$. The working diagram of a pump in the $p–V$ coordinates shown in Fig. 4.5 describes the operation of a pump in a single compression cycle. As is well-known from thermodynamics, for polytropic compression

$$A' = p_1 V_1 \frac{1}{m-1}\left[\left(\frac{p_{out}}{p_1}\right)^{\frac{m-1}{m}} - 1\right],$$ (4.5)

where V_1 is the largest volume of the working chamber of the pump, m is the polytropic exponent, and p_1 and p_{out} are the inlet and output pressures, respectively.

A single compression cycle (the cross-hatched area in the figure) is described as

$$A = A' + p_{out} V_{out} - p_1 V_1,$$ (4.6)

where V_{out} is the volume of the working chamber at the moment of achievement of the output pressure p_{out}. Because for a polytropic process the ratio $p_{out} V_{out}^m = p_1 V_1^m$ is true, then

$$p_{out} V_{out} - p_1 V_1 = p_1 V_1\left[\left(\frac{p_{out}}{p_1}\right)^{\frac{m-1}{m}} - 1\right].$$ (4.7)

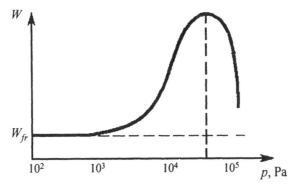

Figure 4.6 Power consumption as a function of pressure.

Substituting Eqs. (4.5) and (4.7) into Eq. (4.6), we obtain

$$A = p_1 V_1 \frac{m}{m-1} \left[\left(\frac{p_{out}}{p_1} \right)^{\frac{m-1}{m}} - 1 \right].$$ (4.8)

Because the useful power consumed by a pump is

$$W_u = \frac{A}{t_c} = nA,$$

where $t_c = 1/n$ is the cycle duration, then, using the expression for A, we obtain

$$W_u = np_1 V_1 \frac{m}{m-1} \left[\left(\frac{p_{out}}{p_1} \right)^{\frac{m-1}{m}} - 1 \right].$$

Because $S_g = V_1 n$, we may write

$$W_u = S_g p_1 \frac{m}{m-1} \left[\left(\frac{p_{out}}{p_1} \right)^{\frac{m-1}{m}} - 1 \right].$$ (4.9)

We differentiate Eq. (4.9) by p_1 and equate it to zero:

$$\frac{\partial W_u}{\partial p_1} = S_g \frac{m}{m-1} \left[\frac{1}{m} p_{out}^{\frac{m-1}{m}} p_1^{\frac{1-m}{m}} - 1 \right] = 0.$$ (4.10)

Analysis of Eqs. (4.9) and (4.10) suggests that $W_u = 0$ at $p_1 = 0$ and $p_1 = p_{out}$, and the maximum power is consumed at

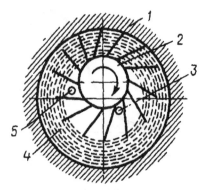

Figure 4.7 Schematic of a liquid-ring pump.

$$p_1 = \frac{p_{out}}{m^{\frac{m}{m-1}}} .$$

The pressure dependence of consumed power is shown in Fig. 4.6. For $m = 1.2$ and p_{out} = 10^5 Pa, the maximum power is consumed at 3.3×10^4 Pa. For low pressures (less than 10^3 Pa), all the power is spent for the friction losses W_{fr}. At high pressures ($\sim 10^5$ Pa) the power consumption decreases, because the required degree of gas compression decreases.

4.3   Design of Displacement Pumps

Displacement pumps can be subdivided into reciprocating and rotary, depending on kinematic scheme. Reciprocating pumps do not provide complete balancing of moving parts. They have lower efficiency than rotary pumps at similar pumping chamber sizes.

In piston pumps, the cylinders may be simple or double-action with water or air cooling. The speed of a piston does not usually exceed 1 m/s. For unlubricated operation the cylinders are covered with polymer materials. One-stage piston pumps with self-acting valves have a limiting pressure of 5×10^3 Pa. Piston pumps with control-acting valves have lower limiting pressures, 3×10^2 Pa for one-stage and 10^1 Pa for two-stage designs. The limiting pressure is improved by by-passing a gas from the dead space at the end of piston cycle to the second volume of the cylinder, in which inflow is completed. In four-stage unlubricated pumps, a limiting pressure of 1 Pa is achieved. A modification of the piston pump are membrane pumps. They can work without lubrication. The speed of modern piston pumps reaches 4000 l/s. Membrane pumps have speeds from 1 to 30 l/s. Pumps are usually started at atmospheric pressure. A disadvantages of piston pumps are the nonuniformity of pumping, incomplete compensation, large friction losses ~200 W/(l/s), and large specific weight, 10–20 kg/(l/s).

Rotary pumps are subdivided into one-rotor and two-rotor pumps. One-rotor pumps have a more simple design, and two-rotor pumps with synchronous rotation of rotors allow

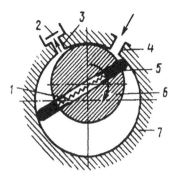

Figure 4.8 Schematic of a rotory vane pump.

one to do without the friction couple between the rotors and the stator. One-rotor pumps are widely used in diverse construction modifications: liquid-ring, rotory vane, multivane, vane-stator, and plunger-type ones.

Liquid-ring pumps, or liquid piston pumps (Fig. 4.7) have off-center driving wheel 2 with stationary fixed blading in cylindrical casing 1. The liquid in the casing is pushed by the centrifugal forces to the casing walls during rotation and forms liquid ring 4. Between the liquid ring and the pump blading cells of different sizes form. Initially, their volume increases and the gas passes to the pump through the inlet aperture 3 in the face cover. Then volume of the cells decreases and the compressed gas through the output aperture 5 leaves the pump. The working liquid for pumping of air mixed with water vapor is water, and for pumping of chlorine, concentrated sulfuric acid. The design and operation conditions of these pumps are more simple than those of piston pumps, because they have not valves and switching devices.

The limiting pressure of these pumps is determined by the pressure of the working liquid vapor. Water-ring pumps have limiting pressures of $(2-3) \times 10^3$ Pa. These pumps can work from atmospheric pressure. In vacuum mode they provide pressures of $(2-3) \times 10^3$ Pa. Their speed is from 25 up to 500 l/s, and the specific power consumption is ~100 W/(l/s) because of the necessity of moving the liquid in the pump. The specific weight of these pumps is about 10 kg/(l/s).

Rotory vane pumps (Fig. 4.8) consist of cylindrical case 7 with inlet 4 and output 3 pipes and off-center rotor 6, in the grooves of which vanes 5 are installed. Due to the centrifugal force the vanes are pressed to the casing and change the volume of the working chamber. Pumps work in oil baths that provide tight sealing of the connections in the pump and reduce the friction losses. To prevent filling the working chamber with the oil, valve 2 is used. Initially plates are pressed to the stator surface by spring 1.

Multivane pumps with pumping speeds of up to 10^3 l/s are made under the scheme shown in Fig. 4.9 with a large number of vanes. These pumps don't have oil baths, and for reduction of friction losses rings 1 are used that are rotated by vanes 2. The apertures of the rings provide passage of the pumping gas. In some designs with vanes made of antifriction materials, the rings can be avoided. The limiting pressure of such pumps is determined, with gas evolution of the pump, by the volume of the dead space and the saturation vapor pressure of the oil.

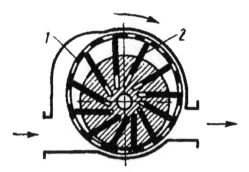

Figure 4.9 Schematic of a multivane rotor pump.

The dead space of a pump is designated in Fig. 4.10 by the letter B. In rotory vane pumps (Fig. 4.8) the dead volume is partially filled with the working liquid. For the pump shown in Fig. 4.9, the dead volume is connected with a passage channel to one of the working chambers that are not connected with the pumped volume. The saturation pressure of the working liquid being neglected, the limiting pressure of such pumps is 1 Pa for the scheme shown in Fig. 4.8 and 2×10^3 Pa for the one in Fig. 4.9. The limit pressure of two-stage rotory vane pumps is 10^{-3} Pa. The specific weight of such pumps is from 10 to 30 kg/(l/s), the specific power consumption from 0.1 to 0.3 KW/(l/s), multivane rotory pumps have smaller values.

Single-rotor cam pumps may be of two types: vane-in--stator and plunger. The plate-stator pump (Fig. 4.11) consists of the following main elements: casing 1, off-center rotor 2, output pipe 3, vane 5, spring 4, and inlet pipe 6. The working chamber of this pump is formed between the off-center rotor and the casing. For clockwise rotation, gas is ingested from the pumped volume during the first round of the rotor and is compressed and let out during the second round. The vane, tightly pressed by the spring, divides the areas of inlet and compression of the pumped gas.

The plunger pump (Fig. 4.12) consists of casing 1, off-center rotor 2, plunger 6, outlet pipe and back valve 3, hinge 7, and inlet pipe 4. Gas passes to the inlet chamber A, which increases during clockwise rotation, from the pumped volume through the inlet pipe and apertures 5 in the plunger. In the same time the volume of the chamber B decreases and the gas in it is compressed and pushed off through the outlet pipe.

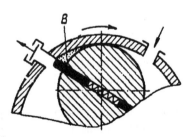

Figure 4.10 Dead space (B) in a rotory vane pump.

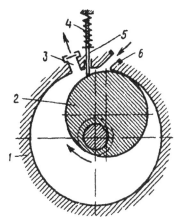

Figure 4.11 Pump wit a vane in the stator.

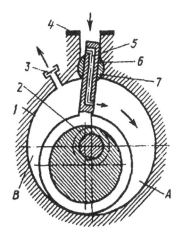

Figure 4.12 Plunger pump.

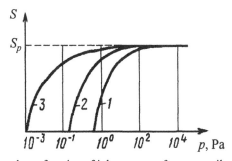

Figure 4.13 Pumping speed as a function of inlet pressure for rotory oil-sealed pumps.

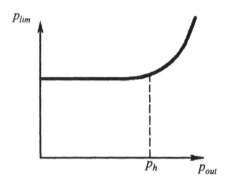

Figure 4.14 Limiting pressure as a function of output pressure.

The stator-vane and plunger pumps work in oil baths, such as the rotory vane pump. The parameters of these pumps are almost identical, but plunger pumps are more rapid. The pressure dependences of the speed of these pumps are shown in Fig. 4.13 and are described by Eq. (4.2). Curve 1 corresponds to one-stage work of the pump, curve 2 to two-stage work without a trap, and curve 3 to two-stage work with a trap. The startup and breaking pressures of these pumps are usually equal to the atmospheric one, but in gas-tight devices they may vary over a wide range. The dependence of limiting pressure on the output one is shown in Fig. 4.14. The highest output pressure is $p_h = 2 \times 10^5$ Pa. At output pressures higher than p_h, the tightness of oil in gaps of a pump is impaired and sharp deterioration of the limiting pressure is observed. At output pressures lower than p_h, the limiting pressure does not practically depend on the output one.

Vacuum oils that are usually used as working liquid of pumps are obtained from usual lubricants by distillation of light and heavy fractions. Characteristics of vacuum oils are listed in Table and Fig. A.4. The flash temperature of the oils should be not less than 200°C, which corresponds to the absence of readily oxidizing fractions in the oil. Semi-lubricated friction that occurs in pumps is accompanied by heating of some microroughnesse contacts to as high as the melting point of the metal. Under these conditions oil cracking occurs with the formation of light hydrocarbons that deteriorate the limiting pressure of pumps. Parameters of industrial one-rotor vacuum pumps of various types are shown in Table A.5 and in Fig. A.5.

Two-rotor vaccum pumps are convenient for high-speed operation at small compression rates. The rotors of these pumps are driven through a synchronizing gears and have such profiles that they do not touch each other during rotation. In some designs of oil pumps, the synchronizing gears are absent and the rotors touch each other. Two-rotor pumps have at the same dimensions considerably larger speed than one rotor pumps, because the absence of friction between the rotor and the stator allows a considerable increase of their rotation frequency.

By the method of gas compression the two-rotor pumps can be subdivided into pumps with external, partially internal, and internal compression. In external compression pumps, gas is compressed only during the pressure stroke. Such pumps are the Roots pumps with lemniscate rotors (Fig. 4.15). During one rotation such rotors twice transfer the cross-hatched volume of gas from the range of high vacuum to that of preliminary rarefaction.

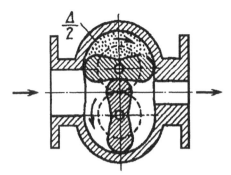

Figure 4.15 Schematic of a Roots two-rotor pump.

The rotors rotate in opposite directions, and their rotation is synchronized by a gear transmission with the gear number equal to unity (shown by dashes in the figure).

The volume of the working chamber of a pump is

$$V_c = 2\pi R^2 l K_\lambda,$$ (4.11)

where R is the maximum rotor radius, l is the rotor length, $K_\lambda = \Delta/\pi R^2$, and Δ is the area of grooves in the circle circumscribed around the rotor. Multiplying the volume of the working chamber of the pump V_c (Eq. (4.11)) by the rotation frequency of the rotor n, one may determine the geometrical pumping speed S_g. The back flow of gas in these pumps is determined by small gaps in the rotor mechanism:

$$Q_b = U_g(p_{out} - p_{in}),$$ (4.12)

where U_g is the conductance of gaps in the rotor mechanism.

Considering gaps to be thin, we write the conductance for air (Eq. (2.48)) in molecular mode in m³/s: $U_g = 116A$, where $A = l(\delta_r + \delta_{rc}) + 4R(\delta_{T_1} + \delta_{T_2})$ is the total area of the gaps, m². Here δ_r is the gap between the rotors, δ_{rc} is the gap between the rotors and the casing, δ_{T_1} and δ_{T_2} are the face gaps between the rotors and the covers. The size of the gaps between the rotor and the stator is usually $0.004R$, and between the rotors and the face covers, $0.006R$.

The back flow of a gas in a pump can also be expressed through the limiting pressure and the maximum speed of the pump: $Q_b = S_{max}p_{lim}$. Solving this equation together with Eq. (4.12) at $p_{in} = p_{lim}$, we find the compression ratio of the pump

$$K = \frac{p_{out}}{p_{in}} = 1 + \frac{S_{max}}{U_g}.$$

The compression ratio of a pump depends on the conductance of gaps that increases with an increase in pressure. In high vacuum $K \sim 50$, and at atmospheric pressure $K \sim 1.5$. To

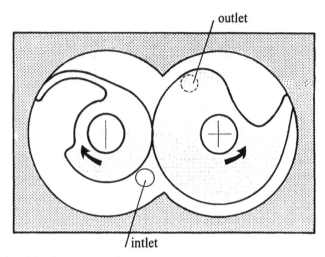

Figure 4.16 Schematic of a two-rotor claw pump.

increase the compression ratio, rotors in Roots pumps may have three or more lobes. For a large number of lobes, the Roots pump transforms to a gear pump. For asymmetrical shafts with claws and grooves (Fig. 4.16), one may obtain high compression at the expense of a decrease in the volume of the working chamber and pumping speed. A single shaft with partitions can provide several sequential stages of pumping. Three-stage pumps provide exhaust directly into the atmosphere.

At high pressures, screw pumps are used. One of the rotors in these pumps is in the form of a multiple-thread screw, and the others have ridges. The length of the working part of the shaft is greater than the thread pitch. Pumping is effected in the axial direction by moving the point of contact between the two shafts. The pumps provide high compression factors, but, because of their large size they are difficult to form into multistage units. The dependence of their pumping speed on inlet pressure for the two-rotor pump (Fig. 4.17) has a characteristic drop in the range of high pressure.

The limiting pressure of these pumps for operation with a roughing pump is 4×10^{-2} Pa with account of the saturation pressure of vacuum oils used for the lubrication of the

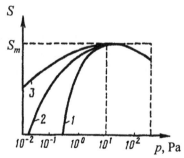

Figure 4.17 $S(p)$ characteristic of a two-rotor pump: (*1*) one-stage, (*2*) two-stage, and (*3*) two-stage with a trap.

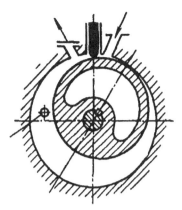

Figure 4.18 Schematic of a ballast gas pump.

bearings. For operation with a trap, the limit pressure is 10^{-3} Pa. The breaking pressure of one-stage pumps is from 10^2–10^3 Pa.

The speed of two-rotor pumps is from 5 to 5000 l/s. The specific characteristics of these pumps are 0.5–3 kg/(l/s) and 6–30 W/(l/s), with high-speed pumps having smaller specific power. The characteristics of two-rotor pumps of various types are shown in Fig. and Table A.6.

The operation of volume vacuum pumps can be accompanied by a number of undesirable phenomena, i.e., penetration of working liquid vapor into the pumped volume, by pollution of the pump by pumped substances with high saturation pressure, and loss of the working liquid through the outlet pipe.

These phenomena can be minimized using special service equipment, which in case of necessity is installed on volume pumps: traps, moisture absorbers, gas purges, condensers, filters, etc.

At high pressures (more than 100 Pa) the back flow of working liquid vapor is swept by the counter flow of the pumped gas, and protection devices are not necessary. At lower pressures for which the free path of gas molecules becomes greater than the diameter of the inlet pipe of the pump, working liquid vapor may move towards the main flow and penetrate into the pumped volume. If the pump temperature is higher than that of the pumped volume, the back flow will exist until all the working liquid in the pump is transferred to this volume. To protect this volume from working liquid vapor, traps, i.e., devices for partial removal of working liquid vapor are used.

Pumping of vacuum systems with vapors of water or other solvents with high saturation pressure at room temperature causes the danger of pollution of the pump with the pumped substances. In this case pumps are equipped with gas-ballast devices that allow one to lower the compression ratio of the pumped vapor and to prevent its condensation in the working chamber. In vane pumps with rolling rotor, the gas-ballast device is in the form of a leak for the supply of atmospheric air to the working chamber.

Example of the device for let-in of ballast gas is shown in Fig. 4.18. The pump casing has aperture A leading to the atmosphere, that is located to the side of the rotor face in such a way that, during the rotation of the rotor, it is opened only during compression in the

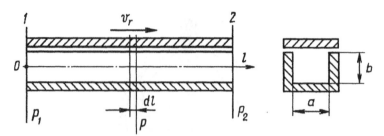

Figure 4.19 Schematic of molecular drag pumping.

chamber. During suction, the face part of the rotor closes the aperture and protects the chamber from filling with atmospheric air. For adjusting the amount of ballast gas let-in, the aperture A is connected to the atmosphere through a leak. If the leak is closed, the pump works in normal mode. Use of a gas-ballast device impairs the limit pressure of a pump because of an increase of gas transfer from the compression chamber to the rarefaction chamber.

For the protection from dust particles, the inlet of a pump may have an inlet filter, which captures dust particle with sizes more than 10 μm. The loss of working liquid through the outlet pipe is reduced by traps. For pumping of large gas flows, suck traps may be inefficient. In this case, external traps and filters for oil mist condensation should be used.

For pumping of oxygen one should use incombustible oils or other flame-proof liquids, for example, phosphate ethers. For radiation-hazardous premises, special radiation proof oils have been developed. Pumping of valuable and radioactive gases requires pumps with gas-tight elements at the output. In such designs, vacuum seals are installed in the oil-feeding pipe, the drain the sight glass, and the pump casing is tested for tightness.

4.4   Molecular Pumping

Removal of a gas from a vacuum system by means of driven surfaces is referred to as molecular pumping. There are two types of molecular pumping. The first-molecular drag one (Fig. 4.19) is pumping through a channel, one of the walls of which moves with the speed v_r parallel to the channel axis. Gas molecules collide with the moving surface and acquire addition impulse in the direction of the roughing pump. The difference of pressure is thus created: $p_2 > p_1$. The maximum pumping speed that can be achieved in such pumps is proportional to the speed of the wall v_r:

$$S_{max} = \gamma F_c v_r , \tag{4.13}$$

where F_c is the cross-section area of the channel and γ is the factor taking into account the ratio of the moving and the stationary parts of the channel perimeter.

Accepting the specific number of molecular collisions with the moving and the stationary surfaces to be identical, we determine the fraction of molecules the speed of which increases permanently:

$$\gamma = \frac{f_m}{f_m + f_s}, \tag{4.14}$$

where f_m and f_s are the moving and the stationary parts of the cross section perimeter of the channel. For a rectangular channel with $a = 2.5$ cm, $b = 1$ cm and $v_r = 165$ m/s, according to Eqs. (4.13) and (4.14), we obtain $S_{max} = 23$ l/s.

The differential equation of gas flow through a channel of constant cross-section in steady-state mode ($Q = $ const) can be written as the difference of the direct and the back flows:

$$Q = S_{max}p - C\frac{dp}{dl}, \tag{4.15}$$

where $C = U_c l_c$, U_c is the conductance of the channel with stationary sides, and l_c is the length of the channel. We write Eq. (4.15) in new notations:

$$\frac{dp}{dl} + Ap - B = 0,$$

$$A = -\frac{S_{max}}{C}; \quad B = -\frac{Q}{C}. \tag{4.16}$$

In view of the boundary condition $p = p_1$, at $l = 0$ the solution of Eq. (4.16) can be written as

$$p = \frac{B}{A}[1 - \exp(-Al) + p_1\exp(-Al)],$$

and then at the end of the channel at $l = l_c$ the pressure is

$$p_2 = \frac{Q}{S_{max}}\left[1 - \exp\left(\frac{S_{max}}{U_c}\right) + p_1\exp\left(\frac{S_{max}}{U_c}\right)\right].$$

Taking into account that $Q = p_1 S_p$, we obtain from the latter expression of pumping speed that

$$S_p = S_{max}\frac{p_2/p_1 - \exp(S_{max}/U_c)}{1 - \exp(S_{max}/U_c)}. \tag{4.17}$$

Equation (4.17) suggests a linear dependence between pumping speed and compression p_2/p_1 (Fig. 4.20). At equal pressures $p_1 = p_2$ the pumping speed is the highest: $S_p = S_{max}$, and at $S_p = 0$, $p_1 = p_{lim}$, and $p_2 = p_{out}$ the compression ratio is the highest:

$$K_{max} = \frac{p_{out}}{p_{lim}} = \exp\left(\frac{S_{max}}{U_c}\right). \tag{4.18}$$

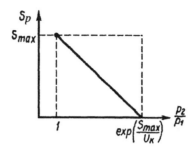

Figure 4.20 Pumping speed as a function of compression ratio.

Because the conductance of the channels U_c in molecular flow mode is proportional to $\sqrt{T/M}$, the maximum compression coeffcient grows with an increase in the molecular mass and a decrease in temperature of pumped gas. The dependence of K_{max} and S_{max} on channel side a at $a = b$, $v_r = 200$ m/s, and $l = 1$ m is shown in Fig. 4.21.

For a rectangular channel with $a = 2.5$ cm, $b = 1$ cm at $l_c = 55$ cm and $v_r = 165$ m/s, according to Eq. (4.18) the maximum compression ratio is $p_{out}/p_{lim} = 10^{10}$. Thus, this scheme of molecular pumping is convenient for obtaining high compression at small pumping speeds. With an increase in the molecular mass of pumped gas, K_{max} grows, and S_{max} remains unchanged. Thus, this pump is more effective for heavy gases.

The second scheme of molecular pumping-turbomolecular (Fig. 4.22, a) uses the dependence of the conductance of an inclined channel, that moves perpendicularly to a gas flow with the speed v_r, on the direction of gas flow. To simplify the problem, we accept that a plate with an inclined channel is bombarded from both sides, perpendicularly to the plate surfaces, with flows of gas molecules q_1 and q_2. Stopping the plate and summing the relative velocity vectors of the molecules v_r with the thermal velocity vectors of the molecules v_{ar}, we obtain changed the resultant direction of molecular movement (Fig. 4.22, b). At $\tan \alpha = v_{ar}/v_r$ the flow q_1 enters the channel along its axis, and the flow q_2, in the perpendicular direction. As a result, the channel has different conductance for the flows q_1 and q_2.

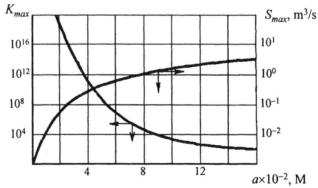

Figure 4.21 Maximum pumping speed and compression ratio as a function of channel side a for $a = b$, $l = 1$ m, and $v_r = 200$ m/s.

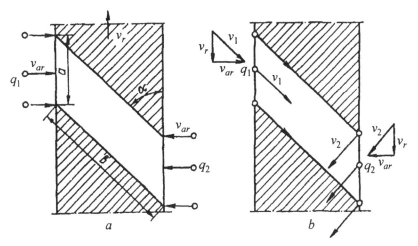

Figure 4.22 Schematic of turbomolecular pumping.

One may approximate that for the flow q_1 the channel is pipe-shaped, and for the flow q_2 it is in the form of a pipe turned through 90° arc.

For steady-state gas flow mode

$$Q = S_p p_1 = U_{12} p_1 - U_{21} p_2, \tag{4.19}$$

where U_{12} and U_{21} are the conductance of the channel, for the flows q_1 and q_2, respectively. The above conductances can be determined using the data on the conductances of straight pipes and pipes with bends or direct mathematical modeling using the Monte–Carlo method. Figure 4.23 presents dimensionless data on the probability of passage of a channel by

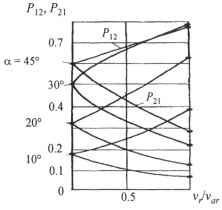

Figure 4.23 Probabilities of transmission the channel P_{12} and P_{21} as a function of velocity ratio v_r/v_{ar} and angle α.

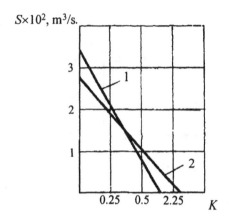

Figure 4.24 Pumping speed as a function of compression ratio K ($a = b = 2\times10^{-2}$ m, $\alpha = 30°$ arc, and $v_r = 200$ m/s): (*1*) N_2 and (*2*) CO_2.

gas molecules as determined by mathematical modeling for channels with a side ratio of $a/b = 1$. From Eq. (4.19), the expression pumping speed is

$$S_p = U_{12} - U_{21}\frac{p_2}{p_1}.$$

The dependence of pumping speed on compression ratio for this pump is similar to that for the first scheme. The maximum pumping speed at $p_2/p_1 = 1$ is

$$S_{max} = U_{12}\left(1 - \frac{U_{21}}{U_{12}}\right) = U_{12}\left(1 - \frac{P_{21}}{P_{12}}\right). \tag{4.20}$$

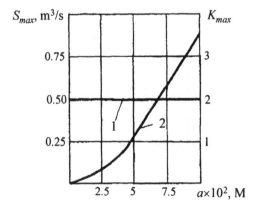

Figure 4.25 Maximum pumping speed and maximum compression ratio as functions of channel side a ($a = b$, $\alpha = 30°$ arc, and $v_r/v_{ar} = 0.5$): (*1*) K_{max} and (*2*) S_{max}.

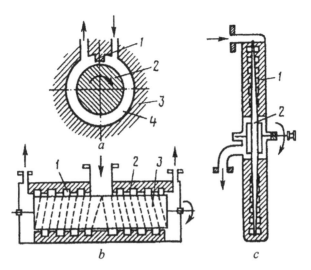

Figure 4.26 Schematics of molecular drag pumps.

where P_{12} and P_{21} are the probabilities of the passage of molecules through a channel in the direct and reverse directions, which is proportional to appropriate conductances. The highest compression ratio is observed at $S_p = 0$. If $p_1 = p_{lim}$ and $p_2 = p_{out}$, then from Eq. (4.19), $K_{max} = p_{out}/p_{lim} = U_{12}/U_{21} = P_{12}/P_{21}$. The p_{out}/p_{lim} ratio for one stage is not large (usually 2–4), and, therefore this scheme is more convenient for obtaining high pumping speed. An increase in compression ratio is achieved by sequential connection of several pumping stages.

The speed of a pump is weakly dependent on the molecular mass of gas (Fig. 4.24), especially in working mode. K_{max} is shown as a function of channel size in Fig. 4.25.

4.5   Design of Molecular Pumps

Molecular drug pumps with the same direction movement of gas and channel walls have many modifications, some of which are shown in Fig. 4.26. The stator of the pump shown in Fig. 4.26, a has three sets of cylindrical grooves, the inlets and outlets of which are separated by partition 1. The rotor 2 rotates with high frequency such that its linear speed is close to the thermal velocity of the gas molecules. The spiral groove 1 (Fig. 4.26, b) on the surface of the stator 2 and the cylindrical surface of the rotor 3 form a working channel. The spiral grooves on the face surfaces of the stator 1, that are at the minimum distance from the rotating disk 2, are used for molecular pumping as shown in Fig. 4.26, c. Through the gap between the stator and the rotor gas passes back from the compression space to the inlet space, which effect impairs the real of these pumps. Normal work of these pumps requires the gap between the rotor and the stator be not wider than 0.1 mm.

These pumps are used at high pressure stages in combination with pumps having mutually perpendicular movement of gas molecules and working surfaces, as well as for

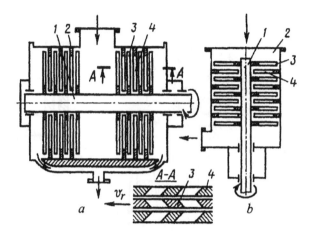

Figure 4.27 Schematics of turbomolecular pumps: (*a*) with horizontal shaft and (*b*) with vertical shaft.

pumping of gases with large molecular mass. The penetration of the oil vapors used for the lubrication of bearing into the pumped volume during pumping is insignificant, but it grows strongly as the pump is stopping.

The speed of pumps is directly proportional to the rotation frequency of the rotor, which reaches 10–40 thousands rpm. The maximum pumping speed does not usually exceed 100 l/s because of the small cross section of the channels. The limiting pressure of these pumps is 10^{-5} Pa and the compression ratio is 10^5–10^6, for air.

Turbomolecular pumps with mutually perpendicular movement of working surfaces and pumped gas flows have found general use. The designs of turbomolecular pumps based on this principle are classified by rotor shaft arrangement into horizontal and vertical, and by the construction and shape of working body, into cylindrical, conical, disc with radial flow, disc with axial flow, and drum-type. The parameters of a pump are greatly affected by the design of the support bearings (lubricated rolling contact bearings, magnetic supports, or gas cushion).

Pumps with horizontal and vertical rotor shaft arrangement are shown in Fig. 4.27, *a* and *b*. The casing 2 has stationary stator wheels 4, between which rotate the wheels 3 that are fixed on the rotor 1. The rotor wheels are in the form of slotted discs. The stator wheels have mirror-image slots of the same shape. For convenience of installation, the stator wheels are cut along the diameter.

For horizontal rotor arrangement, the movement of gas in the pump after entry into the inlet pipe is branched into two flows that are recombined in the outlet pipe.

Because of the small compression ratios at each stage in a turbomolecular pump, working gaps can be increased. For a driving wheel diameter of 200 mm, the axial (between the wheels) and radial (between the casing and the rotor wheel or the rotor and the stator wheels), gaps can be 1–1.2 mm, which allows to considerably increase the reliability of their operation. Increasing of gaps by reducing the compression ratio of a pump affects its speed only slightly.

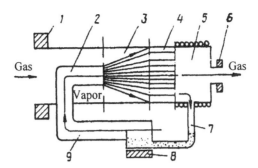

Figure 4.28 Schematic of vapor jet pumping.

The speed of turbomolecular pumps is weakly dependent on the type of gas. The limiting pressure of pumps is 10^{-7}–10^{-8} Pa. With an increase in molecular mass the compression ratio grows because of a decrease in leakage in the gaps and an increase of the ratio of rotor speed to the thermal velocity of molecules v_r / v_{ar}. The highest output pressure of such pumps for air is 1–10 Pa.

The advantages of turbomolecular pumps are the high specific speed, 2 l/s per 1 cm^2 of the inlet aperture area, the reasonably wide range of working pressures, 10^{-6}–10 Pa, fast start of the pump during 5–10 min., and practically oil-free spectrum of residual gases iwhen purging with dry nitrogen during pump starting and stopping.

A disadvantage of these pumps is the high-speed rotor with lubricated fastly worn bearings or complex support scheme. Characteristics of turbomolecular pumps are shown in Fig. and Table A.7.

4.6 Vapor Jet Pumping

During vapor jet pumping (Fig. 4.28), molecules of the pumped gas passing into the pump through the inlet branch pipe 1 interact with the vapor jet that has sound or ultrasonic speed and acquire additional velocity in the direction of the roughing pump that is connected to the outlet pipe 6.

In the chamber 3, the vapor jet mixture from nozzles 2 is mixed with the pumped gas. Isolating channel 4 stops the back flow of gas, thus providing the compression. The pumped gas and the working vapor are separated in chamber 5 during condensation of the working vapor on cooled surfaces, then the pumped gas leaves the pump through the outlet pipe, and the condensed vapor passes through pipeline 7 to boiler 8, where it is evaporated again and, through pipe 9, passes to working nozzle 2, providing continuous pumping.

The interaction of pumped gas with a vapor jet depends on the degree of vacuum. At low vacuum, the molecules in the layer adjacent to the vapor jet entrain other gas layers due to internal friction. Such devices are referred to as ejector pumps.

At high vacuum, all the molecules of the pumped gas move by self-diffusion and directly interact with the moving vapor jet; pumps that use this principle are called diffusion pumps.

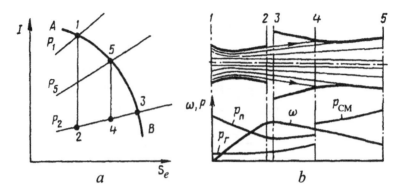

Figure 4.29 (*a*) IS_e diagram of working vapor and (*b*) working parameters of ejector pump nozzle.

We consider the action principle of ejector pumps. The speed of these pumps at given pressures at the inlet p_2 and the output p_5, as well as the throughput of working vapor G_1, are determined from the IS_e diagram of the vapor.

Curve AB in Fig. 4.29, a shows saturated vapor of the working liquid pressure. From the initial condition in boiler (point 1 in curve AB, pressure p_1, section 1 in the nozzle scheme) the working vapor is adiabatically expanded and transfers to the condition (point 2) with pressure p_2 of working vapor jet in the pumped volume (section 2). Adiabatic processes on the IS_e diagram are shown by direct lines parallel to the I-axis. The energy conservation law for adiabatic gas evolution, for which the work of gas expansion is equal to the increase in its kinetic of energy, can be written as

$$G_1(I_1 - I_2) = G_1 \frac{\omega_2^2}{2}, \tag{4.21}$$

where ω_2 is the speed of the vapor jet at the outlet from the nozzle in section 2. The pumped gas (for simplicity we consider, that working liquid vapor is pumped) is in the state corresponding to point 3 and section 3. Mixture of the pumped gas flow G_2 with vapor jet, according to the energy conservation law, will change speed:

$$G_1 \frac{\omega_2^2}{2} = (G_1 + G_2) \frac{\omega_4^2}{2}, \tag{4.22}$$

where ω_4 is the speed of the mixture in section 4 (Fig. 4.29, *b*). In the diffusor between sections 4 and 5, the vapor–gas mixture is adiabatically compressed to the pressure p_5, and the point 5 that correspond to section 5, should be on curve AB. This fact can be used for the location of point 4 by graphical construction. In adiabatic compression the kinetic energy of the jet transforms to enthalpy, for which reason we may write

$$(G_1 + G_2) \frac{\omega_4^2}{2} = (G_1 + G_2)(I_5 - I_4), \tag{4.23}$$

where I_4 and I_5 are the enthalpies in points 4 and 5.

From Eqs. (4.21)–(4.23) one may find the expression for the theoretical speed of a pump:

$$S_t = \frac{G_2}{\rho_2} = \frac{G_1(I_1 - I_2 - I_5 + I_4)}{\rho_2(I_5 - I_4)}, \tag{4.24}$$

where ρ_2 is the density of gas in section 2 and I_1 and I_2 are the enthalpies in points 1 and 2. The speed of a pump depends on the efficiency of nozzles and the properties of the working vapor. In ejector pumps it is in the range from several tens to several thousands of litres per second.

The maximum output pressure cannot be more than p_1 – the pressure of the working vapor in the boiler; therefore for oil vapor pumps it does not exceed $(1-5) \times 10^2$ Pa, and for mercury vapor pumps, $(20-40) \times 10^2$ Pa. It is impossible to increase the maximum output pressure of an oil vapor pump, because the temperature of the vapor in the boiler is limited to the temperature of oil decomposition. In mercury vapor pumps the maximum output pressure can be increased to the atmospheric one, but this is a rare practice because of the large losses and the toxicity of mercury.

The limiting residual pressure of an ejector pump is the pressure of the transition from medium to high vacuum, at which the vapor jet is expanded and the optimum operation mode is violated. The limiting pressure is 10^{-1} to 10^{-2} Pa.

The size of nozzles can be determined, given its efficiency G_1 for the working vapor. Substituting the critical pressure ratio

$$r_c = \frac{p_2}{p_1} = \left(\frac{2}{\gamma+1}\right)^{\frac{\gamma}{\gamma-1}}$$

into Eq. (A4.4), we obtain the expression for the maximum nozzle efficiency

$$P_1 = \sqrt{\frac{p_1}{v_1}\gamma\left(\frac{2}{\gamma+1}\right)^{\frac{\gamma+1}{\gamma-1}}} A_{min}, \tag{4.25}$$

where p_1 is the vapor pressure in the boiler, v_1 is the specific volume of vapor in the boiler, and A_{min} is the minimum nozzle section. From Eq. (4.25) one may find the minimum section as

$$A_{min} = \frac{P_1}{\sqrt{\frac{p_1}{v_1}\gamma\left(\frac{2}{\gamma+1}\right)^{\frac{\gamma+1}{\gamma-1}}}}. \tag{4.26}$$

The outlet section of ultrasonic nozzle A_2 can be found from Eq. (4.21) considering that $\omega_2 = P_1 v_2/A_2$ (is the specific volume of vapor at the nozzle output):

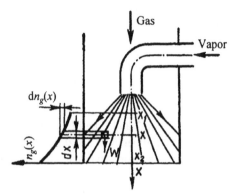

Figure 4.30 Nozzle of a diffusion pump.

$$A_2 = \frac{P_1 v_2}{\sqrt{2(I_1 - I_2)}}.$$

When a vapor jet is evolved into high vacuum it is expanded due to the thermal velocities of its molecules. At a jet speed equal to the speed of sound the jet leaves the nozzle at an angle of 45 deg to its axis (Fig. 4.30). The pressure in the vapor jet is much higher than the pressure of the pumped gas. The best conditions for the trapping of pumped gas molecules are provided if the pressure in a vapor jet corresponds to medium vacuum. Thus, all the molecules of the pumped gas penetrate into the vapor jet at first collision. At greater density of vapor jet, the probability of molecules capturing is reduced. Because of the small amount of pumped gas, the theoretical speed of a diffusion pump is determined not by the thermodynamic characteristics of the vapor jet, as is in ejector pumps, but by the geometrical sizes of the nozzles and the partial pressure of the pumped gas in the vapor jet:

$$S_t = \frac{A}{n}(N_{q1} - N_{q2}), \tag{4.27}$$

where A is the projection of the vapor jet surface accessible for pumped gas molecules onto the plane perpendicular to the x-axis, N_{q1} and N_{q2} are the numbers of pumped gas molecules, incident and and leaving in unit time from unit area A, n is the gas concentration at the pump inlet. According to the expressions for N_q from (1.34) and v_{ar} from (1.18), the speed of a pump (4.27) can be written as

$$S_t = A\frac{\sqrt{8k/(\pi m)}}{4n}(n\sqrt{T} - n_g\sqrt{T_g}), \tag{4.28}$$

where T and T_g are the gas temperatures at the pump inlet and in the vapor jet, respectively, and n_g is the gas concentration in the vapor jet.

Since $n_g/n = p_{lim}/p$ then,

$$S_t = \left(\frac{R}{2\pi}\right)^{1/2} \sqrt{\frac{T}{M}} A \left(1 - \frac{p_{lim}}{p} \sqrt{\frac{T_g}{T}}\right). \qquad (4.29)$$

Thus, the speed of a diffusion pump depends on the temperature and type of gas, with heavy molecules being pumped with smaller speed. The real pumping speed is less than theoretical value calculated from Eq. (4.29), because of partial trapping of pumped gas molecules by the vapor jet. This is associated with the so-called 'vapor splume', which appears as a result of a flow of vapor molecules whose thermal velocities are greater than the speed of the vapor jet.

Introducing the capture coefficient H_0, the expression for the pumping speed of action can be written in the following form

$$S_p = H_0 S = H_0 \sqrt{\frac{RT}{2\pi M}} A \left(1 - \frac{p_{lim}}{p} \sqrt{\frac{T_g}{T}}\right) \qquad (4.30)$$

The mean capture coefficient is $H_0 = 0.3$.

To determine the compression ratio and the limiting pressure we consider in more detail the processes occurring in the vapor jet of a diffusion pump. The gas concentration in the point x of the vapor jet will be denoted as $n_g(x)$ (Fig. 4.30), and the component of the vapor jet speed along the x-axis, as W.

The throughput of pumping is equal to the difference of the forward and back flows:

$$Q = Q_f - Q_b,$$

where $Q_f = n_g(x)W$, $Q_b = D dn_g(x)/dx$, and D is the diffusion coefficient of gas in the vapor jet. The forward flow of pumped gas molecules in a diffusion pump due to collisions with gas and vapor molecules with the vapor jet, and the back flow is due to the thermal movement of the molecules that diffuse in the opposite direction. At the limiting pressure, $Q = 0$, whence

$$n_g(x)W - D\frac{dn_g(x)}{dx} = 0, \qquad (4.31)$$

From Eq. (1.10)

$$\frac{dn_g(x)}{n_g(x)} = \frac{dp_g(x)}{p_g(x)}. \qquad (4.32)$$

Integrating Eq. (4.31) with account of Eq. (4.32) in the limits from x_1 to x_2 and from p_1 to p_2, we have

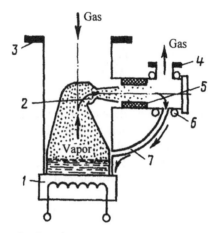

Figure 4.31 Principial schematic of an ejector pump.

$$\ln\frac{p_2}{p_1} = \frac{W}{D}(x_2 - x_1).$$

Designating the length of vapor jet $b = (x_2 - x_1)$, we rewrite the resultant expression as

$$\frac{p_2}{p_1} = \exp\left(\frac{Wb}{D}\right).$$

Thus, the compression ratio for a given jet geometry is determined by the ratio of the projection of the vapor jet speed on the x-axis and the diffusion coefficient of the pumped gas in the vapor jet. To increase the compression ratio one should increase the speed of the

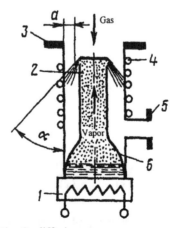

Figure 4.32 Principial schematic of a diffusion pump.

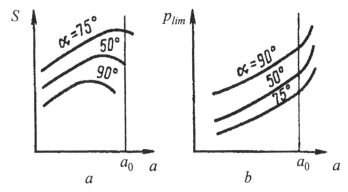

Figure 4.33 (a) S and (b) p_{lim} as functions of ring gap a for different nozzle inclination angle α.

vapor jet and increase its density, because the diffusion coefficient D is inversely proportional to vapor jet density.

We consider the designs of vapor jet pumps. The scheme of an ejector pump is shown in Fig. 4.31. The pump consists of boiler 1, supersonic ejector Laval nozzle 2, mixing chamber 5, and inlet and outlet flanges 3 and 4. The mixing chamber is thermoinsulated from the pump casing. The outlet pipe has cooler 6 cooled by circulating water. Vapors condensing on the cooler walls to the boiler through pipe 7, ensuring continuous working liquid circulation in the pump.

The simplest diffusion pump (Fig. 4.32) consists of boiler 1, expansion nozzle 2 fixed on pipe 6, cooler 4, and inlet and outlet pipes 3 and 5. Vapors of the working liquid pass from the boiler through the pipe and the ambrella-like nozzle and are condensed at the cooled pump walls. During vapor movement from the end of the nozzle to the pump walls the pumped gas diffuses into the vapor jet. After condensation of the formed vapor-gas mixture the evolved gas is pumped through the outlet pipe by the roughing pump, and condensed vapors flow at the pump walls into the boiler through the gap between the pipe and the pump casing.

The effect of dimension a and nozzle inclination angle α on the limit pressure and speed of diffusion pumps is shown in Fig. 4.33, a, b. At greater gaps than the optimum value a_0 the speed of a jet at the pump walls decreases, which results in an increase of the back flow. Similar dependence is true for the angle α, which strongly affects the axial component of the jet speed.

Diffusion pumps intended for work in the pressure range from 10^{-1} to 10 Pa are referred to as booster pumps. In these pumps the power of the heater is increased and thermally stable working liquids are applied, which increases the output pressure and shifts the characteristic of the pump $S_p = f(p)$ to the higher inlet pressures.

The main characteristic of vapor jet pumps is the dependence of pumping speed on the pressure at the pump inlet (Fig. 4.34, a). In the medium range of working pressures the pumping speed is constant and equal to S_{max}. As the working pressure approaches the limiting one, the speed tends to zero because of the back flow of gases and vapors from the pump to the pumped volume. With an increase in the working pressure above the upper limit of molecular flow mode, the pumping speed decreases due to a decrease in the diffusion

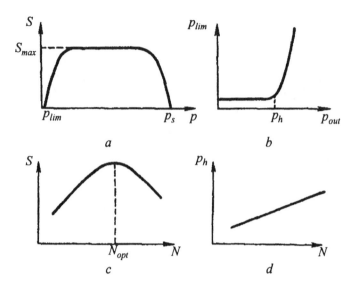

Figure 4.34 Characteristics of vapor jet pumps.

velocities of gas molecules in the vapor jet, and at the maximum inlet pressure p_3 tends to zero.

The limiting pressure depends weakly on the low pressure at the outlet pipe (Fig. 4.34, b). The pump characteristic fails at the equality of the outlet pressure and the pressure in the vapor jet, corresponding to pressure p_h.

With an increase in the power N of pump heating at the expense of an increase in vapor jet speed, the pumping speed first increases (Fig. 4.34, c), reaches a maximum at N_{opt}, and then decreases because of an increase in vapor jet density. The maximum output pressure of a pump p_h grows continuously with an increase in heater power (Fig. 4.34, d).

4.7   Working Liquids

The working liquids used in vapor jet pumps should meet the following requirements: (1) minimum vapor pressure at room temperature and maximum at the working temperature in the boiler, (2) resistance to decomposition at heating, (3) minimum ability to dissolve gases, (4) chemical stability in relation to pumped gases and to the structural materials of the pump, and (5) low heat of vaporization.

The minimum vapor pressure at room temperature is required for obtaining the least limiting pressure of a pump. The maximum vapor pressure at the working temperature increases the output pressure of a pump and reduces the required heater power. The resistance to decomposition of the working liquid at heating affects its service life and the maximum output pressure. High solubility of gases in the working liquid results in an increase of the back flow of gases through the nozzle together with the vapor jet. Chemical stability determines

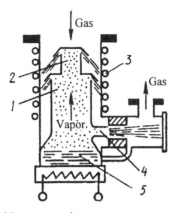

Figure 4.35 Schematic of a multistage vapor jet pump.

the service life of the working liquid and choice of structural materials for pumps. For low heat of vaporization, lower heater power is required.

As a working liquid of vapor jet pumps, mercury, mineral oils, complex ethers of organic spirits and acids, and silicon-organic compounds are used.

Mercury it is not oxidized by air, has a homogeneous structure, does not decompose at working temperatures, dissolves only small amounts of gases, and has a high vapor pressure at working temperatures. The drawbacks of mecury are its toxicity, chemical activity in relation to non-ferrous metals, and high vapors pressure at room temperature (10^{-1} Pa).

Mineral oils used in vapor jet pumps are produced by vacuum distillation of petroleum processing products. They have low vapor pressure at room temperature (10^{-6} Pa), satisfactory thermal stability, but have low thermal oxidation resistance and may form resin deposits on the internal parts of pumps.

The ethers used as working liquids of vapor jet pumps are products of the synthesis of phthalic and sebacic acids with higher alcohols, as well as polyphenyl compounds that consist benzene radicals connected in chains through oxygen atoms. Polyphenyl ethers (PPE) have very low vapors pressure at room temperature (10^{-9} Pa) and high thermal oxidation resistance.

Silicon-organic liquids for vapor jet pumps are polysiloxane polymer compounds consisting of $(CH_3)_2SiO$ functional groups. They have high thermal oxidation resistance and reasonably low vapor pressure at room temperatures (10^{-5} Pa). Hydrocarbon liquids have good oxidation stability and low vapor pressure.

Inexensive mineral oils are often used in pumps, silicon-organic and hydrocarbon liquids are used in systems with frequent inlet of atmospheric air. Ethers, the cost of which is still high, are used in systems in which ultrahigh vacuum is required. Because of its toxicity, mercury is used in vapor jet pumps only for pumping of mercury systems, for example, mercury rectifiers. The main types of working liquids are shown in Table and Fig. A.4.

In pumps working with structurally nonuniform or thermally unstable liquids, fractionating devices are often used that separate heavier fractions with low vapor pressure for the first stage of pumping of a vapor jet pump.

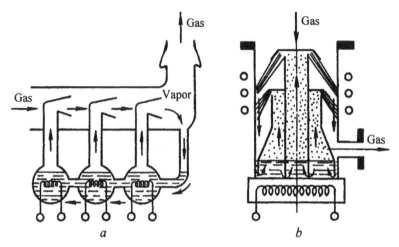

Figure 4.36 Fractionating vapor jet pumps: (*a*) glass and (*b*) metallic.

4.8   Design of Vapor Jet Pumps

Figure 4.35 shows design of a multistage vapor jet pump, in which two diffusion (1, 2) and one ejector (4) pumping stages are used, which are fed by one boiler 5. The pump casing 3 is water-cooled. In Fig. 4.36, *a*, *b* designs of glass and metallic fractionating pumps are shown.

Working liquid condensating on the pump walls because of the large hydraulic resistance reaches the first stage of the pump in a reasonably long time that is necessary for the evaporation of light fractions. The hydraulic resistance that hinders mixing of the working liquid is produced in a glass vessel using thin pipes that consecutively connect the boilers of different pumping stages from the stage working at the maximum pressure of pumped gas. In the metallic design, the same function is performed by the gap between the vapor conducting cylinder and the base of the pump.

The main characteristics of commercial vapor jet pumps are shown in Fig. A.4 and Tables A.8 and A.9.

4.9   Traps

The limiting pressure of oil pumps is determined by the back flows of working liquid vapors from the pump to the pumped volume. These flows can be considerably reduced by installing traps in their course. The traps should meet the following main requirements: maximum protective action for the desired service life and minimum resistance to the main flow of pumped gas. Additional requirements are the capability of regeneration, reliability, simplicity, adaptability to various designs, and convenience of operation. By the principle of action, the traps can be subdivided into condensing, sorbing, and dissociating.

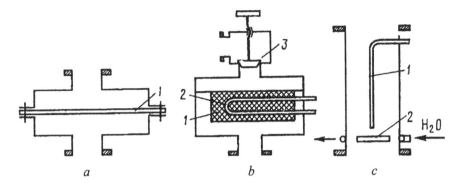

Figure 4.37 Schematics of traps for low-vacuum pumps: (*a*) mechanical, (*b*) adsorption, and (*c*) ion.

Condensing mechanical traps (Fig. 4.37, *a*) are devices that hinder direct flow of working liquid vapors from the pump to the pumped volume. For obtaining of appreciable protective action up to atmospheric pressure conditions of molecular gas flow mode should be provided in the trap, which corresponds to a highest distance between trap elements of 0.1 μm. In protective screens 1 these sizes can be obtained by application of porous materials: porous glass, glass–fiber materials, porous copper, and stainless steel. The most frequently used materials are shown in Table 4.1.

Vapor sorption in oil traps is made by adsorption on walls of capillary channels. The period of continuous work of a trap is several hundreds of hours, and after that the element should be replaced, cleaned by blowing with atmospheric air or heated to high temperatures (~500°C). Constant temperature is maintained in traps during their work by circulating water. Reducing trap temperature additionally increases the protective action of traps, but somewhat reduces the specific conductance.

One of the designs of condensing baffles for vapor jet pumps is water-cooled reflective cap 1 (Fig. 4.38, *a*). The use of reflective cap reduces the back flow of oil vapor through unit area of the gap between the first nozzle and the pump casing from 10^{-3} to 10^{-5} mg/(s cm²). Grating-type and conical disc traps (Fig. 4.38, *b*) reduce the back flow to 10^{-6} mg/(s cm²). The evaporation rate of oil vapor from the surface of such water-cooled traps is considerably smaller than from heated to 200°C top nozzle of a vapor jet pump.

Table 4.1 Parameters of porous materials.

Parameter	Material			
	Fine-pored glass	Coarse-pored glass	Stainless steel	Copper filters
Pore size, μm	3×10^{-3}	1×10^{-1}	1.5×10^{1}	2×10^{1}
Specific conductance, m³/(s cm²)	1×10^{-10}	1×10^{-8}	2×10^{-5}	1×10^{-4}

Note. Trap size depends on the specific conductance of the porous materials used.

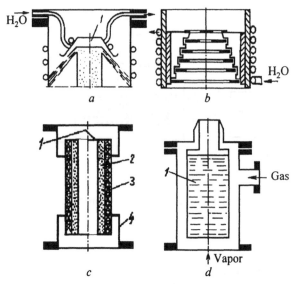

Figure 4.38 Schematics of traps for high-vacuum pumps: (*a*) reflecting cold cap, (*b*) conical disc trap, (*c*) uncooled sorption trap, and (*d*) liquid nitrogen trap.

Cooled traps work below room temperature. The most frequently used are ice with NaCl (−18°C), ice with $CaCl_2$ (−48°C), solid carbonic acid with alcohol (−78°C), freon (−120°C), liquid air (−183°C), and liquid nitrogen (−196°C). Low temperatures (to −70°C) can be produced using semiconductor elements that use the Peltier effect. A decrease in the temperature of the condensing surface reduces the vapor pressure of working liquid. For example, for water at temperatures +15, −78, and −196°C the vapor pressure is 2×10^3, 7×10^{-2}, and 10^{-19} Pa, respectively, for mercury at +18 and −196°C the vapor pressure is 10^{-1} and 10^{-30} Pa, respectively.

The pumping speed per unit surface of condensing traps is determined by the difference of the molecular flows that are incident on the condensing surface N_1 and leave it N_2:

$$S = \frac{N_1 - N_2}{n_1} = \sqrt{\frac{kT_1}{2\pi m}}\left(1 - \frac{n_2}{n_1}\sqrt{\frac{T_2}{T_1}}\right), \tag{4.33}$$

where n_1 and n_2 are the molecular concentrations of working liquid vapor in the gaseous phase and on the condensing surface, respectively, T_1 and T_2 are the temperatures of working liquid vapor in the gaseous phase and on the condensing surface, respectively, and m is the molecular mass of the working liquid, kg. At $T_1 = 298$ K, Eq. (4.33) can be transformed to

$$S = 116\sqrt{\frac{29}{M}}\left(1 - \frac{p_2}{p_1}\sqrt{\frac{298}{T_2}}\right),$$

where p_1 and p_2 are the pressures of working liquid vapor in the gaseous phase and on the condensing surface, respectively, and M is the molecular mass of the working liquid, kg/kmol. The design of metal traps cooled by liquid nitrogen 1 is shown in Fig. 4.38, d.

Adsorption traps adsorb oil vapors by the surface porous adsorbents, such as activated charcoal, zeolites, silicagels, etc.. The adsorbents should be cleaned of foreign substances adsorbed by molecular-size pores under usual atmospheric conditions by heating in vacuum at about 300°C. The adsorption of oil vapor on cleaned surfaces is usually carried out at room temperature until the equilibrum vapor pressure becomes lower than the allowable one. After that a trap should be regenerated by heating. The main components of air, nitrogen and oxygen, are adsorbed in very small amounts at room temperature.

It is possible to roughly calculate the service life of a trap, accepting with a safety margin that it absorbs the entire back flow of oil vapor. Using the equation of polymolecular absorption isotherm (Eq. 1.56)) the equilibrum pressure of oil vapor in a trap can be written in the following form:

$$p = p_T \frac{C(\theta - 1) + \sqrt{C^2(\theta - 1)^2 + 4\theta^2(C - 1)^2}}{2\theta(C - 1)}, \qquad (4.34)$$

where p_T is the saturation pressure of the oil vapor at the working temperature of the trap,

$$C = \exp\left(\frac{Q - E}{RT}\right),$$

θ is the coverage degree of adsorbent surface, and Q and E are the adsorption and condensation heats. Because

$$\theta = \frac{1}{a_m A} \int_0^{t_c} q \, dt,$$

where q is the back flow of oil vapor and A is the active adsorbent surface, then, at a constant back flow of oil vapor ($q = $ const) one may easily determine the service life of the trap:

$$t_c = \frac{a_m A \theta_{max}}{q}.$$

Here θ_{max} is determined from Eq. (4.34) at $p = p_{max}$, where p_{max} is the maximum allowable pressure of oil vapor in the pumped volume.

The service life of traps with porous filters can be increased by adsorbing materials, such as active zeolites with pore sizes of ~9 E and activated aluminium oxide. Traps with adsorbents (Fig. 4.37, b) sould not be exposed to atmospheric air, because they may absorb large amounts of atmospheric water, which will then be released during the operation of the pump. To remove oil and water adsorbed by the adsorbing element 1 the trap is heated by

Table 4.2 Protective elements of traps.

Trap type	Schematic	Optimum size	Specific conductance, $m^3/(s\ cm^2)$
Grating-type		$D/B > 5$ $\theta = 60°$	4.80×10^{-3}
Stripe-type		$D/B > 5$ $\theta = 60°$	3.16×10^{-3}
Conical disc type		$\theta = 60°$	4.50×10^{-3}
Corner-type		$A = B = 3R$	3.27×10^{-3}

heater 2 to 300°C. Valve 3 is placed in one casing with the trap and closed during adsorbent degasing to protect it from polluting the vacuum system.

The constructive scheme of adsorption trap shown in Fig. 4.38, c, consists of casing 4, heater 3, adsorbent 2, and reflectors 1 that provide optical densityof the trap. The materials of the trap should allow heating to 300–400°C. The adsorbent should be placed in such a way as to prevent the migration of oil vapors to the pumped volume through the trap walls. When designing traps, one should constructively reduce their thermal lag, which impair their operation parameters.

The heated surfaces of dissociating traps decompose hydrocarbons into easily pumped gases, such as hydrogen, carbon oxide, carbon dioxide, and solid carbon, which deposits on the trap walls, and light gases are removed by pumps. The dissociating surfaces are heated by electric current. The work of electronic dissociating traps is based on the excitation or ionization of working liquid molecules in a discharge with a cold or a hot cathode. The excitation increases the probability of the molecules to dissociate and, subsequently, polymerize on the trap walls. At sufficient electron energy complex molecules of oil may dissociate due to interaction into light components and carbon. The light components are removed by the pump, and carbon deposits on the trap walls. The efficiency of these traps depends on the density of electric current. Dissociating traps may also use catalytic decomposition of oil vapor on oxidized metal surfaces.

In dissociating ion traps (Fig. 4.37, c), the cylindrical casing serves as a grounded cathode for cold discharge. The anode is in the form of rod 1 arranged along the cylinder axis. The discharge glows at an anode voltage of ~3 kV in the presence of an axial magnetic field produced by external magnets. The electrons emitted by the cathode move over elongated trajectories to the anode and ionize the residual gases. The positive ions that bombard the surface of the casing destroy the surface oil films. This causes hydrogen evolution and the polymerization of hydrocarbons to solid substances. The casing and the protective screen 2 are water-cooled. These traps reduce the partial pressure of oil vapor by 10–100 times.

The common problem of designing traps of any type is to satisfy two mutually excluding requirements: the maximum protective action and the highest specific conductance. The protective action of traps can be evaluated from the average number of collisions of a molecule that passed through the trap with its protective elements. An increase in the protective action of a trap is usually accompanied by a decrease in its specific conductance. Setting the weight factors of the importance of these parameters for an appropriate technological process, one may choose the optimum sizes of a trap. The specific conductance and the optimum size ratios of various protective elements of traps are shown in Table. 4.2.

4.10 Test Questions

4.1 What is the effect of the volume of the dead space on the characteristics of mechanical pumps?

4.2 What is the reason of letting ballast gas into displacement pumps?

4.3 For what purpose multistage mechanical pumps are made?

4.4 Under which conditions does the back flow of working liquid vapor occur in mechanical vacuum pumps?

4.5 What is the purpose of the fractionating device in vapor jet pumps?

4.6 In what does the design of booster pumps differ from other types of vapor jet pumps?

4.7 What is the difference in the action principles of molecular drag and turbomolecular pumps?

4.8 What factors determine the compression ratio of molecular pumps?

4.9 What are the requirements for working liquids of vacuum pumps?

4.10 Which designs of vacuum traps have the maximum conductance without sacrifice in the optical density?

Chapter 5

Physico-Chemical Methods of Vacuum Production

5.1 General

The existing physico-chemical methods of vacuum production permit one to create pumps whose technological parameters are, in a certain range of working pressures, better than those provided by mechanical pumps. The physico-chemical pumps are substantially better because they preclude pollution of the vacuum chamber with vapors of working liquids that are used for lubrication and hermetical sealing in many mechanical pumps.

The general operation principle of ion pumps is based on directed movement of previously charged gas molecules in an electrical field. The general operation principle of evaporation pumps is chemisorption. Physical adsorption and condensation are used to pump out gases by cryosorption pumps that are subdivided in cryoadsorption and cryocondensation pumps.

Because condensation heat is usually less than adsorption heat, condensation of the same amount of gas occurs at lower temperatures than adsorption, for which reason technical implementation of the cryogenic system of a cryocondensation pump more complex than for a cryoadsorption one. The advantage of condesation pumping in comparison to adsorption pumping is a higher pumping speed per unit of cooled surface area.

The general operation principle of ion pumping combined with sorption is used in ion-sorption pumps. Figure 5.1 shows the working pressure ranges of industrial and laboratory pumps whose work is based on physico-chemical pumping.

5.2 Ion Pumping

Constant electrical field can be used to transfer an impetus to gas molecules in the direction of the preliminary vacuum pump, provided neutral gas molecules were previously charged. To ionize gas, one may use α-, β-, and γ-radiation.

Most frequently, β-radiation is used, which is produced by electron bombardment. The efficiency of ionization by medium energy electrons (see Fig. 1.18) that passed a potential

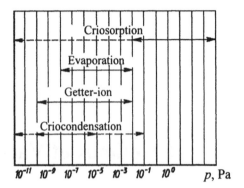

Figure 5.1 Working pressure ranges of vacuum pumps that operate on the basis of physico-chemical pumping techniques: solid line is for industrial specimens and dotted line is for laboratory specimens.

difference of 100 V, is 3 to 25 ion pairs, depending on sort of gas, over a 1 m path at a pressure of 1 Pa. The number of formed ions decreases proportionally with a decrease in pressure.

Figure 5.2 shows schematic of ion pump operation. Gas entering into the pump and ionized in ionization space 1, and it passes to the outlet of the pump under the effect of accelerating electrodes 2, to which potential difference Y is applied. Here the ions are neutralized and pumped out by the preliminary vacuum pump.

This pump generates a direct flow of pumped gas Q_d, which is determined by ion current I^+ in the series of accelerating electrodes 2.

$$Q_d = \frac{kTI^+}{q}, \tag{5.1}$$

where q is the electrical charge of the ions. The back flow of neutral molecules depends on pump conductance U and the difference of pressures at the input p_{in} and output p_{out} of the pump:

$$Q_b = U(p_{out} - p_{in}). \tag{5.2}$$

The pump throughput is determined by the difference of the direct and the back flows:

$$Q = Q_d - Q_b = \frac{kTI^+}{q} - U(p_{out} - p_{in}),$$

and the pumping speed is

$$S_n = \frac{Q}{p_{in}} = \frac{kTI^+}{qp_{in}} - U\left(\frac{p_{out}}{p_{in}} - 1\right). \tag{5.3}$$

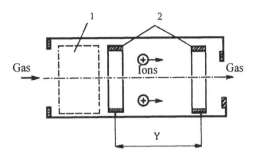

Figure 5.2 Schematic of ion pumping.

At $K = p_{out}/p_{in} = 1$, the pumping speed is the highest:

$$S_{max} = \frac{kTI^+}{qp_{in}}.$$

At $S_n = 0$, the compression factor is the highest:

$$K_{max} = 1 + \frac{S_{max}}{U}.$$

Thus, Eq. (5.3) can be rewritten in the following form:

$$S_n = S_{max} - U(K - 1). \tag{5.4}$$

The pumping factor (5.4) is similar to the same factor of the molecular pump (4.17). However, since effective gas ionization is difficult to achieve and the conductance of pumps for back gas flows is low, the energy consumption per unit of pumping speed of ion pumps is still too high and precludes their general use.

5.3   Chemisorption Pumping

Chemisorption pumping is carried out by sorption of active gases on the surface of a getter. The parameter of gas activity is its adsorption heat on a particular metal. Ti, Zr, Ta, Ba, Mo, W, Hf, Er are most frequently used for chemisorption pumping as getter.

At small degrees of surface coverage adsorption heat Q_a depends strongly on the type of gas. For example, for Ti, which sorbs all the gases except inert ones:

Gas	H_2	CO	N_2	O_2	CO_2	Ar	Kr	Xe
Q_a, MJ/кmol	19.3	419	356	813	461	8.38	16.8	33.5

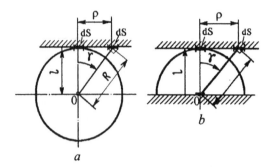

Figure 5.3 Schematic of deposition onto a flat surface from various sources: (*a*) dot source and (*b*) flat source.

To increase the surface of metal that interacts with pumped gases, the metal is deposited in the form of thin films on the electrodes or the casing of a pump. The possibility of multiple continuos replacement of these films increases the service life of pump.

During coating the pressure of metal vapors is higher than the pressure of saturated vapor over the surface, and the number of molecules condensing on unit surface in unit time is greater than evaporating molecules. The rate of condensation from the gas phase can be calculated using Eqs. (1.46) and (1.49).

We now consider the process of deposition of molecules evaporated from an infinitely small source onto a flat substrate in high vacuum. The deposition rate is then

$$dG_c = G_c dP.$$

If a molecule evaporates from a point source, dP is determined according to Eq. (1.36), and

$$dw = \frac{dS \cos\gamma}{R^2},$$

where R is the distance from the source to area dS of the substrate (Fig. 5.3, *a*). The deposition rate can be written as:

$$dG_{cR} = \frac{G_c \cos\gamma\, dS}{4\pi R^2}.$$

The flow of molecules incident on the area dS, locates at the shortest distance l from the source, is

$$dG_{cl} = \frac{G_c dS}{4\pi l^2} \tag{5.5}$$

Therefore, the ratio of the thicknesses of films condensed over the same period of time onto the surface areas at the distances from the source l and R is

$$\frac{h_l}{h_R} = \frac{dG_{cl}}{dG_{cR}} = \frac{R^2}{l^2 \cos\gamma} = \frac{R^3}{l^3} = \left[1 + \left(\frac{\rho}{l}\right)^2\right]^{3/2}. \tag{5.6}$$

It follows from Eq. (5.6) that at $p = l$ the thicknesses of condensed layers in the center and at the edge of the substrate will differ by a factor of 2.8. For the acceptable 5% nonuniformity of coating thickness, the minimum distance between the source and the substrate should be 3.6ρ.

For condensation of molecules from infinitely small flat source (Fig. 5.3, b), the probability that a molecule leaves the surface is determined by Eq. (1.37). Hence, by analogy with Eq. (5.5) and (5.6), we obtain

$$dG_{cl} = \frac{G_c dS}{\pi l^2}; \quad dG_{cR} = G_c \frac{\cos^2\gamma\, dS}{\pi R^2}.$$

The ratio of the thicknesses of condensate films for deposition from a flat source

$$\frac{h_l}{h_R} = \frac{R^4}{l^4} = \left[1 + \left(\frac{\rho}{l}\right)^2\right]^2. \tag{5.7}$$

At $\rho = l$, the film thicknesses in the center and at the edge of the substrate differ by a factor of 4. Therefore, to deposit a film with a 5 % thickness nonuniformity, one should keep the substrate at a distance of 6.3ρ from the source. Flat sources provide for a higher deposition rate than the point sources (by a factor of 4 at $\gamma = 0$), but the coatings become more nonuniform.

Nonuniform thickness coatings, which are equally accessible for pumped out gas molecules, make less use of the active substance in a pump.

The pumping speed of the film S_0 depends on its material, the type of gas and film temperature. For 1 cm^2 of a periodically renewed titanium film, S_0 for H_2 and N_2 at 77 K are 26×10^{-3} and 7×10^{-3}, and at 293 K, 15×10^{-3} and 1×10^{-3} $m^3/(s\, cm^2)$, respectively.

Sorption of gases by films may have surface or bulk character. For surface sorption at small degree of surface coverage, the amount of sorbed gas is proportional to sorption time. Bulk sorption occurs due to diffusion of gases into a film, and the amount of sorbed gas at the initial portion of the sorption curve is proportional to the root square of the sorption time.

Surface sorption is usually observed at low temperatures, and bulk one at high temperatures. Above 150 K titanium films sorb hydrogen by the bulk mechanism. The diffusion coefficient of hydrogen at 220 K is 10^{-15} cm^2/s, and at 300 K, is 2×10^{-11} cm^2/s.

During surface sorption of hydrogen on titanium films at $T > 77$ K, methane CH_4 is formed during the catalytic reactions with free carbon dissolved in the films.

Such gases as N_2, O_2, CO, and CO_2 can be absorbed by titanium only at temperatures above 1000 K. Absorption of water is accompanied by hydrogen evolution.

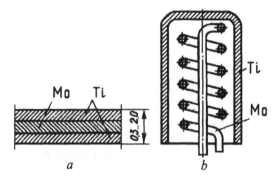

Figure 5.4 Design of electrothermal evaporators: (*a*) straight-channel, and (*b*) heating.

The sorption characteristics of films depend strongly on the conditions of their formation. A film deposited at a low temperature has a porous structure, which results in a significant increase in the absorption rate of gases.

5.4   Evaporation Pumps

The design of evaporation pumps is basically determined by the type of the evaporators used. Evaporators may be of straight-channel, heating, electron-beam and arc types. As a straight-channel evaporator (Fig. 5.4, *a*), bimetallic wire with a molybdenum core is used, on which a titanium layer is deposited using the iodide deposition technique.

Evaporator heater (Fig. 5.4, *b*) is a spherical shell made of an active material, in which a wire heater is embedded. For titanium, the maximum working temperature of such a evaporator is 1150°C, which provides a maximum evaporation rate of 1 mg/s.

Electron-beam evaporator (Fig. 5.5, *a*) is an electron gun with tungsten cathode 1 placed in a transverse magnetic field. This magnetic field allows the gun to be placed outside the deposition zone of the active material. An accelerating voltage of several thousand volts is applied between the gun and the target. The maximum rate of evaporation from the liquid phase is 30 mg/s.

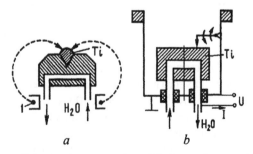

Figure 5.5 Design types of electron-ion evaporators: (*a*) electron beam and (*b*) arc.

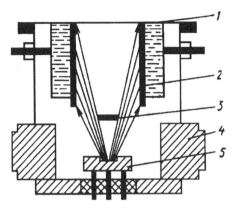

Figure 5.6 Evaporation pump.

In arc evaporators (Fig. 5.5, *b*), the active material is deposited in the cathodic zone of a direct current arc. The cathodic zone moves randomly over the surface of the water-cooled titanium cathode. The current density in the cathodic zone is 10^6 A/cm^2. The arc burns in vapor of a volatile material and allows one to maintain charge even under the conditions of ultrahigh vacuum. An arc can be excited, for example, by shorting of a mobile electrode. The arc is fed from a direct current source with $U = 30$–50 V and $I = 100$–180 A. The maximum start pressure is 10 Pa. At high pressures, the anode zone becomes immobile and can melt the pump wall. The evaporation rate in the arc evaporators may be to 20 mg/s.

In evaporation pumps without means for ion pumping, the limiting pressure may usually be 10^{-7} Pa. Cooling of an active film to the liquid nitrogen temperature reduces the limiting pressure to 10^{-11} Pa.

The maximum working pressure, 10^{-2} Pa, is limited by the formation of oxides, nitrides, and carbides on the surface of active material during operation of the vaporizer, and this hampers the evaporation. The maximum hydrogen pumping speed of such pumps is 2×10^5 l/s. Evaporation pumps are ineffective for pumping of organic products and inert gases.

Evaporation pump (Fig. 5.6) consists of case 4 with evaporator 5. Atoms of active metal evolved from the evaporator and condense onto screens 2 and provide pumping of chemically active gases. Screen 3 protects the chamber, which is connected to the pump through flange 1, from penetration of the evaporated material. To increase the pumping speed, the screens 2 can be cooled by liquid nitrogen.

5.5   Cryocondensation Pumping

Cryocondensation pumping is the removal of gases from a vacuum system by means of their condensation on cooled pump surfaces. According to Eq. (1.47), the condensation rate can be determined as

$$G_c = \gamma p_g \sqrt{\frac{M}{2\pi R T_g}},$$

where p_g and T_g are the pressure and the temperature of gas, respectively, and γ is the probability of condensation on the free surface (1.46). We now use Eq. (1.48) for the evaporation rate:

$$G_e = \gamma p_T \sqrt{\frac{M}{2\pi R T_s}},$$

where p_T is the pressure of saturated adsorbate vapors at surface temperature T_s. Taking into account that the gas density $\rho = p_g M/(R T_g)$, we write the following expression for the rate of condesation pumping:

$$S_c = \frac{\beta}{\rho}(G_c - G_e) = 36.4\beta\gamma\sqrt{\frac{T_g}{M}}\left(1 - \frac{p_T}{p_g}\sqrt{\frac{T_g}{T_s}}\right), \qquad (5.8)$$

where β is the factor that takes into account input resistance.

Cryocondesation pumping can be used if the pressure of pumped gas in the vacuum system is higher than the pressure of its saturated vapors in the pump. Using the dependence of the pressure of a saturated vapor on temperature (1.44) in the form $\ln p = M - N/T_s$, where M and N are constants, and after transformations of the expression, we obtain:

$$S_c = 36.4\beta\gamma\sqrt{\frac{T_g}{M}}\left(1 - \frac{\exp(M-N/T_s)}{p_g}\sqrt{\frac{T_g}{T_s}}\right). \qquad (5.9)$$

At $p_g \gg p_T$ the maximum pumping speed of a cryocondensation pump is equal to the condensation rate. At $\beta = \gamma = 1$

$$S_{max} = 36.4\sqrt{\frac{T_g}{M}}, \qquad (5.10)$$

where S_{max} [m³/(s m²)] varies with temperature and the type of gas and is 118 and 61 for N_2, 110 and 57 for O_2, 139 and 72 for Ne, and 442 and 228 for H_2, at T = 293 and 78 K, respectively.

Figure 5.7 shows the specific pumping speed of nitrogen at 293 K for $\beta = \gamma = 1$ as a function of pressure and temperature of the cryosurface, in accordance with Eq. (5.9). The maximum pressure of cryocondesing pumping depends on the temperature of the surface onto which the pumped gas condenses. At $S_c = 0$ and $p_g = p_{lim}$, we use Eq. (5.9) to determine the limiting pressure:

$$p_{lim} = \sqrt{\frac{T_g}{T_s}}\exp\left(M - \frac{N}{T_s}\right). \qquad (5.11)$$

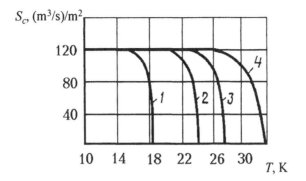

Figure 5.7 Specific speed of nitrogen pumping as a function of cryosurface temperature at pressures of (*1*) 10^{-10} Pa, (*2*) 10^{-6} Pa, (*3*) 10^{-4} Pa, and (*4*) 10^{-1} Pa.

The operation of a pump causes cryolayers deposition on the condensation surface. The thickness of the cryolayers is

$$h = \frac{t(G_c - G_e)}{A\rho_g},$$

where t is the working time of the pump; A is the area of the condensation surface; and ρ_g is the density of gases in the solid state

Gas	N_2	H_2	CH_4	Ar	Kr	Ne
ρ_g, kg/m³	950	80	520	1690	2960	1440

During nitrogen pumping with an efficiency of 10^{-2} m³ Pa/s on an area of 0.1 m² over a period of 10^6 s, a 1.5 mm thick condensed layer is formed.

Thermal flows on the surface of cryolayers produces temperature gradients in the condensed layer

$$q = \lambda \frac{T_s - T_c}{h},$$

here q is the specific thermal flow on the cryolayer surface, T_c and T_s are the temperatures of cryoagent and the external surface of cryolayers, respectively, and λ is the heat conductivity coefficient of the cryolayers:

Gas	N_2	H_2	CH_4	Ar	Kr	Ne
λ, W/(m K)	10^{-1}	1	8×10^{-2}	2	2	1

Solving the thermal equation with respect to T_s, we obtain

$$T_s = T_c + \frac{qh}{\lambda}.$$

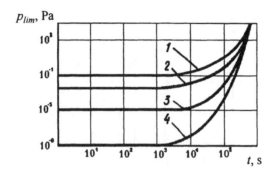

Figure 5.8 Limiting pressure of cryocondensation pump as a function of operation time for nitrogen at $T_g = T_s$, $A = 0.01$ m², $Q = 10^{-1}$ m³; (*1*) 35 K, (*2*) 30 K, (*3*) 25 K, and (*4*) 20 K.

For $q = 50$ Wt/m², after the formation of a 1.5 mm thick nitrogen cryolayers, the excess of cryosurface temperature over the temperature of cryoagent is 0.75 K.

Substituting the expression for T_s in Eq. (5.11), we obtain

$$p_{lim} = \sqrt{\frac{T_g}{T_c + qh/\lambda}} \exp\left(M - \frac{N\lambda}{T_c\lambda + qh}\right). \tag{5.12}$$

Therefore, the limiting pressure of a pump tends to deteriorate by the growing thickness of cryolayers. Figure 5.8 shows the limiting pressure of a pump as a function of pump working time, as calculated using Eq. (5.12).

5.6   Cryoadsorption Pumping

Cryoadsorption pumping is effected by gas adsorption onto cooled adsorbents. The equilibrium pressure of pumped gases is determined by their adsorption isotherms that establish a relationship between the amount of adsorbed gas and its pressure at a constant temperature. The amount of adsorbed gas should be significant compared to the volume of the vacuum system, which is usually observed if the temperature of the adsorbing surface is lower than the temperature of gas boiling at atmospheric pressure.

The mass balance equation of cryosorption pumping using an immobile adsorbent can be written as the equity of the amount of gas in the adsorbed state and in the gas phase before and after the pumping:

$$p_0v + K_{T1}p_0 + \int_0^t q_1 dt = p_1v\frac{T_1}{T_2} + K_{T2}p_1, \tag{5.13}$$

where p_0 and p_1 are the initial and final pressures, respectively, $v = V_c/V_a$ is the volume load of the pump, V_c is the volume of the pumped chamber, V_a is the volume of adsorbent in the

Table 5.1 Characteristics of adsorbents for pumping of nitrogen.

Adsorbent	K_T		H, m^2/s		A, m^2/s		Note
	77 K	298 K	77 K	298 K	298 K	77 K	
Active carbon SKT-2	10^8	10^2	10^{-11}	10^{-9}	10^{-8}	10^{-8}	$p = 10^{-1}$–10^{-5} Pa
Zeolite CaA-4B	10^8	10^3	10^{-10}	10^{-8}	10^{-8}	10^{-8}	$p = 10^{-1}$–10^{-5} Pa
Silica gel KSM-6	10^7	10^2	10^{-11}	10^{-9}	10^{-7}	10^{-7}	$p = 10^{-1}$–10^{-5} Pa
Aluminosilicate catalyzer AS	10^7	10^2	10^{-11}	10^{-9}	10^{-7}	10^{-7}	$p = 10^{-1}$–10^{-5} Pa

pump, q_1 is the gas evolution and inleakage per unit adsorbent volume, T_1 and T_2 are the initial and final temperature of the adsorbent in the pump, respectively ($T_2 < T_1$), and K_{T1} and K_{T2} are the coefficients of gas adsorptivity at temperatures T_1 and T_2 ($a = K_T p$).
If $q_1 = $ const, then

$$p_1 = \frac{p_0(v + K_{T1}) + q_1 t}{v T_1 / T_2 + K_{T2}}. \tag{5.14}$$

If

$$\int_0^t q \, dt \ll p_0 v + K_{T1} p_0,$$

then

$$p_1 = p_0 \frac{v + K_{T1}}{v T_1 / T_2 + K_{T2}}. \tag{5.15}$$

For ordinary vacuum systems ($V_c = 50l$; $V_a = 1l$), $v \sim 50$. For pumping of nitrogen from the atmospheric pressure using active carbon SKT-2 at $T_1 = 293$K, $K_{293} = 10^2$, and $T_2 = 77$ K, $K_{77} = 10^8$ (Table 5.1), we obtain, according to Eq. (5.15), $p_1 = 1.5 \times 10^{-1}$ Pa. Thus, the pressure of gas decreases almost by a factor of 10^6.

Selecting desorption temperature, one may ensure that $K_{T1} \ll v \ll K_{T2}$, and then

$$p_1 \approx p_0 \frac{v}{K_{T2}}. \tag{5.16}$$

If a pumped chamber is connected in parallel with n adsorption pumps, any of which alone and all together satisfy the conditions of Eq. (5.16), a multistep pumping (n is the number of steps), for which each of the pumps is turned on in sequence and then, after achievement of the limiting pressure, is blocked by the valve, ensures that the final pressure will be

Table 5.2 Adsorptivity coefficients of various gases on active carbon at the liquid nitrogen temperature.

Coefficient	He	Ne	H$_2$	N$_2$	Note
K_{77}, (m^3 Pa)/m^3 Pa	2	20	1600	10^8	$p = 10^{-1}$–10^{-5} Pa

$$p_n = p_0 \left(\frac{v}{K_{T2}} \right)^n .$$

For $n = 2$, $v = 50$, and $p_0 = 10^5$ Pa during pumping of nitrogen with $K_{77} = 10^8$ we obtain $p_2 = 2.5 \times 10^{-8}$ Pa.

Much worse results are obtained by pumping of He, Ne and H$_2$ at the liquid nitrogen temperature. The adsorptivity factors of these gases at the liquid nitrogen temperature (77 K) are considerably lower than for nitrogen (Table 5.2). For one-step pumping, one may not reach an accessible reduction in the pressure. For pumping of atmospheric air, residual gases are enriched with weakly adsorbable components He, Ne, and H$_2$ that limit the maximum pressure reduction in a pump. Temperatures of ~20–30 K are required for effective adsorption pumping of all the gases that constitute air.

Adsorption pumping speed is determined by the non-steady-state gas diffusion in the porous structure of the adsorbent. This diffusion proceeds both in the gas phase and on the surface of pores. At low temperatures, surface diffusion in the micropores is the controlling factor. The equation describing this process can be written in the following form:

$$\frac{\partial a}{\partial t} = H \left(\frac{\partial^2 a}{\partial x^2} + \frac{K_1}{x} \frac{\partial a}{\partial x} \right) . \tag{5.17}$$

Given the boundary conditions $a(x, 0) = a_0$; $a(x, \infty) = a_\infty$

$$\frac{\partial a(0, t)}{\partial x} = 0; \quad q_1(R, t) = \frac{H}{K_2 R} \frac{\partial a(R, t)}{\partial x}, \tag{5.18}$$

where a is the amount of adsorbed gas per unit adsorbent volume, $q_1(R, t)$ is the specific efficiency of pumping, x is the current coordinate over the section of the adsorbed layer; R is the characteristic dimension of the adsorbent; H is the unstational diffusion coefficient (Table 5.1), K_1 is the shape factor of the adsorbent, which is $K_1 = 0$ for an unfinite plate, $K_1 = 1$ for an unfinite cylinder, and $K_1 = 2$ for a sphere; $K_2 = V/(FR)$ (F and V are the surface and volume of the adsorbent, respectively) which is $K_2 = 1$ for an unfinite plate, $K_2 = 0.64$ for an unfinite cylinder, and $K_2 = 0.64$ for a sphere.

The solution of Eq. (5.17) with boundary conditions (5.18) can be represented in the following form:

$$a(R, t) = a_0 + \frac{q_1 R^2}{H} \left(K_3 \tau + K_4 - \sum_{n=1}^{\infty} \frac{K_5}{\mu_n^2} \exp(-\mu_n^2 \tau) \right) \tag{5.19}$$

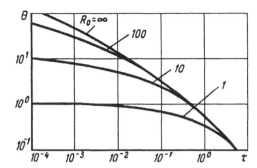

Figure 5.9 Dimensionless specific operation speed θ as a function of dimensionless time τ for spherical adsorbent.

where $\tau = Ht/R^2$, K_3, K_4, K_5 are the shape constants of the adsorbent for an unfinite plate, an unfinite cylinder, and a sphere, respectively ($K_3 = 1$, $4/\pi$, and $6/\pi$; $K_4 = 1/3$, $1/2\pi$; and $2/5\pi$; $K_5 = 2$, $4/\pi$ and $4/\pi$. μ_n are the roots of the characteristic equation $\tan \mu = 0$ (for an unfinite plate), $J_1(\mu) = 0$ (for an unfinite cylinder) and $\tan \mu = \mu$ (for a sphere).

The above solution for $a = K_T p$ can be simplified for small or great dimensionless time $\tau = Ht/R^2$. For an unfinite plate at $\tau \geq 0.42$, an unfinite cylinder at $\tau \geq 0.27$ and a sphere at $\tau \geq 0.19$, we have

$$\theta = \frac{R_0}{1 + R_0(K_3\tau + K_4)}.$$

For an unfinite plate at $\tau < 0.42$, an unfinite cylinder at $\tau < 0.27$, and a sphere at $\tau < 0.19$

$$\theta = \frac{R_0}{1 + 2R_0K_6\sqrt{\tau/\pi}}, \tag{5.20}$$

here

$$\theta = \frac{S_1 R^2}{K_T H}; \quad R_0 = \frac{q_1 R^2}{a_0 H};$$

S_1 is the pumping speed per unit adsorbent volume and K_6 is the shape factor ($K_6 = 1$ for an unfinite plate; $K_6 \approx 0.64(1 + 0.44\sqrt{\tau})$ for an unfinite cylinder, $K_6 \approx 0.64(1 + 0.89\sqrt{\tau})$ for a sphere).

Dimensionless specific cryoadsorption pumping speed as a function of dimensionless time was calculated using Eq. (5.20) for spherical adsorbent (Fig. 5.9).

For a pump with an adsorbent volume of 1 l in the form of a plate with thickness $2R = 3$ mm, at non-steady-state diffusion coefficient 1.4×10^{-11} m²/s, adsorptivity factor 5.3×10^6, and an initial equilibrum pressure of 10^{-4} Pa, the specific pumping speed for $q = 4.23 \times 10^{-6}$ m³ Pa/(s m³) is 0.01 l/(s cm²) and is retained during 10^4 s. The pumping speed depends on gas flow. As the flow increases by two orders of magnitude, the pumping speed increases by the same factor and reaches in this case 1 l/(s cm^2). However, it will not be constant and starts to decrease almost immediately.

The specific adsorption pumping speed is 5–10 times lower than condesation pumping speed and depends much stronger on the amount of absorbed gas.

In adsorption pumps with moving adsorbent the adsorbent is regenerated continuously, this providing for uniform pumping speed irrespective of working time.

The amount of absorbed and desorbed gas under nonequilibrium conditions is determined by the solution of non-steady-state difusion Eq. (5.17) with the following initial and boundary conditions: $a(0, x) = a_0$; $a(t, R) = a_R$ and $\partial a(t, 0)/\partial x = 0$. $a_R = a_m$ (for adsorption, and $a_R = a_0$ for desorption.

Under the specified boundary conditions, the solution of Eq. (5.17) can be written as:

$$a = (a_R - a_0)\left(1 - \frac{6}{\pi^2}\sum_{k=1}^{\infty}\frac{1}{k^2}\exp(-k^2\pi^2\tau)\right) = (a_R - a_0)v .$$

At low $\tau = Ht/R^2 < 2\times10^{-2}$ the degree of adsorbent filling

$$v = 1 - \frac{6}{\pi^2}\sum_{k=1}^{\infty}\frac{1}{k^2}\exp(-k^2\pi^2\tau) = 6\sqrt{\frac{\tau}{\pi}} .$$

The rate of continuous adsorptive pumping speed is

$$S_c = vS_{max}\left(1 - \frac{K_{T1}p_2}{K_{T2}p_1}\right), \tag{5.21}$$

where $S_{max} = v_tFK_{T2}$ is the maximum pumping speed, v_t is the rate of adsorbent transfer, and F is the cross-section area of the adsorption layer. For $v_t = 0.1$ cm/s, $F = 1$ cm^2, and $K_{T2} = 10^7$ we obtain $S_{max} = 10^3$ l/s.

Since the limiting pressure $p_{lim} = p_1$ at $S_c = 0$, we deduce from Eq. (5.21) that

$$p_{lim} = p_2\frac{K_{T1}}{K_{T2}},$$

and, hence, $p_{lim} = 10^{-5} p_2$ for carbon SKT-2 B at $K_{T1} = 10^2$ and $K_{T2} = 10^7$.

Typical dependence $S_c = f(v_t)$ for continuous pumps is shown in Fig. 5.10.

In the first mode at low transfer rates, $v = 1$ until $\tau \geq 0.4$, which corresponds to

$$v_{t1} \leq \frac{Hl_a}{0.4R_a^2},$$

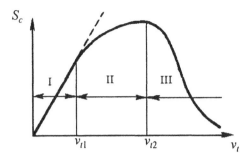

Figure 5.10 Operation speed of a continuous adsorption pump as a function of adsorbent transfer rate.

where l_a is the adsorption chamber length, and R_a is the radius of adsorbent particle. The pumping speed $S_c = S_{max}$ in this mode is proportional to v_t.

In the second mode at average transfer rates, the adsorbent has no time to be saturated with the pumped gas, and, hence, $v < 1$ and the pump speed is proportional to $\sqrt{v_t}$

$$S_c = \frac{3.4 F K_{T2}}{R_a} \sqrt{H l_a v_t} \left(1 - \frac{K_{T1} p_2}{K_{T2} p_1} \right).$$

The onset of the third pump operation mode with high transfer rates is marked by the rate determined by the thermal conductivity factor of adsorbent A and dimensionless cooling time

$$F_0 = \frac{A l_c}{v_{t2} R_l^2} = 0.8, \quad v_{t2} = \frac{A l_c}{0.8 R_l^2},$$

where l_c is the cooling chamber length, and R_l is the typical layer thickness. At transfer rates higher than v_{t2}, the adsorbent has no time to be cooled and the pumping speed decreases drastically. Therefore, v_{t2} is the optimum transfer rate of adsorbent that provides the maximum pumping speed.

5.7   Cryogenic Pumps

We now consider the main types of cryogenic pumps. Open flask pumps are used for operation in low vacuum (Fig. 5.11, a), and pour-in type pumps, for operation in high vacuum (Fig. 5.11, b). Adsorbent 1 that prevents chamber pollution and improves cooling conditions is placed in a porous metal filter 2. Heater 3 serves for adsorbent regeneration after its saturation with pumped gas. The difference in the design of open flask and pour in type pumps is that the Dewar flask 4 for cryoagent 5 in the former pumps is detachable, whereas in the latter pumps, the vacuum produced by the pump acts as a heat insulating layer

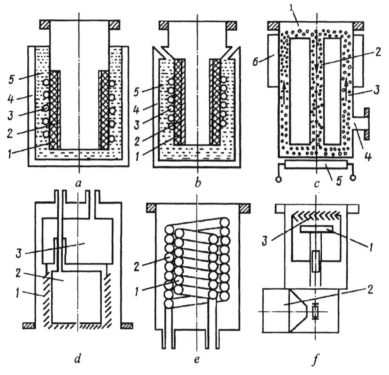

Figure 5.11 Schematic design of cryogenic pumps.

between the pump walls and the flask with cryoagent. Liquefied gases are used as cryoagents (Table 5.3).

Cryoadsorption pumps with immobile adsorbent are simple in design but require additional apparatus and doubling of pumping means for the maintenance of continuous pumping and consume much cryoagent and electricity for changing temperature not only of the adsorbent, but also of the casing.

Schematic of a cryoadsorption pump with mobile adsorbent (Fig. 5.11, *c*) provides stable pumping speed and limiting pressure irrespective of pumping time. The adsorbent moves at a linear speed v_t over a closed contour and passes through the adsorption chamber 1, lock 2, heater 5, adsorption chamber 4, lock 3, cooler 6 and enters again the adsorption chamber.

Table 5.3 The properties of liquefied gases.

Parameters	N_2	Ne	H_2	He
Boiling temperature at atmospheric pressure, K	77.3	27.2	20.4	4.2
Density in the liquid state, g/cm³	0.81	1.18	0.07	0.13
Heat of vaporization, kJ/l	162	102	31.7	2.7
Evaporation rate at thermal load 1 W, l/h	0.021	0.035	0.16	1.4

Cryocondensation pour-in pumps are designed as follows (Fig. 5.11, *d*). A low-temperature cryoagent (liquid helium or hydrogen) and a high-temperature cryoagent (liquid nitrogen) are poured into cavities 2 and 3, respectively. Screens 1 protect the flask surface by the low-temperature cryoagent from the radiation emitted by the pump walls, but do not hinder the penetration of the pumped gas to the cooled surface. The adsorption of non-condensing gases is often used in cryocondensation pumps, for which purpose the flask surface with the low-temperature cryoagent is covered with an adsorbent in the form of a porous oxide film (or, alternatively, well-condensed gases is deposited on it during cryotrapping).

Cryocondensation pumps of the evaporation type (Fig. 5.11, *e*) have cryopanels in the form of coils, in which volatile cryoagent evaporated from the Dewar flask circulate. This circulation may occur due to an excess pressure in the Dewar flask or operation of the mechanical vacuum pump. Volatile cryoagent in cryopanel 1 is used for cooling of external screen 2, which protects the cryopanel from the radiation of the pump walls.

Cryoadsorption pumps can be supplied with autonomous cryogenerators (Fig. 5.11, *f*), in which cryopanel 1 is cooled by autonomous gas machine 2, and screen 3 serves for reduction of heat flow to the cryopanel.

Adsorbents with a large internal surface such as active carbons, zeolites, and silica gels are widely used for pumping.

Activated charcoals are porous carbon adsorbents that are extracted from turf, coal, wood saw dust, and other organic raw materials using thermal processing without access of air for the removal of water and resin with subsequent activation by oxidation at the 900°C in the presence of CO_2. Since carbon surface is electrically neutral, adsorption on activated carbons is basically determined by the dispersion interaction forces. Activated charcoals have pores of various size. The surface area of activated charcoals at a bulk density of 0.5 g/ cm^3 may reach 2000 m^2/ g.

Zeolites are aluminosilicates containing SiO_2 and Al_2O_3, oxides of alkali and alkaline-earth metals, and molecules of crystalline water. After the removal of crystalline water by heat treatment at 400–500°C, a regular pore structure with sizes from 0.3 to 0.9 nm is produced in zeolites of various types. The surface of zeolites is polar, and adsorption in many respects depends on the orientation interaction. Zeolites may be either natural or synthetic, the latter being extracted using crystallization of initial components from solutions at 100°C. The active surface of zeolites may reach 1000 m^2/g at a bulk density of 0.7 g/cm^3.

Silicagel is the amorphous form of hydrated silica (SiO_2 nH$_2$O), which is extracted by interaction of silicates of alkaline metals and mineral acids. During drying of silicon acid hydrogel, a structural lattice of spherical particles is formed, in which the pore surface may reach 500 m^2/g at a bulk density of 0.7 g/cm^3. The size of pores varies from 1 to 7 nm depending on the processing mode. Characteristics of cryoadsorption pumps are listed in Table A.10.

5.8   Getter-Ion Pumping

Getter-ion pumping uses two methods of gas absorption, i.e., introduction of ions into solids using electric fields and chemical interaction of pumped gases with thin films of active metals.

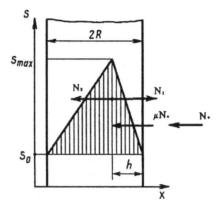

Figure 5.12 Steady-state concentration distribution in an unfinite plate bombarded by high-energy ions.

High-energy ions or neutral particles bombard a solid and penetrate into it to a depth sufficient for their dissolution. This method of gas removal is a modification of ion pumping. Figure 5.12 shows the equilibrum distribution of concentration for ion pumping in the bulk of an unfinite plates with thickness $2R$ placed in a vacuum chamber. The maximum specific geometrical ion pumping speed can be calculated using the formula

$$S_i = \frac{\mu N^+}{n} = \frac{\mu i^+}{nq},$$ (5.22)

where μ is the factor of ion penetration, $N^+ = i^+/q$ is the specific frequency of bombardment, i^+ is the ion current density, q is the elementary electric charge, and n is the molecular concentration of the gas.

The penetration factor takes into account the partial reflection and dispersion caused by ion bombardment. The penetration factor depends strongly on temperature of the body and slightly on current density and accelerating voltage. The value $\mu \to 1$ is observed for Ti and Zr at 300–500 K. The maximum dissolved gas concentration for ion pumping can be determined from the condition of gas flow balance

$$\mu N^+ = N_1 + N_2;$$

$$N_1 = D\left(\frac{ds}{dx}\right)_1; \quad N_2 = D\left(\frac{ds}{dx}\right)_2,$$ (5.23)

where D is the coefficient of gas diffusion in a solid. The concentration gradients are determined by the following ratio:

$$\left(\frac{ds}{dx}\right)_1 = \frac{s_{max} - s_0}{h}; \quad \left(\frac{ds}{dx}\right)_2 = \frac{s_{max} - s_0}{2R - h}.$$

Here $h = CE$ is the ion penetration depth (E is the accelerating voltage) and s_{max} and s_0 are the maximum and initial concentrations of the absorbed gas, respectively.

Since h is minor in comparison with $2R$ (the constant C does not exceed 1 nm/kV even for light gases), N_2 in Eq. (5.23) can be neglected:

$$\mu N^+ \approx D\frac{s_{max} - s_0}{CE}.$$

Hence, the expression for the maximum concentration of dissolved gas is as follows:

$$s_{max} = s_0 + \frac{\mu N^+ CE}{D}.$$

If s_{max} as calculated using the above formula is higher than the maximum possible gas solubility in metal under these conditions, absorbed gas starts to coalesce into gas bubbles that cause destruction of metal. This phenomenon is referred to as the *blister effect*.

In stainless steel the hydrogen blister effect is observed at an absorption of 3×10^{-2} m^3 Pa/ cm^2, which corresponds to continuous work for approximately 300 h at pumping speed of 10^{-2} m^3/(s cm^2) and a pressure of 4×10^{-8} Pa.

Using calculated s_{max}, one may determine the total amount of gas that is adsorbed by unit surface area: $a = (s_{max} - s_0)R$.

Ion bombardment causes material sputtering which is accompanied by deposition of thin films on the electrodes and pump casing. The gettering activity of these films is used in chemisorption pumping.

Spraying of the active material can be controlled irrespective of pumping process, for example, by adjusting heater temperature. The active material of these pumps is consumed irrespective of pumped gas flow.

Active metal is consumed more economically in pumps with self-adjusted sputtering. In these pumps, sputtering is effected by pumped gas ions that bombard the cathode, which is made from an active material. Sputtered material deposits on the casing and the anode, where chemisorption pumping occurs.

5.9 Getter-Ion Pumps

By way of example, Figure 5.13, *a* shows the design of a pump with independently sputtered active material. The pump consists of casing 6, cooler 5, getter 4, control grid 3, ioniziting anode grid 2, and cathode 1. The electrons emitted from the thermocathode are directed to the getter and the anode grid, to which a positive voltage of several hundred Volts is applied. Electron bombardment heats the getter to the evaporation temperature of the active metal inside. The electrons make several oscillations before they reach anode grid 2 thereby ionizing the molecules of residual gases. Grid 3 serves to maintain the emission current at a constant level during temporary cathode contamination.

Ionized gas is well-sorbed by the active metal, which is deposited onto the surface of the casing and removed by ion pumping due to the incorporation of positive ions the casing

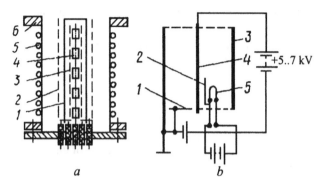

Figure 5.13 Schematic design of getter-ion pumps.

walls. The main parameters of the pumps of this type are shown in Fig. A.11 and Table A.11.

The electrode system that provides effective ionization of the molecules of residual gases during ion pumping, is used in Orbitron pumps (Fig. 5.13, *b*). Electrons emitted by small tungsten cathode 5 are directed by the positive potential of anode 4 and grids 1 to the ionization space between the concentrically located anode 4 and ion collector 3. At the same time, special electrode 2 attached to the cathode imparts the electrons tangential velocities providing their circular movement. As a result, the electron path increases by several thousand times in comparison with the distance between the cathode and the anode. Positive ions, that form as a result of the ionization of residual gases are introduced by the accelerating field into ion collector, which may be the pump casing. At the same time, active metal is evaporated from the anode, which is heated by electron bombardment to the evaporation temperature, onto the ion collector, and, hence, chemisorption of the pumped gas and "dissolution" of adsorbed molecules occurs.

Disadvantage of these pumps is that the evaporation rate of the active metal and the efficiency of pumping do not depend on each other. This often results in inefficient consumption of the active metal.

Self-adjustment of evaporation rate is implemented in magnetic discharge pumps (Fig. 5.14, *a*, *b*), which consist of two cathodes 1 and wire or cylindrical anode 2 placed in a magnetic field with induction *B*. The magnetic field is directed along the axis of the anode,

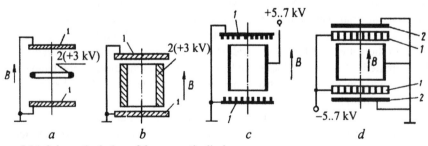

Figure 5.14 Schematic design of the magnetic discharge pumps.

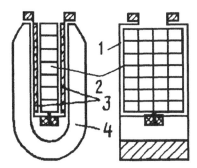

Figure 5.15 Magnetic discharge pump.

which has a positive potential of 3 to 7 kV with respect to the cathodes. In highly efficient diode magnetic discharge pump, (Fig. 5.15), anode 2 is in the form of a cellular framework, each cell of which, together with cathodes 3, which are attached to casing 1, and with magnet 4, correspond to scheme in Fig. 5.14, *b*.

Pumped gases are ionized by electrons that are produced by autoelectronic emission from the cathode and by secondary electrons produced by the bombardment of the cathode by ions of the pumped gas. Magnetic field intensity is chosen in such a way that the radius of the electron trajectory be less than the anode radius. Thus, the total length of the electron trajectory before its collision with the anode increases greatly and, hence, increases the probability of ionization of residual gases.

Positive ions, which are affected by the magnetic field only slightly, bombard the cathode and sputter the active metal that is deposited onto the anode. On the average, one ion knocks out one atom of active metal, thus providing for the self-adjusted sputtering rate during pumping.

Active gases chemically interact with sputtered atoms of cathode material and deposit onto the anode in the form of chemical compounds. Inert gases are pumped out during ion pumping: positive ions by introduction into the cathode material, and negative ions and high-energy neutral particles deposit onto the anode. The most part of a gas is pumped at the anode, because absorbed gases evolve from cathodes during their sputtering.

To increase the efficiency of active material sputtering, circuits of diode magnetic discharge pumps with ribbed cathodes (Fig. 5.14, *c*) and the triode circuit (Fig. 5.14, *d*) with grid cathode 1 can be used. Sputtering of the active material of these pumps occurs from large surfaces at small ion incidence angle. In triode pumps, sputtering is additionally performed onto collector 2 (the pump casing), which is not bombarded by positive ions. Magnetic discharge pumps are appreciably selective during pumping. The hydrogen pumping speed of these pumps is higher by a factor of 3, and the pumping speed of oxygen, is lower by a factor of 2, than the pumping speed of nitrogen. For inert gases, the speed of diode pumps is 10% for He, 4% for Ne, and 1–2% for Ar, Kr and Xe as compared to the pumping speed of nitrogen. In triode-type pumps and pumps with ribbed cathodes, the argon pumping speed is higher by 25 and 10 %, respectively, than the nitrogen pumping speed. During long-term pumping of argon, an argon instability may occur in the pump, which is accompanied by periodic fluctuations of pressure caused by argon evolution.

By the reasons of economic efficiency and reliability, the magnetic system of pumps is made of barium ferrite (2BA, 3BA), iron-cobalt alloy (YuNDK35G5), and cobalt-samarium alloy permanent magnets. The maximum baking temperatures of these magnets are 150, 500, and 150°C, respectively.

The maximum pressure of magnetic discharge pumps ranges from 10^{-8} to 10^{-10} Pa. Easy ignition of discharge in these pumps during operation in ultrahigh vacuum is achieved using a trigger device on the basis of radioactive isotopes with an electronic multiplier.

The upper limit of working pressures is determined by gas evolution due to overheating of pump electrodes. At pressures higher than 5×10^{-3} Pa, long-term operation of a pump is possible only if its electrodes are additionally cooled. Short-term operation of a pump when starting is allowed at a pressure of about 1 Pa.

Contamination of pump electrodes, especially organic one, reduces the speed of a pump and deteriorates the maximum pressure. Hence, preliminary pumping should be made by oil-free pumps.

Problems are faced when a high pumping speed is required in the magnetic discharge pumps, because of the small conductance of the pump casing in the magnetic gap. For this reason, multisection pumps were elaborated, in which magnetic discharge sections are mounted around the vacuum pump casing. The pumping speed of these pumps is proportional to the number of sections or the diameter of the pump inlet tube.

5.10   Test Questions

5.1   How can the operation speed of magnetic discharge pumps be increased at pumping of inert gases?

5.2   Explain the phenomenon of the self-adjusting rate of cathode sputtering in of the magnetic discharge pumps?

5.3   On what factors does the limiting pressure of cryosorption pumps depends?

5.4   Why getter pumps do not pump inert gases?

5.5   Specify the methods to improve the upper limit of working pressures in magnetic discharge pumps.

5.6   What types of adsorbents are used in adsorption vacuum pumps?

5.7   What emission techniques of active materials are used in ion-sorption pumps?

5.8   What is the advantage of adsorption pumps over condesation ones?

5.9   Why there is the optimum rate of adsorbent transport in continuously-operated pumps?

5.10   What materials are used as getters in getter-ion pumps?

Chapter 6

Analysis of the Vacuum Systems

6.1   Typical Vacuum Systems

The variety of vacuum systems that are used in industry and scientific research can be sub-divided into several types, which are intended to produce low, medium, high, and ultrahigh vacuum. For the basic schematic of vacuum systems, one may use certain designations of elements of the vacuum systems listed in Table A.13.

To produce low vacuum, a vacuum system (Fig. 6.1) provides working pressure, from 10^5 to 10 Pa in a vacuum chamber. The type of pump 1 is chosen on the basis of the limiting pressure (Table 6.1) and the composition of residual gases (Fig. 6.2).

To increase the reliability, as well as if a cryosorption pump is used that may absorb a limited quantity of pumped gas, the elements 1, 2, and 5 of the schematic are duplicated. The additional elements are shown by dotted lines.

Valve 2 equalizes the pressures at the inlet and outlet pipes of the pumps with a working liquid during their stop or lets out gas that is desorbed during heating of the cryosorption pump. These valves are sometimes inserted into pumps.

Protective chamber 3 is used in mechanical pumps and is a trap for vapor of the working liquids or a ballast volume preventing the migration of the working liquid from the pumps into the vacuum chamber. It is not used in pumps of other types. If necessary, manometer 4 permits one to determine the serviceability of a vacuum pump. Valve 5 switches out the

Table 6.1 Main characteristics of low-vacuum pumps.

Pump type	Designation	Limiting pressure, Pa	Pumping speed, m^3/s
Centrifugal	CP	$10^5 - 10^4$	$10^{-3} - 10^{-2}$
Air jet	AJP	2×10^3	$10^{-3} - 10^{-2}$
Water jet	WJP	2×10^3	$10^{-3} - 10^{-2}$
Watering	WRP	$2 \times 10^3 - 10^4$	$10^{-2} - 8 \times 10^{-1}$
Multivane	MP	5×10^2	$5 \times 10^{-2} - 10^0$
Oil rotary	ORP	$5 \times 10^0 - 2 \times 10^2$	$10^{-3} - 10^{-1}$
Cryosorption	CRP	$10^5 - 10^0$	$10^{-3} - 10^{-1}$

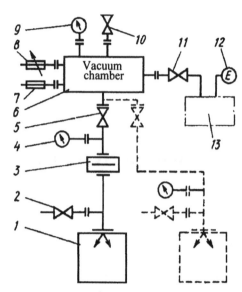

Figure 6.1 Low-vacuum system (10^5–10 Pa).

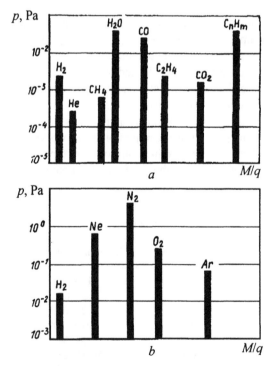

Figure 6.2 Composition of residual gases in low-vacuum systems with various pumps: (*a*) low-vacuum mechanical oil pumps; (*b*) adsorption pumps.

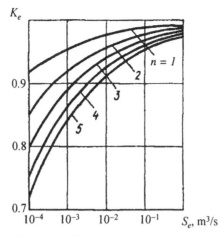

Figure 6.3 Recommended utilization coefficients K_e of rotary pumps in low-vacuum systems as a function of effective pumping speed S_e and the number n of elements n between the pump and the pumped object.

vacuum chamber from the pump when the latter achieves the working pressure and permits one to throttle the pumping, and, as a result, prevents overheating of the mechanical pumps during their operation at high pressures.

Vacuum chamber 6 may have the required quantity of movement inputs 7 and electrical inputs 8. Manometer 9 measures the pressure in a vacuum chamber. With valve 5 opened, this manometer can be used to determine the effective pumping speed, and, with valve 5 closed, it measures the speed of total gas evolution and leakage in the vacuum chamber. Valve 10 is intended for letting air into the vacuum chamber. During adjustment of the system, valve 10 can be connected to a leak detector.

If the partial pressures of all gases in a vacuum chamber should be measured, gas analyzer 12 is put in operation through valve 11. Auxiliary vacuum system 13 is necessary to reduce the pressure of the analyzed mixture of gases, because modern vacuum gas analyzers can work at a maximun total pressure of 10^{-2} Pa.

The cost of the fragment of a vacuum system between the pump and the pumped object depends on the number of sequentially connected parts. Figure 6.3 shows the optimum utilization coefficient of the pump as a function of the required effective pumping speed for various numbers of elements of the vacuum system between the pump and the pumped object.

Table 6.2 Main characteristics of medium vacuum pumps.

Pump type	Designation	Limiting pressure, Pa	Pumping speed, m^3/s
Two-rotor	TRP	5×10^{-3}	5×10^{-3}–5
Vapor jet booster	VJBP	5×10^{-1}	0.45–15
Adsorption	ASP	10^0–10^{-3}	10^{-3}–5

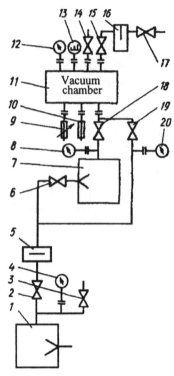

Figure 6.4 Vacuum system for low and medium vacuum (10^5–10^{-2} Pa): (*1*) roughing pump; (*2, 3, 6, 14, 15, 17, 18, 19*) vacuum valves; (*5*) oil trap; (*7*) pump for production of medium vacuum; (*9*) electric input; (*10*) mechanical input; (*11*) vacuum chamber; (*13*) gas analyzer; (*16*) moisture trap.

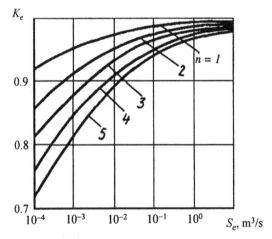

Figure 6.5 Optimum utilization coefficients K_e of two-rotor pumps as a function of effective pumping speed S_e and the number of elements n between the pump and the pumped object.

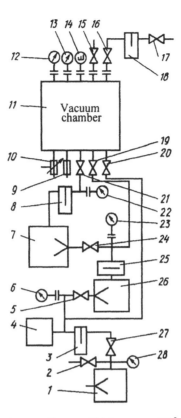

Figure 6.6 Vacuum system for low, medium, and high vacuum (10^5–10^{-5} Pa): (1) medium-vacuum pump; (2, 5, 15, 16, 17, 19, 20, 21, 24, 27) vacuum valves; (3, 8, 25) traps; (4) preliminary vacuum cylinder; (6, 12, 13, 22, 23, 28) manometers; (7) high-vacuum pump; (10) electric input; (11) pumped oblect; (14) gas analyzer; (18) hydroscopic trap; (26) medium-vacuum pump.

Vacuum system for producing of low and medium vacuum (Fig. 6.4) have two vacuum pumps. Pump 7 produces medium vacuum and pump 1 provides preliminary rarefaction. As a medium vacuum pump, one may apply two-rotor, vapor-jet, or adsorption pumps (Table 6.2). Pumps 1 should provide low vacuum. The speed and the limiting pressure of the preliminary rarefaction pump should be in accordance with the characteristics of the medium vacuum pump. The optimum efficiencies of two-rotor pumps depending on the effective pumping speed and the number of sequentially connected elements are shown in Fig. 6.5.

When pump 1 is off, one may equalize the pressure at the inlet and outlet pipes by closing valve 2 and opening valve 3. Manometer 4 is necessary to test the serviceability of pump 1. Trap 5 prevents penetration of vapor of the working liquid from pump 1 into pump 7 and vacuum chamber 11. Valves 6, 18, and 19 permit one to work in the 'direct' and 'by-pass' pumping modes that imply passage of pumped gas through both pumps or only the preliminary pump, respectively. In this latter case, low vacuum is produced in the pumped object.

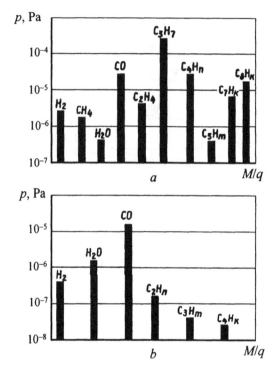

Figure 6.7 Composition of residual gases in vacuum systems with various pumps: (*a*) vapor jet pumps and (*b*) turbomolecular pumps.

Manometer 20 permits one to determine the efficiency of trap 5, and manometer 8 controls the limiting pressure of pump 7 with valve 18 closed.

The vacuum chamber is equipped with electrical 9 and mechanical 10 inputs, manometer 12, gas analyzer 13, valve 14 connecting a leak detector, and valves 15 and 17 with hygroscopic trap16 for letting air into the vacuum chamber. The trap protects the vacuum chamber from water vapor of atmospheric air and hence essentially shortens the time of pumping the chamber to the working pressure.

Vacuum systems for low, medium and high vacuum (Fig. 6.6) contain all the elements of the previous schematic and additional ultrahighvacuum pump 7 (Table 6.3). As a high-vacuum pump, one may use steam-jet, turbomolecular, or getter-ion pumps. The composition of residual gases for steam-jet and turbomolecular pumps is shown in Fig. 6.7. Trap 8

Table 6.3 Main characteristics of high-vacuum pumps.

Pump type	Designation	Limiting pressure, Pa	Pumping speed, m^3/s
Vapor jet	VJP	10^{-4}–10^{-5}	5×10^{-3}–40
Turbomolecular	TMP	10^{-6}–10^{-7}	10^{-2}–10
Getter-ion	GIP	5×10^{-7}	5×10^{-1}–20

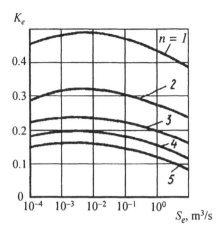

Figure 6.8 Optimum utilization coefficients K_e of vapor-jet pumps as a function of effective pumping speed S_e and the number of elements n between the pump and the pumped object.

protects the vacuum chamber from penetration of working liquid vapor of the steam-jet pumps. Trap 8 is not installed in turbomolecular and getter-ion pumps. Valve 21 connects the pump with the vacuum chamber. Manometer 22 permits one to control the efficiency of pump 7 and trap 8 with valve 21 closed. Manometers 12 and 13 measure the pressure in the vacuum chamber at medium and high vacuum, respectively. To maintain high vacuum in the vacuum chamber, all the pumps are connected sequentially. If medium vacuum should be produced in the vacuum chamber, pump 26 is connected to the pumped object through valve 19. In this case valve's 21 and 24 must be closed.

Similarly, a low-vacuum pump is directly connected to the pumped object or pump 26 through valves 20 and 5. If the vacuum system operates in high-vacuum mode, the pumping gas flows are very small. This fact permits one to disconnect pump 1 for a long time and to operate with preliminary vacuum cylinder 4.

The schematic may be simplified by excluding commutation equipment 19 and 20, but this will considerably reduce the efficiency of the system in non-steady-state modes.

If the system is intended to operate only at high vacuum, pump 26 with manometer 23 and valves 19 and 24 may be excluded. In this case, the time of pumping the vacuum chamber to the working pressure increases. Some recommendations for the selection of the optimum operating efficency of high-vacuum pumps are shown in Fig. 6.8.

Table 6.4 Main characteristics of ultrahigh vacuum systems.

Pump type	Designation	Limiting pressure, Pa	Pumping speed, m^3/s
Vapor jet with a trap	CJP	10^{-8}	$5 \times 10^{-3} – 5 \times 10^{-1}$
Magnetic discharge	MDP	5×10^{-8}	$6 \times 10^{-3} – 1.2 \times 10^{0}$
Orbitron	OP	10^{-8}	$10^{-1} – 5 \times 10^{-1}$
Cryocondensation	CP	10^{-11}	$5 \times 10^{-1} – 4 \times 10^{1}$

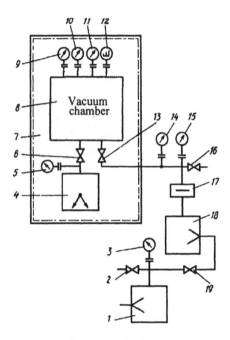

Figure 6.9 Vacuum system for ultrahigh vacuum: (*1*) low vacuum pump, (*2, 6, 13, 16, 19*) valves; (*3, 5, 9, 10, 11, 14, 15*) manometers; (*4*) ultrahigh vacuum pump; (*7*) 200°C heated vacuum unit; (*8*) pumped object; (*12*) gas analyzer; (*17*) trap; (*18*) high vacuum pump.

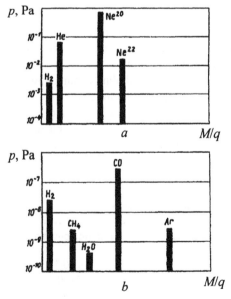

Figure 6.10 Composition of resudual gases in (*a*) cryosorption and (*b*) magnetic discharge ultrahigh vacuum systems.

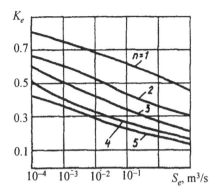

Figure 6.11 Optimum utilization coefficients of ultrahigh vacuum pumps as functions of their effective pumping speed S_e and the number of elements n between the pump and the pumped object.

Vacuum systems for the production of ultrahigh vacuum (Fig. 6.9) contain heated vacuum block 7. Heating to 200°C reduces gas evolution of all parts of the vacuum system that are directly connected to sorption pump 4. The composition of the residual gases in ultrahigh vacuum systems is shown in Fig. 6.10. The main characteristics of ultrahigh vacuum pumps are listed in Table 6.4.

The chamber is supplied with manometers 9, 10, and 11 that measure pressures from the atmospheric one to ultrahigh vacuum. Manometer 5 controls the efficiency of pump 4. The high vacuum part of the system consists of two pumps, i.e., 18 and 1. As high vacuum pump 18, one may use a steam-jet pump with trap 17 or a turbomolecular pump without a trap. The valve 16 connects a leak detector to the vacuum system, and manometers 14 and 15 measure the pressure of medium and high vacuum. A rotary pump is more often used as low-vacuum pump 1. The schematic of high-vacuum pumping is greatly simplified, because it is only used in non-steady-state mode on starting the system.

When the system should be more reliable to protect the vacuum chamber from working liquid vapor, one may use systems with cryosorption pumps for preliminary pumping.

The optimum utilization coefficients of ultrahigh vacuum pumps as a function of their pumping speed in the pumped object and the number of parts between the pump and the pumped object are shown in Fig. 6.11.

The vacuum systems in which the outlet pressure is provided by one preliminary rarefaction pump for several vacuum pumps connected in parallel are referred to as centralized vacuum systems. These systems are used for a large number of similar vacuum equipment, in automatic pumping devices, etc.

Schematic of the centralized three-pipeline pumping system is shown in Fig. 6.12. The vacuum chambers V_1–V_n are operated by the high vacuum pumps P_1–P_n through the automatic valves K_1–K_n. The high vacuum pumps are connected to the preliminary rarefaction pipeline *III* through the automatic valves B_1–B_n. The pipelines *I* and *II* serve for preliminary pumping of air from the vacuum systems. The pipeline *I* reduces the pressure from the atmospheric one to 10^2 Pa, and with the pipeline II, from 10^2 to 5 Pa.

Each pipeline is operated by one mechanical pump. The pipelines *II* and *III* have two-rotor pumps. A reserved pump can be connected to any of the three pipelines. Switching

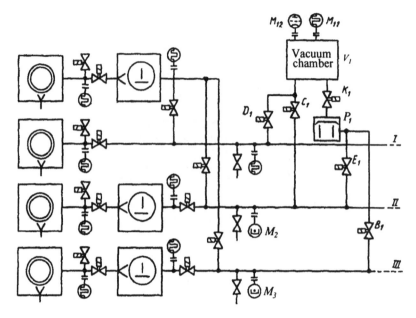

Figure 6.12 Centralized tree-pipeline pumping system.

and operation of the centralized vacuum system for pumping of vacuum chamber V_1 are performed as follows:

1. Pumping for 2–3 minutes from the atmospheric pressure through the pipeline I, with valve D_1 opened, and C_1, K_1, and E_1 closed, valve B_1 opened, high-vacuum pumps P_1 are prepared for operation;
2. A time relay switches from pipelines I to pipelines II; at this time valve D_1 is closed, and valve C_1 opens, and the pumping is continued to a pressure of ~5 Pa;
3. Manometer transducer M_{11}, in the schematic of blocking vacuum gauge at a pressure of 5 Pa connects oil–vapor pump P_1 to the system, in this case valves B_1, and C_1 are closed, and valves E_1 and K_1 are opened;
4. When the working pressure is produced in the volume, the pumps are switched to the pipeline III by manometer M_{12}; valve E_1 is closed and B_1 is opened.

To provide continuous operation of a centralized vacuum system in case of a failure, one may use two methods:

1. protection of the centralized pipeline from the vacuum system, and
2. protection of the vacuum systems from the centralized pipeline.

The former protection operates from the vacuum relays M_{11}–M_{1n} if the pressure in the vacuum system is higher than allowed (6–8 Pa); valves K_1–K_n are automatically closed. The latter protection comes into operation from vacuum relay M_2 or M_3 installed in the centralized pipeline, if the pressure in it is above the highest allowable value, and in this case the valves B_1–B_n or E_1–E_n are automatically closed.

As relays M_2, M_3, and M_{11}–M_{1n}, one may apply manometer transducers that show relay characteristics.

The above centralized pumping systems are designed for high vacuum pumps that require continious preliminary rarefaction. For high vacuum sorption pumps the centralized pumping system can work without pipeline *III*.

If the number of vacuum systems is not too large or their total volume is commensurable with the volume of the centralized pipeline, one may work without pipeline *I*. Thus, two- and one-pipeline centralized pumping systems can be used.

Economic reasons permit one to determine the expediency of using the failure protection systems. If the probability of failure is minor and its consequences are easily recoverable, the operation of both or one of protection systems is useless.

6.2   Calculation of Gas Load

A steady-state gas load that is pumped during the operation of a vacuum system consists of several integral parts:

$$Q = Q_g + Q_l + Q_t + Q_p,$$

where Q_g is the diffusive gas evolution from the material, Q_l is the leakage through the casing of the vacuum chamber, Q_t is the steady-state process gas evolution, and Q_p is the permeability of the materials. Below we consider each of the specified integral parts in detail.

Gas evolution is of a non-steady-state nature, but for the majority of gases and materials the time constant of these processes is so large that they can be considered steady-state. A simplified method for determining gas evolution is based on experimentally determined coefficients of specific gas evolution (Table C.19) that depend on the type of gas, material and its preliminary outgasing, and working temperature. The gas flow is

$$Q_g = \sum_{i=1}^{N} q_i F_i = q \sum_{i=1}^{N} F_i, \tag{6.1}$$

where F_i is the area of the ith material that is present in the vacuum system; N is the number of materials; and q is the average specific gas evolution of the materials of the vacuum system:

$$q = \frac{\sum_{i=1}^{n} q_i F_i}{\sum_{i=1}^{n} F_i}.$$

Quantitative estimation of the processes of steady-state gas penetration through the walls of vacuum systems that are made of various materials or have various thickness can be made with account of the permeability constants K_{0i} and Q_{pi} (Table C.21) using Eq. (1.68)

$$Q_0 = \sum_{i=1}^{N} K_{0i} F_i \frac{p_2^{1/n} - p_1^{1/n}}{2h_i} \exp\left(\frac{Q_{pi}}{nKT}\right),$$ (6.2)

where K_{0i} and Q_{pi} are the permeability constant and the activation energy for a material of the ith wall of the vacuum system, respectively, F_i and h_i are the area and half-thickness of the ith wall, respectively, p_1 and p_2 are the pressures at the inner and outer sides of the walls, respectively, n is the number of atoms in a gas molecule that penetrates through the wall, T is the absolute temperature of the wall, $R = 8.31$ kJ/(kmol K), and N is the number of walls of the vacuum chamber, fittings and pipelines that are made of various materials or have various thicknesses. Gas permeation increases with reduction of the vacuum chamber wall thickness. This can especially be seen for such parts as bellows, membranes, etc., where the small thickness of a part is determined by its work conditions. Gas permeation can be reduced apart from the choice of materials, using such methods as 'double' vacuum and cooling of parts during operation directly in the vacuum chamber.

The gas flow through a leak varies with pressure in a vacuum chamber:

$$q_L = U_L(p_1 - p_2),$$

where U_L is the conductance of the leak; p_1 and p_2 are the pressures outside and inside the vacuum chamber.

A decrease in the pressure in the chamber increases the leakage flow. The conductance of a leak may decrease in accordance with pressure if several gas flow modes are present.

The effective diameter of a leak for a cylindrical capillary with length l that is equal to the wall thickness at molecular flow mode can be calculated using the formula

$$d_e = \sqrt[3]{\frac{q_L l}{121(p_1 - p_2)}}.$$ (6.3)

The leakage mode is determined by the Knudsen criterion for average pressure in the capillary $(p_1 + p_2)/2$. If molecular flow mode is not the case, d_e can be recalculated using the formulas for molecular-viscous and viscous modes.

Leakage through the casing of a vacuum chamber usually occurs in detachable and undetachable connections that are not absolutely tight. Leakage can also occur through the structural defects of the wall material. Therefore, leakage in a vacuum system can be estimated using the formula:

$$Q_l = \frac{K_b Q_{lm} N}{m},$$ (6.4)

where Q_{lm} is the minimum flow registered by a leak detector (Table 6.5), K_b is the probability that the leak is smaller than the sensitivity of the leak detector, N is the number of connections, and m is the number of simultaneously tested connections.

Table 6.5 Gas flows that are registered by leak detectors

Type of leak detector	Smallest registered flow, m^3 Pa/s	Purpose
Mass spectrometer	10^{-11}	Testing of objects allowing evacuation
Ditto, with a throttle of pumping out	10^{-12}	Ditto
Ditto, with accumulation	10^{-13}	Testing of small-sized objects allowing evacuation
Halogen detector with an atmospferic transducer	10^{-7}	Testing of cavities filled with halide-containing substances
Halogen detector with a vacuum transducer	10^{-9}	Testing of objects allowing evacuation
Spark leak detector	10^{-5}	Testing of vacuum systems with glass parts

Possibility of reducing Q_l for the calculation using Eq. (6.4) is the reduction of the number of tightness tests. The extreme case is one test when $m = N$ and the entire system is tested for tightness. For operation with helium leak detector, such test is made by placing the entire system in a helium atmosphere with the use of polyethylene covers or other auxiliary system sized means.

Process gas evolution depends on the type of treated object and the method by which the technological process is implemented. In design calculations its value is taken constant during the whole time of steady-state mode.

6.3   Equations of Steady-State Pumping

For steady-state pumping mode the stability of gas mass flows and pressures in time is typical in all the sections of a vacuum system. When the gas mass flow is also constant in all the sections, this is a vacuum system with concentrated parameters (Fig. 6.13, a) and the main equation of vacuum engineering (2.110) is valid for it.

If $dQ/dx = f(x)$ depends only on x, the sources of gas evolution are distributed over the length of the vacuum system. Such sources may be specially installed leaks, leaks through looseness of connections, gas permeation and gas evolution of materials (Fig. 6.13, b). If $dQ/dx = f(p)$ depends on pressure, there are usually sorbing walls in a vacuum system (Fig. 6.13, c).

The equation of balance of forces is written from the condition that the difference of pressures is the driving force, which at low vacuum is equal to the resistance force due to the internal friction between the layers of pumped gas, and at high vacuum, it is equal to the force of external gas friction with the walls of the vacuum system: $dp/dx = \varphi(x, p, Q)$.

If dp/dx is constant, the gas is in a pipe of constant cross section at molecular flow. For $dp/dx = f(x, Q)$ the pipe has a cross section variable (Fig. 6.13, d), and if the dependence is $dp/dx = f(p, Q)$, the gas flow is viscous or molecular-viscous.

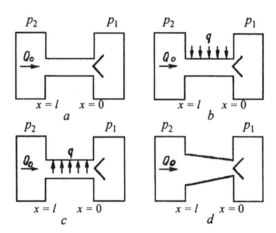

Figure 6.13 Calculated schematics of vacuum pipelines.

The equations of mass balance and the balance of forces, with the exception of Q, can be rewritten as an ordinary second order differential equation: $d^2p/dx^2 = \psi(x, p)$. The boundary conditions can be given as constant pressures at both ends of the pipeline flow and pressure at one of the ends of the pipeline. The solution is a distribution function equation.

To calculate the distribution of pressure for the schematic in Fig. 6.13, b, we write the differential equation of the balance of forces

$$Q = C\frac{dp}{dx},$$

where C is the proportionality coefficient.

We consider that the pipeline has a constant cross section shape with perimeter P along its total length, and the gas evolution from unit surface is q. Therefore, the differential equation of the mass balance can be written in the form

$$\frac{dQ(x)}{dx} = -qP.$$

Excluding Q from the above equations, we obtain the differential equation of steady-state pumping:

$$\frac{d^2p}{dx^2} = -\frac{q}{C}P, \tag{6.5}$$

given the boundary conditions $x = 0$, $p = p_1$; $x = l$, $dp/dx = Q_0/C$.

We now consider molecular mode of gas flow when C is constant. Integrating Eq. (6.5) we obtain

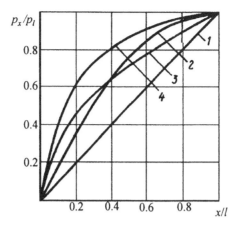

Figure 6.14 Pressure distribution over the pipeline length: in molecular mode: (1) $q = 0$; $Q_0 \neq 0$; (2) $Q_0 = 0$; $q \neq 0$; in viscous mode: (3) $q = 0$; $Q_0 \neq 0$; (4) $Q_0 = 0$; $q \neq 0$.

$$p(x) = p_1 + \frac{qPl + Q_0}{C}x - \frac{qP}{2C}x^2 . \qquad (6.6)$$

Examples of this distribution are shown in Fig. 6.14 (curves 1 and 2).

The pressure p_2 at $x = l$ can be written as

$$p_2 = p_1 + \frac{l}{C}\left(Q_0 + \frac{qPl}{2}\right).$$

If $Q_0 > qPl/2$, the effect of internal gas evolution on the distribution of pressure can be neglected. Denoting $C/l = U$ and taking into account the above assumption, we may transform the equation to the form

$$Q_0 = U(p_2 - p_1),$$

which is similar to Eq. (2.35). In this case the distribution of pressure becomes linear, and the gas flow is constant in all the sections of the vacuum system:

We consider a viscous gas flow through a pipeline whose conductance is directly proportional to gas pressure, i.e., $C = Ul = C_0 p$.

Using the same transformations for Eq. (6.5), we may determine the pressure in any section:

$$p^2(x) = p_1^2 + \frac{2(qPl + Q_0)}{C_0}x - \frac{qP}{C_0}x^2 . \qquad (6.7)$$

Examples of pressure distribution along the total length of a pipeline with a viscous gas flow are shown in Fig. 6.14 (curves 3 and 4) . At $x = l$ the pressure at the end of the pipeline connected to the pumped object is

$$p_2^2 = p_1^2 + \frac{2l}{C_0}\left(Q_0 + \frac{qPl}{2}\right). \tag{6.8}$$

If $Q_0 \gg qPl/2$, the dependence of pressure on the pipeline length in viscous flow mode is parabolic. Denoting $U = C_0(p_1 + p_2)/(2l)$, we again obtain Eq. (2.35).

Pressure p_1 is determined by the characteristics of a pump. If we accept the theoretical dependence of pumping speed of S_p on pressure in the form of function

$$S_p = S_m\left(1 - \frac{p_{lim}}{p_1}\right),$$

where S_m is the nominal speed of the pump; and p_{lim} is the limiting pressure of the pump, then the working pressure of the pump is

$$p_1 = \frac{Q_0}{S_p} = \frac{Q_0}{S_m(1 - p_{lim}/p_1)}.$$

Solving this equation relative to p_1, we obtain

$$p_1 = p_{lim} + \frac{Q_0}{S_m}. \tag{6.9}$$

6.4   Connections of System Elements and Pumped Objects

If ducts (switching elements, traps, pipelines) and pumped objects are connected to one another, they form complex element of vacuum systems. A connection of two or more parts is referred to as a joint of a vacuum system. In a joint, the sum of net gas flows passing through the connected channels is equal to zero, i.e.,

$$\sum_{i=1}^{K} Q_i = 0. \tag{6.10}$$

If a vacuum pump is in the system, Eq. (6.10) can be rewritten as

$$\sum_{i=1}^{K-1} Q_i = S_p p_p \tag{6.11}$$

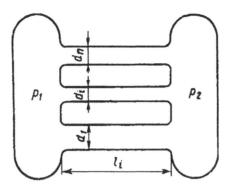

Figure 6.15 Parallel-connected pipelines.

where S_p and p_p are the speed and pressure of the pump, respectively. If there are several joints in a vacuum system, for each of them one may write an equation such as (6.10). Solving the resultant set of algebraic equations, one may determine the pressure in all the joints of the vacuum system.

Parallel connection of pipelines (Fig. 6.15) increases the gas flow between two adjacent elements of a vacuum system. The pressures p_1 and p_2 at the ends of all pipelines are identical, and the total flow between the nodes is in terms of Eq. (2.35)

$$Q = \sum_{i=1}^{K} Q_i = (p_1 - p_2)\sum_{i=1}^{K} U_i,$$

where Q_i is the flow through the ith pipeline, K is the number of pipelines.

The calculation of a vacuum system can be simplified if its nodes are connected by one equivalent pipeline with conductance U:

$$U = \sum_{i=1}^{K} U_i. \tag{6.12}$$

For generalized molecular-viscous gas flow mode at constant temperature and molecular mass the conductance of a pipeline is a function of pressure and geometrical size of the pipeline:

$$U_i = \frac{A d_i^3}{l_i} + \frac{B d_i^4}{l_i} p_{av}, \tag{6.13}$$

where A, B are constant, d and l are the diameter and length of the pipeline, respectively, and p_{av} is the average pressure in the pipeline: $p_{av} = (p_1 - p_2)/2$. We rewrite (6.12) in terms of (6.13):

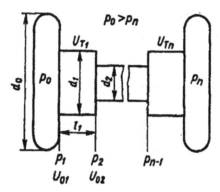

Figure 6.16 Series-connected pipelines.

$$\frac{Ad^3}{l} + \frac{Bd^4}{l} p_{av} = A \sum_{i=1}^{K} \frac{d_i^3}{l_i} + B p_{av} \sum_{i=1}^{K} \frac{d_i^4}{l_i}. \tag{6.14}$$

Equalizing the terms in the left-hand and the right-hand sides of Eq. (6.14) at similar p_{av}, we obtain two equations

$$\frac{d^3}{l} = \sum_{i=1}^{K} \frac{d_i^3}{l_i}; \ \frac{d^4}{l} = \sum_{i=1}^{K} \frac{d_i^4}{l_i},$$

from which one may determine the diameter d and length l of the equivalent pipeline.

Serial connection of elements (holes and pipelines) is shown in Fig. 6.16. For similar gas flow in all the parts one may write

$$\frac{Q}{U_i} = p_{i-1} - p_i. \tag{6.15}$$

Summing all the n equations of set (6.15), we obtain

$$Q\left(\frac{1}{U_1} + \frac{1}{U_2} + \dots + \frac{1}{U_i} + \dots + \frac{1}{U_n}\right) = p_0 - p_n.$$

For the entire system of connected in series parts one may also write

$$\frac{Q}{U} = p_0 - p_n. \tag{6.16}$$

Comparing Eqs. (6.15) and (6.16), we obtain the general conductance of connected in series parts in the form

$$U = \frac{1}{\sum_{i=1}^{n} 1/U_i} . \tag{6.17}$$

For calculation of complex vacuum systems using the methods of continuum mechanics, we will make the following assumptions:

1. for serial connection of parts with various diameters, there are additional resistances to the gas flow in the places of pipe narrowing and;
2. the places of pipe broadening do not resist the gas flow.

For molecular flow mode, the system of connected in series orifices and long pipelines can be replaced by one equivalent orifice connected to an infinitely large object. For the system shown in Fig. 6.16, taking into account that the conductance of a round orifice from Eq. (2.50) is

$$U_{0i} = \frac{C d_i^2}{(1 - d_{i-1}^2 / d_i^2)} ,$$

where C is a constant, d_i is the diameter of a orifice, d_{i-1} is the diameter of the previous part, and the conductance of the long pipeline $U_p = A d_i^3 / l_i$, according to Eq. (6.17), we obtain

$$\frac{1}{C d^2} = \frac{1 - d_0^2 / d_1^2}{C d_1^2} + \ldots + \frac{1 - d_{i-1}^2 / d_i^2}{C d_i^2} + \ldots$$

$$+ \frac{1 - d_{m-1}^2 / d_m^2}{C d_m^2} + \frac{l_1}{A d_1^3} + \ldots + \frac{l_j}{A d_j^3} + \ldots + \frac{l_n}{A d_n^3},$$

where m is the number of narrowings of the pipeline; and n is the number of pipelines in the system. Solving the above equation, one may find the diameter d of the orifice, the conductance of which is equal to the conductance of a complex system of connected in series elements.

For viscous flow mode one may also determine the size of the equivalent pipeline. Neglecting the resistance of orifices, set of equations (6.15) can be rewritten with account of the general formula of pipeline conductance at a constant temperature in the following form:

$$Q \frac{2 l_i}{B d_i^4} = p_{i-1}^2 - p_i^2 .$$

Combining all the equations, we obtain

$$\frac{2Q}{B} \left(\frac{l_1}{d_1^4} + \ldots + \frac{l_1}{d_i^4} + \ldots + \frac{l_n}{d_n^4} \right) = p_1^2 - p_n^2 ,$$

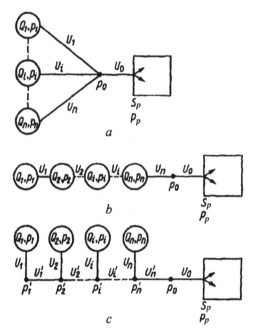

Figure 6.17 Connections of pumped objects: (*a*) parallel; (*b*) series; (*c*) combined.

whence it follows that the diameter *d* and length *l* of the equivalent pipeline should satisfy the condition

$$\frac{l}{d^4} = \frac{l_1}{d_1^4} + \dots + \frac{l_1}{d_i^4} + \dots + \frac{l_1}{d_n^4}.$$
(6.18)

Accepting the length of the equivalent pipeline *l*, one may determine from Eq. (6.18) its diameter *d*.

Parallel connection of several pumped objects to one pump (Fig. 6.17, *a*) is frequently used for pumping of small chambers. Writing Eq. (6.10) for a joint with pressure p_0, we determine

$$p_0 = \frac{\sum_{i=1}^{n} Q_i(S_p + U_0)}{S_p U_0}.$$
(6.19)

The pressure in any pumped object p_i depends on p_0. According to Eq. (2.35), we obtain

$$p_i = \frac{Q_i}{U_i} + p_0.$$
(6.20)

For serial connection of pumped objects (Fig. 6.17, *b*) the pressure p_0 is determined using Eq. (6.19), and the pressure p_i in any pumped object, according to Eq. (6.20).

Serial-parallel connection of pumped objects (Fig. 6.17, *c*) differs from parallel connection (Fig. 6. 17, *a*) by the presence of a collector the conductance of which is commensurable with the conductance of the pipelines that are directly attached to the pumped objects. The pressure p_0 is determined using Eq. (6.19). Writing Eq. (6.10) for a joint, the pressure in which is

$$p'_n = p_0 + \frac{\sum_{i=1}^{n} Q_i}{U_n} \tag{6.21}$$

The pressure in any pumped object:

$$p_i = p'_i + \frac{Q_i}{U_i} = p'_{i+1} + \frac{\sum_{j=1}^{i} Q_j}{U_i} + \frac{Q_i}{U_i}. \tag{6.22}$$

6.5 Connection of Pumps

In vacuum systems with large gas loads parallel connection of pumps is used (Fig. 6.18, *a*). We write Eq. (6.10) for unit *A*, which is connected through elements with conductances $U_1 - U_i - U_n$ to all the inlet branch pipes of parallel-connected pumps:

$$Q = \sum_{i=1}^{n} U_i(p_A - p_i) = \sum_{i=1}^{n} S_{pi} p_i, \tag{6.23}$$

where Q is the total flow in the system of pumps.

For the inlet sections of the pumps, Eq. (6.11) can be written as equality of the flows passing through a part with conductance U_i and through a pump with pumping speed S_i:

$$U_i(p_A - p_i) = S_{pi} p_i. \tag{6.24}$$

Solving Eq. (6.24) with respect to p_i, we will obtain

$$p_i = p_A \frac{U_i}{S_{pi} + U_i}. \tag{6.25}$$

The pumping speed in joint *A* is determined by substituting Eq. (6.25) into (6.23):

$$S_A = \frac{Q}{p_A} = \sum_{i=1}^{n} \frac{S_{pi} U_i}{S_{pi} + U_i} = \sum_{i=1}^{n} S_{Ai}. \tag{6.26}$$

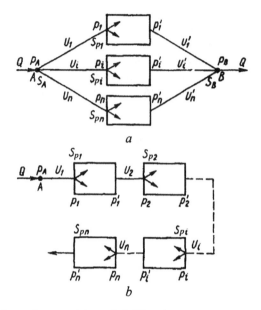

Figure 6.18 Connections of pumps: (*a*) parallel; (*b*) series.

It follows from Eq. (6.26) that the pumping speed in joint A is equal to the sum of the effective pumping speeds of all the pumps. Hence, a system of parallel-connected pumps can be replaced by one pump that is connected to the joint A and whose pumping speed is (6.26).

If all pumps are identical, then $S_A = nS_{Ai}$, and to maintain pressure p_A in joint A, the following number of pumps is required:

$$n = \frac{Q}{p_A S_{Ai}}.$$

We will determine the limiting pressure in joint A at $Q = 0$ for a system consisting of pumps with various limiting pressures $p_{lim\,i}$. In this case we write equation of flow (6.10) for joint A as

$$\sum_{i=1}^{n} S_{pi} p_i = 0. \tag{6.27}$$

Using the dependence of S_{pi} on pressure according to Eq. (4.1) at $p \ll p_s$, one may transform Eq. (6.27) to

$$\sum_{i=1}^{n} S_{mi}(p_i - p_{lim\,i}) = 0. \tag{6.28}$$

For of each of the pumps the equation

$$U_i(p_A - p_i) = S_{mi}(p_i - p_{lim\ i})$$ (6.29)

is true, from which we find

$$p_i = \frac{U_i p_A + S_{mi} p_{lim\ i}}{S_{mi} + U_i}.$$ (6.30)

Substituting Eq. (6.30) into Eq. (6.28) and solving the obtained expression relative to pressure p_A we will obtain the equation of the limiting pressure of a system with parallel-connected pumps:

$$p_A = \frac{\sum_{i=1}^{n} S_{ei} p_{lim\ i}}{\sum_{i=1}^{n} S_{ei}},\ \text{where } S_{ei} = \frac{S_{mi} U_i}{S_{mi} + U_i}.$$ (6.31)

At the maximum gas flow for a system of parallel-connected pumps, the pressures in the outlet pipes of the pumps are their maximum outlet pressures that are specified in the catalogues. Equation of flows (6.10) in joint B is written in the asumption that $p'_i = p_{mi}$:

$$\sum_{i=1}^{n} (p_{mi} - p_B) U_i = Q.$$ (6.32)

Solving Eq. (6.32) with respect to the pressure in note B, we obtain

$$p_B = \frac{\sum_{i=1}^{n} U_i p_{mi} - Q}{\sum_{i=1}^{n} U_i}.$$ (6.33)

The effective pumping speed of a preliminary rarefaction pump in note B

$$S_B = \frac{Q}{p_B} = \frac{\sum_{i=1}^{n} U_i}{\sum_{i=1}^{n} (U_i p_{mi}/Q) - 1}.$$ (6.34)

The pumping speed is determined by Eq. (6.34) with a safety margin, because it is improbable that all the pressures at pumps outlets in a system of different pumps simultaneously reach the maximum level. S_B may be specified for a particular system of pumps by the condition $p'_i \leq p_{mi}$.

A system of series-connected pumps (Fig. 6.18, *b*) is frequently used in high-vacuum devices. If the internal gas evolution of the pumps and connecting pipelines is neglected, the gas flow Q for steady-state operation mode in all the pumps remains constant, and one may write the condition of flow constancy as

$$S_{pi}p_i = S_{p(i+1)}p_{i+1} = Q. \tag{6.35}$$

It follows from Eq. (6.35) that the required pumping speed of series-connected pumps decreases proportionally with increasing pressure:

$$\frac{S_{p(i+1)}}{S_{pi}} = \frac{p_i}{p_{i+1}}. \tag{6.36}$$

The pumping speed of the first pump can be written using Eq. (6.35)

$$S_{p1} = \frac{Q}{p_1}. \tag{6.37}$$

To describe the main characteristic of a pump, we accept Eq. (4.1) at $p \ll p_s$. Using Eq. (6.37) one may determine the maximum pumping speed:

$$S_{m1} = \frac{Q}{p_1 - p_{lim1}}. \tag{6.38}$$

We will write the equation of flows (6.11) for the inlet section of the first pump as

$$Q = U_1(p_A - p_1) = S_{p1}p_1. \tag{6.39}$$

Solving Eq. (6.39) with respect to pressure p_1, we will obtain, in terms of expressions (2.111), the utilization coefficient of pump K_{e1}

$$p_1 = \frac{U_1}{U_1 + S_{p1}}p_A = K_{e1}p_A. \tag{6.40}$$

Substituting Eq. (6.40) into Eq. (6.38), we transform the expression for the maximum pumping speed of the first pump to

$$S_{m1} = \frac{Q}{K_{e1}p_A - p_{lim1}}. \tag{6.41}$$

Similarly, for any of the pumps in series we obtain

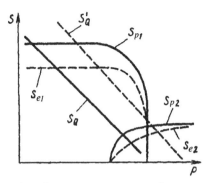

Figure 6.19 Combined operation of two pumps at constant flow.

$$S_{mi} = \frac{Q}{K_{ei}p'_{i-1} - p_{limi}} . \tag{6.42}$$

The maximum outlet pressure of a previous pump can be accepted as p'_{i-1}. When the characteristic of a pump is given graphically and is not described by Eq. (4.1), the compatibility of the pumps can be determined graphically (Fig. 6.19). Using the graphic characteristics of pumps S_{p1} and S_{p2} one may calculate and plot their effective pumping speed

$$S_{e1} = \frac{S_{p1}U_1}{S_{p1} + U_1}; \; S_{e2} = \frac{S_{p2}U_2}{S_{p2} + U_2}, \tag{6.43}$$

where U_1 is the conductance of the vacuum system from the first pump to the pumped object, and U_2 is the conductance of the vacuum system from the second pump to the first one.

In calculations using Eq. (6.43) one should take into account that the conductance U in general case is a function of pressure.

Figure 6.19 also shows the leakage and gas evolution speed as a function of pressure in the pumped object: $S_Q = Q/p$.

The intersection point of curves S_{e1} and S_Q corresponds to steady-state operation mode of the first pump. The pressure in the intersection point should be equal to the working pressure of the first pump. Similarly, the intersection point of curves S_{e2} and S_Q determines the working pressure of the second pump. If it is lower than the maximum outlet pressure of the first pump, the pumps may work together in serial connection.

Using Fig. 6.19, one may easily determine the conditions required for starting a system. This condition can be considered as the absence of double intersection of curves S_Q and S_{e1} in the first pump range of the working pressures. The system cannot be started for S_Q'.

In systems with variable gas evolutions and leakages, it is of interest to find the area of combined work of two series-connected pumps in the curve $Q = f(p)$. This range can be conveniently found graphically (Fig. 6.20). We consider that the condition of flow continuity $Q_1 = Q_2$ is satisfied, where Q_1 is the throughput of the first pump in the vacuum chamber, and Q_2 is the throughput of the second pump in the section of the outlet pipe of the first pump.

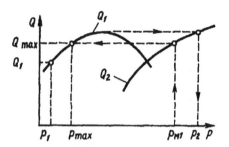

Figure 6.20 Compatibility range of two pumps.

Q_1 and Q_2 as a function of pressure can be calculated using the curves of effective pump-ing speed, of the pumps (see Fig. 6.19), by the formulas

$$Q_1 = S_{e1}p; \ Q_2 = S_{e2}p.$$

If at the maximum throughput of a previous pump the next one provides such as pressure at its outlet pipe that is lower than the maximum outlet pressure of the previous pump, these pumps allows work together.

Pressure p_2 corresponding to the maximum throughput of the system is determined from the equation

$$(Q_1)_{max} = Q_2(p_2),$$

which is convenient to solve graphically by plotting a straight line parallel to axis p in Fig. 6.20 (it is shown by an arrow), from the maximum of curve $Q_1(p)$ to the intersection with curve $Q_2(p)$.

If $p_2 < p_{m1}$ (p_{m1} is the maximum outlet pressure of the first pump), compatible work of the pumps is provided. At $p_2 \geq p_{m1}$ a combine work of pumps is only possible in the range of pressures from p_1 to p_{max} and flows from Q_1 to Q_{max}. p_{max} and Q_{max} are determined by the inverse plotting that is shown by arrows in Fig. 6.20, beginning from the pressure p_{m1}. If the total flow is higher than Q_{max}, the pressure in a vacuum chamber increases drastically from p_{max} to p_{m1}. The drastic change in pressure may be by several orders of magnitude.

When vacuum pumps are connected in series, several additional pumped objects that are then referred to as ballast vacuum cylinders may be placed between them.

A ballast vacuum cylinder permits one to switch off a preliminary rarefaction pump for a long time during operation of the system. The gas flow that is pumped by a pump passes through a ballast vacuum cylinder and increases the pessure in it from the minimum one p_n, that is produced by the preliminary vacuum pump, to the maximum one p_k that is close to maximum outlet pressure of the high-vacuum pump. It is convenient to choose the period of time during which the preliminary vacuum pump can be switched off to be equal to the period of the steady-state mode of system operation.

If the total gas flow passing through a high-vacuum pump is spent to increase the pres-sure in a ballast vacuum cylinder, one may write the mass conservation equation as

$$Qt' = V_b(p_k - p_n),\qquad(6.44)$$

where V_b is the volume of the ballast vacuum cylinder; and t' is the operating time after switching-off of the preliminary vacuum pump.

From the above equation, the volume of the ballast vacuum cylinder is

$$V_b = \frac{Qt'}{p_k - p_n}.\qquad(6.45)$$

Here

$$p_k = (0.3{-}0.5)p_h;\; p_n = \frac{Q}{S_p K_e}$$

where p_h is the maximum outlet pressure of the high-vacuum pump, S_p is the pumping speed of the preliminary vacuum pump, and K_e is the efficiency of rotary pump for pumping of the ballast vacuum cylinder to pressure p_n.

If an adsorbent is placed in a ballast vacuum cylinder, the quantity of absorbed gas, in which

$$A = K_T p\, G,$$

where G is the volume of the adsorbent, m^3, p is pressure, Pa, K_T is the adsorption factor at temperature T (for activated carbon Grade SKT, $K_{293} = 10^2{-}10^3$; $K_{77} = 10^6{-}10^7$), then the equation of a ballast vacuum cylinder with an adsorbent is as follows:

$$Qt'' = V_b(p_k - p_n) + K_T G(p_k - p_n),$$

whence

$$V_b = \frac{Qt'' - K_T G(p_k - p_n)}{p_k - p_n}.\qquad(6.46)$$

where t'' is the operating time of the preliminary vacuum cylinder having the adsorbent. If the whole preliminary vacuum cylinder is filled with adsorbent, $G = V_b$. With account of adsorbent porosity ε the mass conservation equation is

$$Qt'' = \varepsilon V_b(p_k - p_n) + K_T V_b(p_k - p_n),$$

whence the volume of the ballast vacuum cylinder is

$$V_b = \frac{Qt''}{(p_k - p_n)(K_T + \varepsilon)}.\qquad(6.47)$$

Comparing Eqs. (6.45) and (6.47), one may determine the order of magnitude by which the duration of work of the ballast vacuum cylinder will be increased due to filling with an adsorbent

$$\frac{t'}{t''} = K_T + \varepsilon$$

where t' and t'' are the operating times of the ballast vacuum cylinder with and without adsorbent, respectively. For nitrogen $T = 293$ K, $K_{293} = 10^2$, and $e = 0.5$, $t'/t'' = 10^2$. For $T = 77$ K, $K_{77} = 10^6$, and $\varepsilon = 0.5$, $t'/t'' = 10^6$. Therefore, the use of adsorbent in ballast vacuum cylinders at room temperature reduces its volume by a factor of 100, and at the temperature a liquid nitrogen, by a factor of 10^6.

6.6   Time of Pumping

At the initial stage of gas removal from a vacuum system or at the change of working conditions, the pressure in different parts of a vacuum system always changes in time, i.e., the operation is non-steady-state. At this mode, the concept of pipeline conductance that is accepted for constant difference of pressure at its ends cannot be used and the calculations are substantially complicated.

In some cases the non-steady-state operation of a vacuum system can be reduced to quasi-steady-state mode, for which the period of time during which the pressure in the pipeline becomes equilibrum is considerably shorter than the time for the pumped object. The pressure in the pipeline changes much faster than in the pumped object. In this case, the distribution of pressure over the vacuum system in every moment of time can be calculated using the same methods as those for the steady-state mode.

Mathematically, the condition of quasi-steady-state mode can be written through a ratio of time factors

$$\frac{\tau_1}{\tau_2} = \frac{VU}{V_{pl}S_e} \gg 1, \tag{6.48}$$

where $\tau_1 = V/S_e$, V is the volume of the pumped object; S_e is the effective speed for the pumped object, $\tau_2 = V_{pl}/U$, and V_{pl} and U are the volume and conductance of the pipeline, respectively.

Since S_e is always less than U, condition (6.48) is satisfied with a safety margin if

$$\frac{V}{V_{pl}} \gg 1. \tag{6.49}$$

If condition (6.48) is satisfied, the pipeline can be considered as part of a pumped object with volume $V + V_{pl}$.

By way of example of the quasi-stady-state operation mode of a vacuum system with a large pumped object, one may consider pumping of an object from atmospheric pressure to the working one. In this case, the pressure in the pumped object is a function of time, but in every moment of time the gas flow at the inlet and outlet of the pipeline is almost constant.

Thus, one may distinguish three operation modes of a vacuum system: steady-state, non-steady-state and quasi-steady-state. For the steady-state mode, the constancy of the pressures and the flows in all sections of a vacuum system in time is typical; in this case the condition of flow continuity is fulfilled. Then non-steady-state mode is characterized by the dependence of pressures and flows in various sections of a vacuum system on the time of pumping. The quasi-steady-state mode is a variant of the non-steady-state mode, for which condition (6.48) is fulfilled.

We now determine the pumping time under the conditions of the quasi-stady-state operation mode of a vacuum system. We consider that condition (6.48) is fulfilled and assume that there are no adsorbed gases on the internal surface of the vacuum system, and the gas flow is an isothermic process, i.e. pV = const. Differentiating this equation, dividing it by dt and taking into account that $dV/dt = S$, we obtain

$$dt = -\frac{V dp}{Sp}. \qquad (6.50)$$

Intergrating Eq. (6.50) from t_1 to t_2 and p_1 to p_2, we obtain:

$$\Delta t = t_2 - t_1 = -V \int_{p_1}^{p_2} \frac{dp}{Sp}. \qquad (6.51)$$

In general case the pumping speed of an object in the integrated expression of Eq. (6.51) is a function of pressure:

$$S = S_p K_e - \frac{Q}{p}; \quad K_e = \frac{U}{S_p + U}, \qquad (6.52)$$

where S_p is the pumping speed of the pump, K_e is the utilization coefficient of the pump, U is the conductance, and Q is the total throughput of gas evolution and leakage.

We consider particular cases:

1. Let S = const. This case occurs in high-vacuum systems at a molecular mode of gas flow in a pipeline when S_p = const, U = const, and $Q = 0$. Then, using Eq. (6.51), we write the pumping time of object from pressure p_1 to p_2 as

$$\Delta t_1 = 2.3 \frac{V}{S_p K_e} \log \frac{p_1}{p_2}. \qquad (6.53)$$

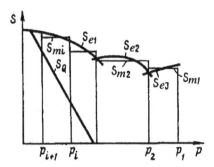

Figure 6.21 Graphical analytic calculation of pumping time in vacuum systems with a centralized pumped object.

2. Let S = const, $U = Cp$, and $Q = 0$. This case is typical rotary pumps in viscous gas flow mode. Similarly to Eq. (6.51), we obtain

$$\Delta t_2 = -V \int_{p_1}^{p_2} \frac{S_p + Cp}{S_p Cp^2} dp = \frac{V}{C}\left(\frac{1}{p_2} - \frac{1}{p_1}\right) + 2.3 \frac{V}{S_p} \log\left(\frac{p_1}{p_2}\right). \tag{6.54}$$

In general case, when the simplifications are inadmissible and Q cannot be taken to be zero, one should use the graphic-analytical method of determining pumping time. We plot the graphs of the effective pumping speed of vacuum pumps in a pumping volume and the speed of evolution and leakage of gas $S_Q = Q/p$ in Fig. 6.21, where it is supposed that three series-connected pumps operate in a vacuum system.

We now subdivide the entire range of pressures from the atmospheric pressure to the working one into several ranges and determine the average value in each of them:

$$S_{mi} = \frac{(S_{ei} - S_{Qi}) + (S_{e(i+1)} - S_{Q(i+1)})}{2}.$$

In this case the pumping time can be calculated using Eq. (6.53), and for all the ranges we obtain

$$\Delta t_g = 2.3 V\left(\frac{1}{S_{m1}} \log\frac{p_1}{p_2} + \dots + \frac{1}{S_{mi}} \log\frac{p_i}{p_{i+1}} + \dots + \frac{1}{S_{mn}} \log\frac{p_n}{p_{n+1}}\right), \tag{6.55}$$

where n is the number of ranges into which the entire range of working pressures is subdivided.

If the walls of a vacuum system were not preliminarily heated, then, determining pumping time, one should take into account desorption of water vapor due to change in pressure. Determining the time of pumping for a vacuum system that contains a two-component mixture of dry air and water vapor, we write two equations of isothermal pumping process:

$$p_1 V_k = \text{const}, \tag{6.56}$$

$$p_2 V_k + a F_g = \text{const}, \tag{6.57}$$

where p_1 is the air pressure without account of the pressure of water vapor, Pa, p_2 is the pressure of vapor water in the air, Pa, a is the amount of water vapor that is adsorbed by unit surface of the chamber, m^3 Pa/m^2, F_g is the total area of the chamber surface, m^2, and V_k is the volume of the chamber, m^3.

We now differentiate Eqs. (6.56) and (6.57) and divide them by dt. Considering that the adsorption balance is established instantaneously, we obtain

$$V_c \frac{dp_1}{dt} + S_e p_1 = 0, \tag{6.58}$$

$$V_c \frac{dp_2}{dt} + S_e p_2 + F_g \frac{da}{dt} = 0, \tag{6.59}$$

where S_o is the effective pumping speed, m^3/s.

Substituting the expression of a into Eq. (6.59), in the agreement with the Freindlich equation (1.58) after integration in the condition $S_e = \text{const}$, we obtain

$$t = 2.3 \frac{V_c}{S_e} \log\left(\frac{p_{2i}}{p_{2f}}\right) + \frac{F_g d}{S_e(1-m)}\left(p_{2i}^{\frac{1}{1-m}} - p_{2f}^{\frac{1}{1-m}}\right), \tag{6.60}$$

where p_{2i} and p_{2f} are the pressures of water vapor before and after pumping, respectively d and m from Eq. (1.58). Using Eq. (6.60), one may find the time of pumping that is required to reduce the pressure of water vapor in a chamber from the initial pressure to the final one.

Upon integrating Eq. (6.58) at $S_e = \text{const}$ we obtain, similarly to Eq. (6.53),

$$t = 2.3 \frac{V_c}{S_e} \log\left(\frac{p_{1i}}{p_{1f}}\right), \tag{6.61}$$

where p_{1i} and p_{1f} are the initial and final air pressures without account of the pressure of water vapor. Equations (6.60) and (6.61) allow one, using the known value of pumping time, to determine the final pressure of a two-component mixture that is equal to the sum of the pressures of dry air and water vapor.

According to experimental data, the times of pumping of wet and dry air in the range of pressures from 10^5 to 10^0 Pa are almost equal, the effect of water vapor adsorption on the time of pumping of vacuum systems becomes significant at pressures below 10 Pa, and at pressure lower than 10^{-3} Pa the time of pumping of a two-component mixture is completely determined by desorption gas evolution of water vapor from the inner walls of a vacuum chamber.

If one neglects the presence of water vapor contained in residual gases, a great error is involved in the determination of the pumping time.

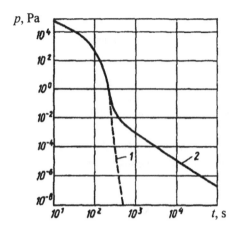

Figure 6.22 Pumping curves of a vacuum system with a volume of 0.1 m³, an inner stainless surface area of 1 m², and a pumping speed of 5×10^{-3} m³/s (*1*) without account of water desorption; (*2*) with account of water desorption.

By way of example, Fig. 6.22 shows the curve of pumping of a vacuum chamber at $V_c = 0.1$ m³, $F_g = 1$ m², and $S_e = 5 \times 10^{-3}$ m³/s. During pumping from the atmospheric pressure to 10^{-4} Pa the real time of pumping is 10 times that calculated without account of the desorption of water vapor, and for pumping to 10^{-6} Pa, this factor is 100.

The effect of temperature on pumping time is shown in Fig. 6.23, where it can be seen that for vacuum systems with limiting pressures up to 10^{-11} Pa, heating of a vacuum system to 500 K allows substantial reduction of the pumping time. Moreover, determining pumping time, one may not take into account the presence of water vapor in a mixture of residual gases.

We will determine pumping time in non-steady-state mode for a pipeline with distributed volume without account of gas evolution from the walls. One end of this pipeline is closed by a plug, and the other end is connected to a pump with a very high pumping speed, i.e., the pressure in an opened section of this pipeline can be considered to be zero.

The difference of gas flows through the sections of the pipeline that are located at distance dx determines the speed of gas removal from these sections:

$$F \mathrm{d}x \frac{\partial p}{\partial t} = -Ul \frac{\partial p}{\partial x} + Ul\left(\frac{\partial p}{\partial x} + \frac{\partial^2 p}{\partial x^2}\mathrm{d}x\right),$$

where F is the cross section area of the pipeline, and U and l are its conductance and length, respectively.

The latter expression can be reduced to a differential equation such as

$$\frac{\partial p}{\partial t} = D \frac{\partial^2 p}{\partial x^2},$$

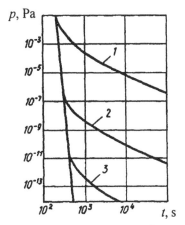

Figure 6.23 Pumping curves of a vacuum system ($V = 0.1$ m³; $F = 1$ m²; $S_e = 5 \times 10^{-3}$ m³/s) at various temperatures of the vacuum system and the pumped object: (*1*) 298 K; (*2*) 400 K; (*3*) 500 K.

that has at $D = Ul/F$ the following initial and boundary conditions: $t = 0$, $p = p_1$; $x = 0$, $dp/dx = 0$; $x = l$, $p = 0$.

The solution of this equation is:

$$p = p_1 \frac{4}{\pi} \sum_{k=1}^{\infty} \frac{(-1)^{k-1}}{2k-1} \cos\left[\frac{\pi}{2}(2k-1)\frac{x}{l}\right] \exp\left[-\frac{\pi^2}{4}(2k-1)^2\tau\right], \qquad (6.62)$$

where $\tau = Dt/l^2 = Ut/V$, and $V = Fl$ is the volume of the pipeline.

For $\tau \geq 0.1$ and $x = 0$, Eq. (6.62) can be simplified:

$$p = p_1 \frac{4}{\pi} \exp\left(-\frac{\pi^2}{4}\tau\right). \qquad (6.63)$$

From here the expression of time required to reduce pressure from p_1 to p_2 in a closed section of a pipeline is as follows:

$$\Delta t = \frac{V}{U} \log\frac{p_1}{p_2}. \qquad (6.64)$$

If U depends on pressure,

$$\Delta t = V\left(\frac{1}{U_1}\log\frac{p_1}{p_2} + \ldots + \frac{1}{U_i}\log\frac{p_i}{p_{i+1}} + \ldots + \frac{1}{U_n}\log\frac{p_n}{p_{n+1}}\right), \qquad (6.65)$$

where $U_1, \ldots U_i, \ldots U_n$ are the average conductances in appropriate pressure ranges.

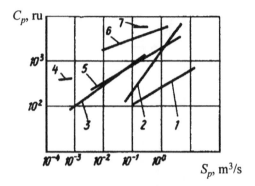

Figure 6.24 Cost of various vacuum pumps as a function of their pumping speed: (*1*) vapor jet; (*2*) getter-ion; (*3*) rotary (old model); (*4*) rotary (new model); (*5*) two-rotor; (*6*) magnetic discharge; (*7*) turbomolecular.

6.7   Cost of Pumping

The cost of pumping of a chamber with a speed of 1 m³/s during 1 hour is

$$C = \frac{(C_p + C_f)/t_o + (Z_p + Z_f)}{S_e},\qquad (6.66)$$

where t_o is the representative period of depreciation, h; C_p is the cost of a pump in relative units (ru) , for which we take the exponential dependence $C_p = C_1 S_p^{K1}$, C_f is the cost of pipelines, valves, traps and other auxiliary fittings which we take as $C_f = C_2 U^{K2}$, Z_p is the charges per one hour for operation of a pump, which we take as $Z_p = C_3 S_p^{K3}$, Z_f is the operational charges of the fittings, we take $Z_f = C_4 U^{K4}$, where C_1, C_2, C_3, C_4 and K_1, K_2, K_3, K_4 are the constants that depend on the conditions and operation of the vacuum equipment (Figs. 6.24–6.28).

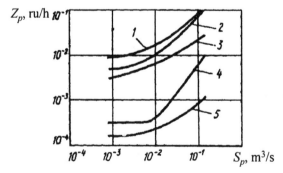

Figure 6.25 Operating cost of rotary vacuum pumps as a function of their pumping speed: (*1*) total cost; (*2*) power cost; (*3*) salary cost; (*4*) material cost; (*5*) maintenance cost.

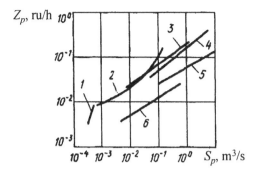

Figure 6.26 Operating cost of vacuum pumps as a function of their pumping speed: (*1*) rotary (new model); (*2*) rotary (old model); (*3*) two-rotor; (*4*) getter–ion; (*5*) vapor jet; (*6*) magnetic discharge.

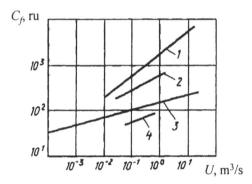

Figure 6.27 Cost of vacuum fittings as a function of conductance: (*1*) cut-off fitting for high vacuum; (*2*) sorption traps; (*3*) cut-off fittings for low vacuum: (*4*) nitrogen traps.

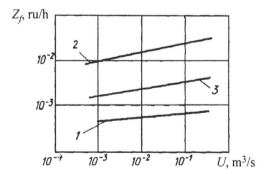

Figure 6.28 Operating cost of vacuum fittings as a function of conductance: (*1*) cut-off fitting; (*2*) nitrogen traps; (*3*) sorption traps.

Table 6.6 Cost coefficients of vacuum pumps.

Pump type	C_1, ru/(m^3/s)$^{-K1}$	K_1	C_3, ru/(m^3/s)$^{-K3}$	K_3
vapor jet	260	0.478	5.40×10^{-2}	0.330
getter-ion	1680	0.900	1.52×10^{-1}	0.517
rotary	3130	0.500	3.17×10^{-1}	0.500
two-rotor	2120	0.427	1.85×10^{-1}	0.451
magnetic discharge	4650	0.206	3.16×10^{-2}	0.417

Taking into account the above expressions, one may write Eq. (6.66) as follows

$$C = A_1 K_e^{-K1} + A_2 (1 - K_e)^{-K2} + A_3 K_e^{-K3} + A_4 (1 - K_e)^{-K4}, \tag{6.67}$$

where

$$A_1 = \frac{C_1 S_e^{K1-1}}{t_o}; \quad A_2 = \frac{C_2 S_e^{K2-1}}{t_o}; \quad A_3 = C_3 S_e^{K3-1}; \quad A_4 = C_4 S_e^{K4-1}.$$

The cost of vacuum pumps depends on their working principle and pumping speed. In regards modern vacuum pumps, turbomolecular pumps are the most expensive, and vapor-jet pumps are the cheapest. The costs per unit of pumping speed may vary for various pumps by a factor of 100 or more. Expensive pumps prevent residual gas environment from contamination with working liquid vapor and are used if the use of inexpensive pumps does not provide the sufficient quality of products or the required experimental conditions.

The operational charges of vacuum pumps include the costs of energy carriers, materials, staff salary, routine inspections, and repair. The depreciation term t_o can be taken to be 10 years that corresponds to 53,520 hr of all-day-long work and a 0.85 factor of the efficient use of working hours.

The cost and the operation costs can be calculated using the coefficient C_1–C_4 and K_1–K_4 that are listed in Tables 6.6 and 6.7.

Table 6.7 Cost coefficients of vacuum fittings.

Fitting type	C_2, ru/(m^3/s)$^{-K2}$	K_2	C_4, ru/(m^3/s)$^{-K4}$	K_4
low vacuum valves	125	0.117	8×10^{-4}	0.077
high vacuum valves	2160	0.500	8×10^{-4}	0.077
nitrogen traps	100	0.200	3.4×10^{-2}	0.200
adsorption heated traps	558	0.250	5×10^{-3}	0.160

6.8   Verifying Calculation of a Vacuum System

Verifying calculation at steady-state made aims at determining the pressure distribution over the vacuum system in hand. The verification is carried out for the determination of parameters of existing vacuum systems or performing more accurate project calculations. The initial data are as follows:

1. schematic of the vacuum system,
2. parameters of the vacuum system (including the pumped objects),
3. specifications and specific data on gas evolution of the materials used in the vacuum system,
4. the minimum flow Q_{ld} registered by a leak detector,
5. the number, m, of the tested connections, and
6. the process gas evolution Q_t.

A verifying calculation can be subdivided into several stages carried out in the following order:

1. determination of the internal gas evolution,
2. calculation of the pressure distribution,
3. verification of the choice of pumps,
4. test of the possibility of starting the system,
5. test of the combined work of the vacuum pumps, and
6. calculation of the cost of pumping.

Internal gas evolution of a vacuum system is combined from gas permeation, gas evolution from structural materials, and leakage through the casing of a vacuum system. It determines the limiting pressure, that can be produced in the vacuum chamber.

Gas evolution is determined using Eq. (6.1) and Table C.19, and gas permeation, by Eq. (6.2) and Table C.21. To determine possible leakage into the vacuum system Q_i, we use Eq. (6.3), in which K_b is the probability of a leak that is lower than the sensitivity of a leak detector; it can be taken to be 0.2, and the number of testings of the pipelines for tightness is equal to the number of detachable and undetachable connections, m.

The minimum detected flow Q_{ld} is determined by the type of the leak detector (Table 6.5). In the calculations, one should separately determine the internal gas evolution in the system and the pumped object. If the amount of internal gas evolution in the system is maximum 20% of this value for the pumped object, one may simply add these amounts and consider that gas evolution occurs only in pumped object. In this case, the pressure distribution over the length of the vacuum system will be calculated with a safety margin, and one may consider this vacuum system as one with localized parameters, which is a simpler case.

The total gas evolution in a vacuum system in steady-state operation mode is combined by the internal and the process gas evolution:

$$Q = Q_i + Q_t.$$

We write the pressure at the inlet section of a vacuum pump according to Eq. (6.9):

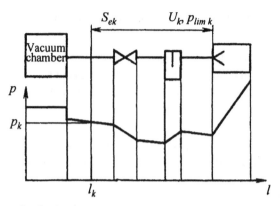

Figure 6.29 Pressure distribution in vacuum system.

$$p_p = p_{lim} + \frac{Q}{S_m},$$ (6.68)

where p_{lim} and S_m are the limiting pressure and maximum pumping speed of the pump, respectively.

An increase in the pressures at the subsequent parts of the ith section of a vacuum system can be determined by the flow Q and conductance U_k of the kth part:

$$\Delta p_k = \frac{Q}{U_k},$$ (6.69)

The pressure leaps appear in the narrow places of the pipelines:

$$\Delta p = \frac{Q}{U_a}$$ (6.70)

where U_a is the conductance of the orifice.

The pressures in the cross-sections between elements of a vacuum system can be calculated using the formula

$$p_k = p_{lim\,k} + \frac{Q}{S_m} + \frac{Q}{U_k}$$ (6.71)

The limiting pressure p_{lim} may decrease only at a trap. The second term in Eq. (6.71) is constant, and the third term increases. At a trap, the pressure may either increase or decrease depending on what term in Eq. (6.71) gives the greatest contribution to the change in pressure. Approximate graph of pressure distribution between a pump and a pumped object is shown in Fig. 6.29.

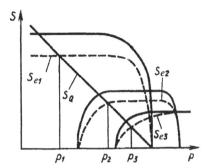

Figure 6.30 Graphical verification of the selection of vacuum pumps.

The calculated points are connected by straight lines, which agrees with the real pressure distribution for molecular gas flow mode. For other flow modes this graph may be considered as a simplified one. For the exact description of pressure distribution between the ends of a vacuum system in viscous flow mode, one should use Eq. (6.7).

It is convenient to plot pressure distribution separately for internal gas evolution of the system and the total gas evolution. The former is a parameter of the system itself, and the latter is a parameter of the process the system uses.

Graphical test of choice of vacuum pumps allows one to determine their working pressures. Using the graphic characteristics of the pumps, one may determine the effective pumping speed of the first pump for the pumped object, the speed of the second pump for the outlet of the first pump, etc., according to Eqs. (6.43). The intersection of these curves with the curve of leakage speed $S_Q = Q/p$ determines the working pressure of the pumps (Fig. 6.30).

The possibility to start a vacuum system is tested graphically. The start of the system is possible, if at any pressure higher than the working one the leakage speed in the pumped object is lower than the effective pumping speed of the pumps.

Series- or parallel-connected pumps may work jointly if the pressure at their outlet pipes is lower than the maximum allowed outlet pressures of these pumps.

The method of determining the range of possible joint operation of pumps on the basis of flows or working pressures is described above (see Section 6.5). Either partial or complete joint operation of pumps is possible. At constant gas evolution, one may use partial joint operation of pumps. In vacuum systems with variable gas evolution, e.g., in vacuum metallurgy, complete joint operation of pumps should be provided.

Using the above data (see Section 6.7), one may calculate the cost of vacuum system operation per one hour $C_h = CS_e$, and one year of two-shift work, $C_y = 4000CS_e$, where S_e is the effective pumping speed.

Verifying calculation of vacuum systems at non-steady-state operation mode is frequently used for the determination of pumping time.

Veryfying calculations require the following initial data:

1. schematic of the vacuum system;
2. parameters of the pumps, i.e., the pumping speed, the limiting pressure, and the starting pressure;
3. parameters of the fittings and pipelines, their sizes and conductance;

4. parameters of the pumped objects, i.e., their sizes and volume;
5. the total gas evolution and leakage for non-steady-state operation mode, and
6. the working pressure.

The order of calculations is as follows:

1. test of the quasi-steady-state condition,
2. plotting of graphs of the effective pumping speed and the leakage speed in the pumped object, and
3. timing of pumping of the object to the working pressure.

Quasi-steady-state condition (6.48) is tested using the parameters of the pumped object and the pipelines, which is the initial data. As a result of test of the quasi-steady-state condition, the character of non-steady-state operation mode is determined more exactly, and the choice of the calculation formulas for determining pumping time is thus corrected. For quasi-steady-state pumping of localized volumes, one may use Eq. (6.55) and (6.60), and for pumping of vacuum systems with distributed volume, Eq. (6.65) is suitable.

To determine the pumping time, one should plot effective pumping speed of all the pumps

$$S_{ei} = \frac{S_{pi}U_i}{S_{pi} + U_i}$$

as a function of pressure at the inlet sections of the pumped object; where S_{pi} is the pumping speed of the pump (it is given in the form of the graph plotted *vs* inlet pressure) and U_i is the conductance of the vacuum system from the ith pump to the pumped object (it is determined by the parameters of the fittings and pipelines using the initial data, and, in general case, it is also a function of pressure).

Then one plots the graph of gas evolution and leakage $S_Q = Q/p$, which is a straight line in logarithmic coordinates if Q does not depend on pressure.

Under the quasi-steady-state conditions, the calculation is performed using Eq. (6.55) with subdivision of grate into parts with constant S_m, in accordance with Fig. 6.21.

If the quasi-steady-state condition is not satisfied, then for molecular-viscous and viscous flow modes one may use the graph $U(p)$ by subdividing it into portions p_i-p_{i+1}, in which the average values U_i are used; the calculation is then performed using Eq. (6.65).

The working time of ballast vacuum cylinders without an adsorbent is calculated using Eq. (6.45), which can be presented as

$$t' = \frac{(0.5p_hS_pK_e - Q)V_b}{QS_pK_e} .$$

If the ballast vacuum cylinder is completely filled with adsorbent, then, using Eq. (6.47), we obtain its operating time as:

$$t'' = \frac{(0.5p_h - Q/(S_pK_e))(K_T + \varepsilon)V_b}{Q} ,$$

where K_T and ε are the constants of adsorptivity and porosity of the adsorbent, respectively.

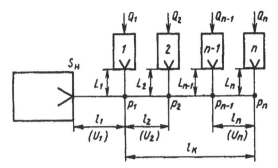

Figure 6.31 Schematic of the centralized vacuum system.

In centralized pumping systems (Fig. 6.12), one pump simultaneously operates several objects connected to the common collector. An example of a centralized system is connection of several high-vacuum pumps to one preliminary vacuum pump. Calculated schematic of a centralized pumping system is shown in Fig. 6.31. The vacuum pump is connected to a collector with length l_k through a pipeline with length l_1 and conductance U_1. Pumps with efficiencies Q_1-Q_n are connected to the collector through connecting pipelines with lengths L_1-L_n and are characterized by the coefficient of synchronous work, m, and the instabilities of the working pressure g_1-g_n.

For the parts of the collector operating in steady-state mode, the equations of gas flows are as follows:

$$m \sum_{i=2}^{n} Q_i g_i = U_2(p_2 - p_1);$$

$$m \sum_{i=3}^{n} Q_i g_i = U_3(p_3 - p_{2(1)});$$

$$\cdots\cdots\cdots\cdots\cdots\cdots\cdots\cdots\cdots\cdots$$

$$m Q_n g_n = U_n(p_n - p_{n-1}) \qquad (6.72)$$

Set of equations (6.72) can be rewritten in another form:

$$p_2 = p_1 + \frac{m}{U_2} \sum_{i=2}^{n} Q_i g_i;$$

$$p_3 = p_1 + m\left(\frac{1}{U_2} \sum_{i=2}^{n} Q_i g_i + \frac{1}{U_3} \sum_{i=3}^{n} Q_i g_i \right);$$

$$\cdots\cdots\cdots\cdots\cdots\cdots\cdots\cdots\cdots\cdots$$

$$p_n = p_1 + m\left(\frac{1}{U_2} \sum_{i=2}^{n} Q_i g_i + \ldots + \frac{Q_n g_n}{U_n} \right). \qquad (6.73)$$

Given collector sizes that determine the conductances U_2–U_n, the flows Q_2–Q_n and the pressure at one of the connection points of the pumped objects (usually p_n), set (6.73) allows one to determine the pressure at all the other sections of the collector including the pressure p_1.

The pressure p_n for centralized pumping is always lower than the maximum outlet pressure of a high-vacuum pump. The maximum speed of preliminary pump is

$$S_m = \frac{Q_s}{K_e p_1 - p_{lim}},$$

where Q_s is the total pumping efficiency,

$$Q_s = m \sum_{i=1}^{n} Q_i g_i,$$

K_e is the utilization coefficient of the pump, p_1 is the pressure at the connection of the first pumped object, and p_{lim} is the limiting pressure of the pump. The conductance of a pipeline is

$$U_1 = \frac{K_e S_m}{1 - K_e}$$

Using the conductance U_1, the pipeline length l_1, and the previously determined gas flow mode, one may calculate the diameter of the first pipeline d_1. The diameters of the connecting pipelines with lengths L_1–L_n can be determined from the conductance using the given difference of pressures between the collector and the pumped objects $U_i' = Q_i/\Delta p_i$.

6.9 An Example of Verifying Calculation

For this verifying calculation, we will choose a simple vacuum system that is shown in Fig. 6.32, *a*. The pumped object and the pipelines are made of stainless steel. The vacuum system is in a nitrogen atmosphere at a pressure of 10^5 Pa and a temperature of 300 K. The tightness test is carried out using a halogen leak detector with an atmospheric transducer. The number of simultaneously tested connections is $m = 1$. Process gas evolution is absent.

We will determine the internal gas evolution. Using Table C.19, we obtain the specific gas evolution q_g, and, using Table C.21, the gas permeation q_p. The total surface of the pumped object and the pipeline, that are shown in Fig. 6.32, can be calculated using the formula

$$F = \sum_{i=1}^{4} F_i = 4.7 D^2 + 2.4 D_b^2,$$

where D and D_b are measured in meters.

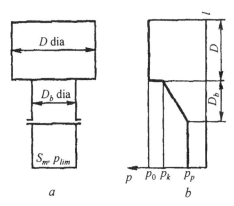

Figure 6.32 Verifying calculation of the vacuum pumps in steady-state mode: (*a*) vacuum schematic; (*b*) pressure distribution over the pipeline length.

Gas evolution and gas permeation of the structural materials, m³ Pa/s, for the entire system are

$$Q_g + Q_p = (q_g + q_p)F = (q_g + q_p)(4.7D^2 + 2.4D_b^2).$$

From the type of a leak detector (Table 6.5) we obtain the minimum flow registered by the detector: $Q_{lm} = 10^{-7}$ m³ Pa/s. The total number of tested connections is ten (two in the pipeline and eight in the chamber). The number of tightness tests should be equal to the number of connections. Using Eq. (6.3), we obtain $Q_l = 0.2 \times 10 Q_{lm} = 2 \times 10^{-7}$ m³ Pa/s. The internal gas evolution of the system is

$$Q_s = Q_g + Q_{pl} + Q_l = (q_g + q_p)(4.7D^2 + 2.44D_b^2) + 2 \times 10^{-7} \text{ m}^3 \text{ Pa/s}.$$

For $q_g + q_{pn} = 2 \times 10^{-4}$ m³ Pa/(m² s), $D = 1$ m, and $D_b = 0.5$ m we obtain $Q_s = 2 \times 10^{-3}$ m³ Pa/s.

Now we calculate the distribution of pressure. Using schematic in Fig. 6.32, we calculate the two sections of the vacuum system: the first section is between the pumped object and the pipeline, and the second one is between the pump and the pipeline. We calculate the pressure in the section between the pump and the pipeline of the vacuum system with localized parameters using Eq. (6.68):

$$p_p = p_{lim} + \frac{Q_s}{S_m}.$$

The conductance of the pipeline of round cross-section at molecular mode of flow for $l = D_b$ is $U_{pl} = 121D_b^2$. According to Eq. (6.69), the pressure at the end of the pipeline is

$$p_k = p_p + \Delta p_k = p_p + \frac{Q_s}{121D_b^2}.$$

The conductance of the inlet orifice from the pumped object in the pipeline is for molecular mode

$$U_{or} = \frac{91D_b^2}{1 - D_b^2/D^2}.$$

Taking into account Eq. (6.70), we write the expression for pressure in the pumped object:

$$p_o = p_k + \Delta p_{or} = p_k + \frac{Q_s(1 - D_b^2/D^2)}{91D_b^2} = p_{lim} + Q_s\left(\frac{1}{S_m} + \frac{1}{121D_b^2} + \frac{1 - D_b^2/D^2}{91D_b^2}\right).$$

Figure 6.32, b shows the distribution of pressure over the pipeline length. At $p_{lim} = 10^{-5}$ Pa and $S_m = 1$ m³/s, $Q_s = 2\times10^{-3}$ m³ Pa/s, $D = 1$ m, $D_b = 0.5$ m, the pressure in the pumped object $p_o = 2\times10^{-3}$ Pa.

6.10   Test Questions

6.1   Which forms of gas load occur in pumping of vacuum systems?

6.2   How can one reduce the pumping time of a vacuum chamber on the walls of which water vapor is adsorbed?

6.3   Which assumptions are taken for the calculation of vacuum systems with localized parameters?

6.4   How does the conductance of a pipeline depends on its geometrical sizes in various gas flow modes?

6.5   What are conditions of joint operation of series-connected vacuum pumps?

6.6   Under which conditions a trap increases the total pressure in a vacuum system?

6.7   What gas source is taken into account only for non-steady-state operation of vacuum system?

6.8   In which case centralized vaccuum systems should be used?

6.9   What is the purpose of heating elements of ultrahigh vacuum systems?

6.10 Why does filling with adsorbent increase the efficiency of ballast vacuum cylinders?

Chapter 7

Design of Vacuum Systems

7.1 Database of Vacuum System Elements

Database of the elements of vacuum systems is a set of records that can be reviewed, edited, sorted, and deleted. The database contains conditionally permanent and variable parts of information. The permanent information contains standards, regulations, and archives. The variable information includes technical and commercial characteristics of elements of vacuum systems.

The basic principle of the database is the priority related model of data. The first significant level includes machines, units, pumps, traps, valves, elements of pipelines, vacuum gauges, analyzers, etc. The second level classifies elements by operation principle, the third level, by size and type, and the fourth one, by type of information. The last relevant level contains general information, the operation principle, photos, tables of technical and commercial characteristics, applications and assembly schematics (Fig. 7.1). The applications are six flat projections of an element without scale lines that are suitable for automatic design.

A special program allows one to synthesize parametric elements of pipelines, for example, during response to a query, and to considerably reduce the volume of stored information. The database can work in query and the program modes. Local databases are used to work in the program mode with reduced volume of information. To create them, direct access files are used. Local databases do not require considerable resources and allow one to create effective programs, but they are not enough flexible for frequent updating of information.

The properties of gases and materials are represented in the database by the tables of sorption and diffusion properties of structural materials, i.e., steels, non-ferrous metals, rubber, ceramics, glass, polymers, etc. for various gases. Users of different types may operate with the database by means of a dialogue using the multilevel menu with diagnostics of error situations. Software developers have access to the database format and can create their own limited local bases. Designers of vacuum equipment can edit, delete, or add new records. Users operating with the database as an information retrieval system cannot change the contents of the database.

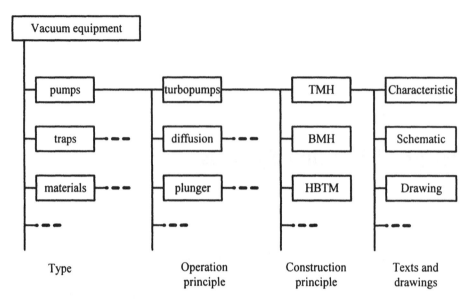

Figure 7.1 Database of vacuum system elements.

7.2   Structural Design of Vacuum Systems Using the Method of Selecting Variants

The structural design of vacuum systems is the initial stage that determines in their quality in many respects. The initial data are determined by the technical project. The results of structural design can be represented as a block diagram that contains information on the types of the elements used and the method of their connection. The elements of vacuum systems in the block diagram correlate with the symbolic designations in Table A.14. The functional connections are shown by thin lines.

The simplest and most frequently used method of structural design is the method of selecting variants. For manual designing without a computer, one may usually consider 2 to 4 variants. The quality of solution depends on the quantity of selected variants. An increase in the number of variants simplifies the procedure of their selection. The use of computer increases selection efficiency by many times. Consideration of all available variants allows one to select the best one of them, but not elaborate a fundamentally new vacuum system. Figure 7.2 shows a set of variants to choose a vacuum system.

The selection method begins with the formulation of the choice criterion of the best variant. Single-criterion selection implies specification of the most important parameter of a vacuum system. For scientific units with small gas evolution this can be the limiting pressure. In industrial units the most important parameter is working pressure, while for special objects it is reliability or pumping time. The universal selection criterion can be economic efficiency, the simplest form of which is minimum cost price. The problem can be presented either with or without restrictions.

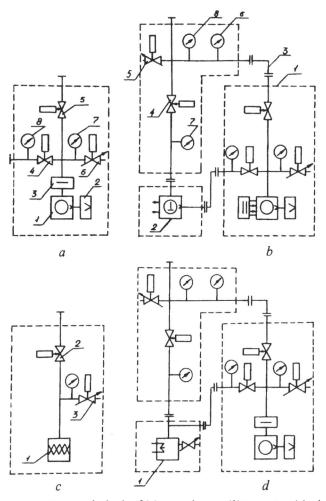

Figure 7.2 Vacuum systems on the basis of (*a*) vane plunger, (*b*) two-rotor, (*c*) adsorption, and (*d*) cryogen pumps.

We will consider a representation of the problem of vacuum schematic design using the criterion of minimum cost price with restrictions of limiting pressure and pumping speed. In the technical project the required pumping efficiency Q and working pressure, p, are specified. The structure of the chosen database imposes restrictions on the limiting pressure and pumping speed. Check of fulfilment of restrictions for the ith variant can be made for each of the parameters

$$p_{limi} < p; \quad S_{ei} > \frac{Q}{p - p_{limi}}. \tag{7.1}$$

From condition (7.1) it is obvious that the restrictions are related to one another by a functional dependence. The problem of selection appears unsoluble if the base of variants does not include one satisfying to condition (7.1). From the variants that satisfy condition (7.1) one may choose one requiring minimum costs.

For multi-criterion selection, optimization one may use additive or multiplicative goal functions. The minimizing additive goal function of selection with account of cost and reliability can be written in the following form:

$$Z_a = \frac{b_1}{Z_{n1}} Z_1 + b_2 \left(1 - \frac{Z_2}{Z_{n2}} \right), \tag{7.2}$$

where Z_1 and Z_2 are the price and reliability, respectively; b_1 and b_2 are the weight factors, and Z_{n1} and Z_{n2} are the normalizing factors. There are some methods of setting the normalizing factors: directional (under the technical project of the customer), maximum (maximum values of the parameters of the elements from the database), and differential (by the difference of maximum and minimum values of the same parameters). The additive goal function is based on the principle of absolute compensation of the optimization criteria.

For the multiplicative minimized goal functions we have

$$Z_n = \frac{(Z_1 / Z_{n1})^{b_1}}{(Z_2 / Z_{n2})^{b_2}}, \tag{7.3}$$

where designation are the same as in Eq. (7.2). The multiplicative function is based on the principle of relative compensation of the optimization criteria. The principle of criteria compensation and the form of goal function are selected by the designer. When selection is difficult, both design solutions are used at the initial stage of designing, the best of which is chosen with account of additional criteria.

The weight factors are set directly by the designer and are normalized to unity using the ratio

$$\sum_{k=1}^{n} b_k = 1,$$

where n is the number of normalized factors. If the number of criteria is large, it is convenient to use the method of pairwise comparison in its simplest modification with the use of binary matrixes. Let 5 optimization criteria Z_1–Z_5 be given. The designer chooses pairwise relation

$$a_{kj} = \begin{cases} 1 \\ 0 \end{cases}.$$

if the kth criterion is more important than the jth one, then $a_{kj} = 1$, otherwise $a_{kj} = 0$. The results of selection are written in a table, an example of which is presented below.

In the last column of Table 7.1, using the formula

Table 7.1 The binary matrix.

k	Z_1	Z_2	Z_3	Z_4	Z_5	b_k
			j			
Z_1	–	0	1	0	0	0.1
Z_2	1	–	1	1	0	0.3
Z_3	0	0	–	0	1	0.1
Z_4	1	0	1	–	0	0.2
Z_5	1	1	0	1	–	0.3

$$b_k = \frac{\sum_{j=1}^{5} a_{kj}}{\sum_{k=1}^{5} \sum_{j=1}^{5} a_{kj}}. \tag{7.4}$$

as a result of additive normalization, the weight factors b_k are calculated, the sum of which is equal to unity. For multicriterion choice there is a problem to use the factors that have no numerical values, for example, if the noise level, on which the database contains no information, is included in the optimization criteria. One may set numerical values using the method of expert evaluations by averaging the results on the number of experts. Expert evaluations that use strict logics lead to binary matrices (Table 7.1). The use of fuzzy logic permits one to take into account the features of compared variants more completely. In this case evaluations 1, 3, 5, 7, 9 are set (Table 7.2). Evaluations 2, 4, 6, and 8 may be used as intermediate.

All the evaluations are written in a table similar to Table 7.1. The normalized values of the criteria are calculated using the equation

$$\frac{Z}{Z_n} = \frac{\sum_{j=1}^{n} a_{kj}}{\sum_{k=1}^{n} \sum_{j=1}^{n} a_{kj}}. \tag{7.5}$$

Table 7.2 Expert evaluations.

a_{kj}	Value	Note
1	Similar	The kth is not better than the jth
3	A little better	The kth is a little better than the jth
5	Better	The kth is better than the jth
7	Substantially better	The kth is substantially better than the jth
9	Far better	The kth is far better than the jth.

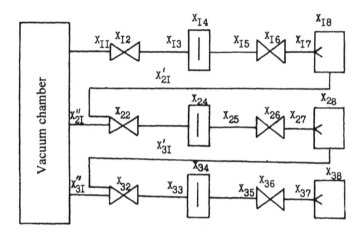

Figure 7.3 Maximum number of elements for designing of a three-stage vacuum system using the truncation method.

To determine the reliability of expert evaluations one may use the solubility criterion of the matrix of pairwise comparisons:

$$CI = \frac{\lambda_{max} - n}{n},$$

where λ_{max} is the maximum inherent matrix number and n is the number of variantes. If CI < 0.1, one may consider that the evaluations are soluble in the sense that the resulting vector is insensitive to small changes of the initial data. If $CI > 0.1$, the reference matrix is unsoluble, and the expert should change the evaluations.

7.3   Structural Design Using Typical Pattern

The structural design of vacuum systems can be performed using the method of typical patterns. This method is widely used for the design of various technical objects. The method has two versions, i.e., truncation and enrichment. For the truncation method, the typical pattern is chosen with the maximum set of elements, and in the enrichment method, with the minimum set.

Figure 7.3 shows the maximum set of elements for designing the schematic of three sections using the method of typical pattern truncation. Each section contains 8 elements, e.g., a pump, a trap, two valves, and four pipelines. The pump is the necessary element, whereas any of the other seven elements may be absent. Some types of pumps do not require a trap. The absence of a trap results in rejection of one valve and several pipelines. The design begins with the first section ensuring the lowest working pressure. The pump is chosen using the above-described selection method. Then the remaining elements of the first section are chosen in the same manner.

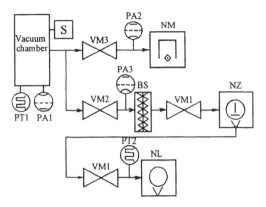

Figure 7.4 Structural schematic of vacuum system.

If the maximum outlet pressure of the pump in the first section is less than the atmospheric pressure, one should choose the second section, etc.. Sections can be connected in series or in parallel, depending on the type of pump in the previous section. For compression pumps the sections are connected in series, and for sorption pumps, in parallel. From each pair of pipelines X'_{21} and X''_{21} ; X'_{31} and X''_{31} (Fig. 7.3) only one is chosen depending on the method of sections connection.

The result of schematic structural design of a vacuum system can be represented as a matrix (X_{ij}), where $i = 1–3, j = 1–8$. The linear number of the matrix corresponds to the number of the section. The numerical values of the matrix elements are codes of the types of chosen elements. By way of the example, Fig. 7.4 shows block diagram automatically obtained by computer using the truncation method when gas load $Q = 6 \times 10^{-5}$ m^3 Pa/s, working pressure 10^{-6} Pa, vacuum chamber diameter 0.5 m.

7.4   Parametric Design of a Vacuum System Using the Utilization Coefficient of Vacuum Pump

Parametric design differs from the structural one by additional determination of trademark or size of the element. For pipelines, the inner diameter and length are determined. The results of parametric design are presented as a basic schematic. The design rules of the basic schematics require designation of the elements by an alphabetic code containing one or several capital letters of the latin alphabet. Ordinal numbers are assigned in the directions from top to bottom and from left to right. The list of elements with indication of their types and sizes can be presented in a separate sheet or in a free place together with the basic schematic.

For parametric design the following data are required: structural schematic, working pressure, gas load, size of the pumped object, length of the pipelines, and other special requirements.

The optimization methods at selection of types and sizes of the elements can vary in accuracy and requirements to computer resources. At early stages of design one may use less accurate but high-speed methods. The optimization criteria should remain the same as at the stage of structural synthesis.

We now consider criterial minimization of the goal function, i.e., the cost price of a vacuum system, using the utilization coefficient of a vacuum pump as an optimization criterion. This approximate method is widely applied to designing without using a computer. Continuous optimization is conducted at the first stage, and discrete optimization, at the second stage. Discrete optimization is necessary because the set of elements in the database is limited. To simplify the problem, independent optimization should be performed for each section of the vacuum system. The optimum utilization coefficient of a pump, the type and size of the pump, traps, and valves, and diameters of the connecting pipelines are determined, and then economic efficiency is evaluated. Parametrical design begins with determination of the optimum utilization coefficient of a pump

$$K_e = \frac{S_e}{S_p},$$

where S_e is the effective pumping speed for the pumped object and S_p is the pumping speed of the pump. The use of basic equations of vacuum engineering permits one to write the equality

$$\frac{U}{S_p} = \frac{K_e}{1 - K_e}. \tag{7.6}$$

It follows from Eq. (7.6) that selection of the utilization coefficient of a pump is identical to setting a ratio between the general conductance of the entire vacuum system and the pumping speed of the pump. This optional character of designing is due to the fact that for various values of this ratio one may obtain the same pumping speed of a chamber, and, consequently, the same working pressure. One should take into account that the cost of purchase and operation of a pump decreases with increasing K_e, whereas the cost of purchase and operation of fitting increases. The capital and working costs of separate elements of a vacuum system will be approximated using exponential functions of pumping speed or conductance of a element according to Eq. (6.66).

We now consider continuous optimization without restrictions. We differentiate Eq. (6.67) by K_e and equalize the resulting expression to zero at $K_1 = K_2 = K_3 = K_4$.

$$(1 - A_5)K_e^2 - 2K_e + 1 = 0, \tag{7.7}$$

where

$$A_5 = \frac{C_2 + C_4 t_0}{C_1 + C_3 t_0}$$

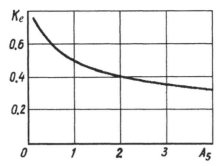

Figure 7.5 Optimum values of pump utilization coefficient K_e versus the ratio of apparatus cost to the cost of the pump.

The coefficient A_5 represents the ratio between the costs of apparatus and the costs of pumps. At $A_5 = 1$, it is obvious that $K_e = 0.5$. Taking into account that K_e is always less than unity, we write the solution of Eq. (7.7) as

$$K_e = \frac{(1 - \sqrt{A_5})}{1 - A_5}. \qquad (7.8)$$

For illustration, dependence (7.8) is shown in Fig. 7.5. If the optimum value K_e cannot be calculated for certain reasons, its selection can be made by analogy with existing systems (Table 7.2).

For analytical description of the main characteristics of a vacuum pump we will accept the following expression:

$$S_p = S_m\left(1 - \frac{p_{lim}}{p_p}\right), \qquad (7.9)$$

where S_m is the nominal pumping speed of the pump, and p_{lim} and p_p are the limiting and working pressure of the pump. Equation (7.9) in steady-state operation mode can be transformed to

$$S_m = \frac{Q}{K_e p - p_{lim}}, \qquad (7.10)$$

where Q and p are the gas evolution and the working pressure in the pumped object, respectively (initial data for designing). The limiting pressure of a pump can be obtained from tables for the known type of the pump, which was determined at the stage of structural synthesis. The type and size of the pump are chosen from the tables so that its nominal pumping speed be not less than S_m determined from Eq. (7.10).

The total conductance of the system section can be represented, using the main equation of vacuum engineering, as

Table 7.2 Pump utilization coefficients for different vacuum installation.

Type of installation	Pump utilization coefficient, K_e	
	high vacuum, K_{e1}	low vacuum, K_{e2}
Automatic machines for pumping electronic devices	0.05–0.001	0.1–0.2
Stationary exhaust system for electronic devices	0.1–0.2	0.8–0.9
Chamber pumping coefficient for electronic devices	0.8–0.9	0.8–0.9
Vacuum coater	0.5–0.6	0.8–0.9
Space simulator	0.5–1.0	0.5–0.9
Vacuum furnace	0.5–0.7	0.7–0.8
Electron beam welding plant	0.3–0.5	0.5–0.8

$$U_o = \frac{Q}{p(1 - K_e)}. \tag{7.11}$$

We will simplify the problem accepting the equality of the conductances of all the in-series connected elements of the fixture and pipelines:

$$U_e = U_1 = U_2 = \dots = U_{n-1} \tag{7.12}$$

where n is the number of elements of the section including the pump. In this case, the conductance of an element is determined from the expression

$$U_e = (n-1)U_o = \frac{(n-1)Q}{p(1 - K_e)}. \tag{7.13}$$

This expression is used to choose valves and traps. The conductance of elements chosen from the table of fixture elements may differ from the value determined using Eq. (7.13), because of the parametrical numbers are discrete. Diameter of the pipelines is selected from the standard list by the known conductance and length of the pipeline. The lengths of pipelines are preset in the initial data and may be corrected at the stage of configuring the vacuum system. When the elements are selected, one should seek to minimize the deviation of the total conductance of the section from that required by Eq. (7.11). The cost price of a multisection pumping unit is determined as the sum of the cost prices of separate sections determined using Eq. (6.67).

7.5   Multiparametric Design of a Vacuum System

Multiparametric design of vacuum systems includes the following stages: determination of the vector of optimum parameters for continuous optimization, quantification of the results, and calculation of the goal function. As a goal function, we accept the cost price of pumping. Unlike Eq. (6.66), where the cost of apparatus was considered as common for all the elements, we take them into account separately. In this case, the cost price of pumping for one section of a vacuum system containing n elements is

$$C = \frac{(K_1 + K_2 + \dots + K_n) + t_0(E_1 + E_2 + \dots + E_n)}{t_0 S_e} \qquad (7.14)$$

where C is determined in relative units (ru) in one hour of pumping with a speed of 1 m³/s, $K_1, K_2, \dots K_n$ are the costs of elements, ru, $E_1, E_2, \dots E_n$ are the exploitation charges, ru/hr, t_0 is the certain payback time, $t_0 = 10^4$ hours for one-shift work during 5 years, and S_e is the effective pumping speed in the working chamber, m³/s. We approximate the cost and operating costs using the exponential functions of the following form

$$K_i = a_i X_i^{k_i}; \quad E_i = b_i X_i^{m_i}, \qquad (7.15)$$

where $a_1, \dots, a_n, b_1, \dots, b_n, k_1, \dots, k_n, m_1, \dots, m_n$ are the constant factors, n is the total number of elements, and $X = (X_1, X_2, \dots, X_n)$ is the vector of parameters: for a pump it is the pumping speed and for a apparatus it is the conductance. We will optimize each section of the vacuum schematics independently from one another.

Multiparametric optimization of goal function (7.14), with given restrictions, is reduced to determination of vector X^* that corresponds to the minimum goal function in the allowed range of parameters:

$$C(X^*) = \min C(X)$$
$$X \in D \qquad (7.16)$$

The allowed for the optimization of parameters space D satisfies the set of restrictions. Parametrical restriction for the parameter X_i is

$$X'_i \le X_i \le X''_i, \qquad (7.17)$$

where X'_i and X''_i are the minimum and maximum allowed values of the ith parameter.

The database imposes the quantification restriction

$$X_{ij} = \{X_{i1}, X_{i2}, \dots, X_{iN}\}, \qquad (7.18)$$

where $j = 1, 2, \dots, N$, and N is the number of sizes of the ith element in the database. Functional restriction as that of a main equation of vacuum engineering, written for n in-series connected elements, is

$$G(x) = \sum_{i=1}^{n} \left(\frac{1}{X_i}\right) - \frac{p}{Q} = 0. \qquad (7.19)$$

where p is the working pressure and Q is the gas load.

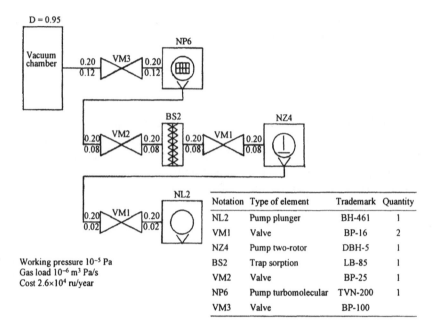

Figure 7.6 Basic scheme of vacuum system.

At the first stage we will only take into account restriction (7.19). This problem can be solved by the Lagrange multipliers method. We will compose the Lagrange function as

$$L(X, \lambda) = C(X) - \lambda G(X), \; G(x) = \sum_{i=1}^{n} \left(\frac{1}{X_i} \right) - \frac{P}{Q},$$ (7.20)

where λ is the Lagrange multiplier.
The required conditions of the extremum

$$\frac{\partial L}{\partial X_i} = 0 \; ; \; \frac{\partial L}{\partial \lambda} = 0$$

allow one to write a set of nonlinear equations

$$\begin{cases} A_i X_i^{k_i - 1} + B_i X_i^{m-1} - \lambda X_i^{-2} = 0; \\ \sum_{i=1}^{n} X_i^{-1} - \frac{P}{Q} = 0; \end{cases}$$ (7.21)

where $A_i = a_i/(t_o S_e)$; $B_i = b_i/(t_o S_e)$; $i = 1, 2, ..., n$.
This system includes $(n + 1)$ nonlinear algebraic equations and the same number of unknown values $X_1, X_2, ..., X_n, \lambda$. As a result of solution of set (7.21), for example using the iteration method, we determine the optimum vector X_p^*. The possibility of choosing an

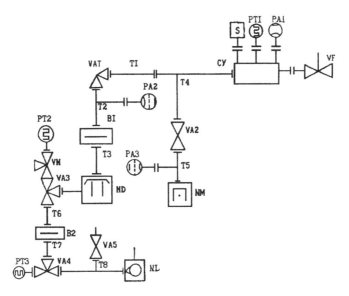

Figure 7.7 Schematic of vacuum conduits.

element from the database corresponds to fulfillment of restrictions (7.17) and (7.18). Thus,we obtain the solution of the optimization problem as $X^* = \{X_1^*, X_2^*, ..., X_n^*\}$.

The results of parametric design can be represented as matrix Y_{ji} containing the codes of types and sizes of the elements (j is the section number and i is the element number) and can be represented on a computer in graphical form. Fig. 7.6 shows basic vacuum schematic designed on a computer using the above method. As the initial data, the structural scheme shown in Fig. 7.4 was used.

Generalizing the results of designing for various working pressures and gas flows, one may determine the range of the effective use of separate elements of a vacuum system that are included in the database. The application field in terms of economic efficiency depends on relative, but not of the absolute cost, this making the results independent of economic inflation.

The presence of this field indicates a high technical level of the element. Its absence specifies inexpediency of further manufacture of this element.

7.6   Schematic Connections and Assembly

The next stage of designing after the design of the basic scheme is the arrangement of connections. It includes data on the mutual arrangement of the elements in a vacuum system. The feature of elements of vacuum systems is the possibility of their rotating during assembly about the axis passing through the connection point, up to 90°. The rotation angle can be measured from the basic position in the database along the clockwise direction from the side of the connected element.

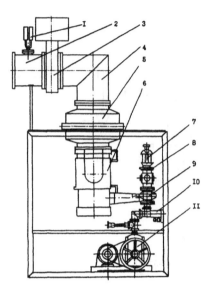

Figure 7.8 Computer aided layout of a vacuum system.

Some elements, for example, diffusion pumps, can be rotated about the axis passing through the inlet pipe, but not about the axis passing through the outlet pipe. Thermal manometer transducers should be placed vertically. However, most elements of vacuum systems allow arbotrary spatial arrangement.

The algorithm of connection of elements should provide flat or volume configuration. Figure 7.7 shows schematic of element connection in a planar vacuum system. A universal problem of connection is to minimize the total volume occupied by the vacuum system. The results of designing connection schematics can be written analytically as matrix of rotation angles Z_{ji}, where j is the section number and i is the element number.

The layout schematic is drawn on the basis of connection schematic. The configuring is performed on a computer using special graphical editors. Flat configuration requires preliminary preparation of the applications of element projections to all planes of the Cartesian system of coordinates. Flat configuration of complex designs is unsufficiently descriptive and is difficult to formalize. Volume assembly is considerably more convenient, but it requires larger computer resources.

Each element of a configuration is characterized by a set of input and output parameters. The input parameters are as follows: the point number of connection to the previous element, the coordinates of the connection point, the rotation angle of the element and the scale multiplier. The output parameters are the point number of connection to the subsequent element and its local coordinates.

During automated configuration the user handles in every moment of time two elements: the base (previous) and the current (subsequent). The elements are selected according to the schematic of connections from the menu of the graphical database. A multi-window interface is applied for convenience of the user. Figure 7.8 shows example of designing the layout schematic on a computer.

7.7   Projecting Calculation of a Vacuum System

We will consider the non-automated method of projecting calculation with approximate optimization using the economic criterion. The cost of a vacuum system is minimized by selecting the utilization coefficient of vacuum pump.

The following initial data should be given:

1. gas load Q,
2. working pressure in the vacuum chamber p
3. time of operation in stationary regime t_s
4. dimensions of the pumped volume and length of the pipelines, and
5. additional data.

Gas load is preset roughly in accordance with existing designs. The working pressure in a vacuum chamber is determined by technological process requirements. The operating time of a system in steady-state mode depends on technical process duration.

The additional conditions may contain operational and design requirements, such as combined work of vacuum pumps in a given range of flows or pressures, duplication of systems to increase their reliability, adjustment of pressure in the pumped volume, use of "double" pumping, etc.

Projecting calculation without computers includes the stages of structural and parametric synthesis. The vacuum circuit is chosen by analogy with existing installations of similar type. Choice of the type of vacuum pump may take into account the features of the technical of process, for example, mercury rectifiers are suitable for pumping by mercury vapor pumps, equipment for the impregnation of transformer coil by oil vapor pumps, installations for the study of properties of materials at low temperatures by cryosorption pumps, installations with own magnetic fields by magnetic discharge pumps, etc.

The condition of steady-state mode in vacuum systems is the equality of the effective pumping speed S_e and the speed of gas evolution inside the vacuum chamber S_Q. The speed of gas evolution can be determined using the formula $S_Q = Q/p$. The neccessary pumping speed in vacuum chamber can be determined using the formula

$$S_e = \frac{S_Q}{p} \tag{7.22}$$

On the basis of the resultant S_e, the vacuum circuit, and the type of pump, using the method described in Section 7.4, one may determine the optimum utilization coefficient K_{e1} of a vacuum pump that directly pumps a vacuum chamber.

The rated pumping speed of a pump for a vacuum chamber in steady-state mode can be determined using Eq. (7.10), which will be in the following form:

$$S_{m1} = \frac{Q}{K_{e1}p - p_{lim1}}, \tag{7.23}$$

where Q, p and K_{e1} is already determined, and p_{lim1} is a characteristic of the chosen pump type. Choice of the type of pump is determined by the condition $p_{lim1} \ll K_{ei}p$. A vacuum pump of nominally correct pumping speed is chosen from catalogues as determined by Eq. (7.23). If no pump with necessary pumping speed can be found, one may use parallel connection of pumps, for which the effective pumping speeds of the pumps are summarized in the pumped volume.

It is also possible in the same way to choose sequentially connected pumps. Then the working pressure p_i of the ith compressing pump should be accepted as the maximum output pressure of the previous pump, and for sorprtion pumps this should be the highest working pressure $p_{h(i-1)}$.

$$p_i = \frac{p_{h(i-1)}}{\eta},$$

where η is the safety margin; one may accept $\eta = 2$.

The pumping efficiency Q for sequentially connected pumps remains constant, and p_{limi} and K_{ei} are found from tables. Any sequentially connected pump working in steady-state mode can be chosen by analogy with Eq. (7.23):

$$S_{mi} = \frac{Q}{K_{ei}p_i - p_{limi}}. \tag{7.24}$$

The overall conductance of the stage of a vacuum system from the pump to the pumped volume or between two sequentially connected vacuum pumps can be determined using the main equation of vacuum engineering by writing it in the following form:

$$U_{pi} = S_{mi}\frac{K_{ei}}{1 - K_{ei}}, \tag{7.25}$$

where U_{pi} is the conductance of the stage of a vacuum system from the ith pump to the previous pump or to the pumped volume. The conductance of a stage of a vacuum system can be expressed through the conductances of separate elements, such as valves, traps, and pipelines for their sequential or parallel connection. If a portion of a vacuum system has several sequentially connected elements, the number of which is $(n-1)$, then the condition should be satisfied

$$\frac{1}{U_{pi}} = \frac{1}{U_{1i}} + \frac{1}{U_{2i}} + \ldots + \frac{1}{U_{ji}} + \ldots + \frac{1}{U_{(n-1)i}}. \tag{7.26}$$

When designing vacuum systems one should aim at making the conductances of all the elements of a vacuum system be equal:

$$U_{1i} = U_{2i} = \ldots = U_{ji} = \ldots = U_{(n-1)i}$$

The conductance of any element of a vacuum system should be $(n-1)$ times greater than the conductance of the entire part. Complex elements, such as traps and valves, are chosen from the catalogues (Table A.13). In the absence of the necessary element in the catalogue, the problem of its design arises. An equivalent calculation circuit is plotted, the complex element being subdivided into more simple parts, connected sequentially or in parallel. In molecular flow mode, the transmission probability that gas molecules pass through an element is

$$P_i = \frac{U_i}{U_h} = \frac{U_i}{91\,d_h^2}, \qquad (7.27)$$

where U_i is the conductance of the element; U_h is the conductance of the inlet hole; and d_h is the diameter of the inlet hole. The probability is a function that is only dependent on the size ratio of the element,

$$P_i\left(\frac{d_1}{l_1}, \frac{d_2}{l_2}, \dots, \frac{d_1^2}{d_h^2}, \frac{d_2^2}{d_h^2}, \dots\right)$$

and the value of which remains constant at proportional change of the element sizes. Presetting the basic circuit of an element, one may determine its relative sizes, and, hence, the value of the function P_i. As the value U_i is known, one may find from Eq. (7.27) the diameter of inlet hole d_h and all other sizes of a element.

Using the layout circuit of a vacuum system, the lengths of all pipelines are determined. The mode of gas flow can be determined from the diameter of the inlet branch pipe of the pump and the working pressure. Using the above formulas for the conductances of pipelines in various flow modes, the diameter of a pipeline can be determined from its known conductance and length.

The task of projecting calculation of a vacuum system in non-steady-state mode is the choice of pumps, fixture and pipeline sizes from the required time of pumping to the working pressure, at which steady-state operation mode is obtained.

The initial data for this calculation are

1. the initial and final pressures in the pumped volume p_1 and p_f,
2. characteristics of the pumped object, such as volume V, diameter d, and length l,
3. time of pumping in non-steady-state mode t_n;
4. additional conditions.

Vacuum schematic is chosen depending on the required degree of vacuum, determined from the working pressure and the diameter of the pumped object. If several pumps are required to achieve the working pressure, the total time of pumping needs to be distributed between them.

The rated pumping speed of a pump will be chosen in the assumption that quasi-steady-state operation mode of the vacuum system occurs. At low vacuum, using Eq. (6.53), we obtain that to reduce pressure from p_1 to p_2 in a chamber of volume V in time t_1, one needs a pump with the rated pumping speed

$$S_{m1} = 2.3\frac{V}{\Delta t_1 K_{e1}}\log\frac{p_1}{p_2}, \qquad (7.28)$$

where p_1 is the initial pressure (usually atmospheric) and p_2 is the final pressure in the pumped volume between viscous and molecular-viscous flow modes.

In the range of low vacuum, the utilization coefficient of a pump K_{e1} is variable due to the pressure dependence of the conductance of a vacuum system. For calculations with a safety margin, we accept for K_{e1} the minimum value corresponding to steady-state flow mode at pressure p_2. Approximately, one may take $K_{e1} = 1$.

In medium vacuum, the pressure changes from p_2 to p_3. The coefficient K_{e2} is determined with a safety margin at the pressure p_3 and corresponds to molecular gas flow mode. We determine the rated pumping speed of a pump using the formula analogous to Eq. (7.28):

$$S_{m2} = 2.3 \frac{V}{\Delta t_2 K_{e2}} \log \frac{p_2}{p_3}. \tag{7.29}$$

At high and ultrahigh vacuum pumping is performed over the time Δt_3 from the pressure p_3 to p_f, which is equial to the working pressure preset in the initial data. The rated pumping speed of a pump can be determined from the equation

$$S_{m3} = 2.3 \frac{V}{\Delta t_3 K_{e3}} \log \frac{p_3}{p_f}, \tag{7.30}$$

where K_{e3} is the utilization coefficient of a high-vacuum pump.

The overall conductance between the pump and the pumped object is

$$U_{oj} = S_{mj} \frac{K_{ej}}{1 - K_{ej}}, \tag{7.31}$$

where j is the number of a corresponding pump.

The overall conductance U_{o1} determined using Eq. (7.31) corresponds to the lower pressure limit of viscous flow mode, and U_{o2} to the lower pressure of molecular-viscous flow mode in the pumped volume.

Further choice of elements of a vacuum system does not differ from the above consideration for steady-state mode. The overall conductance is expressed through the conductances of sequentially connected elements. A layout schematic is developed, from which the lengths of pipelines are determined. Standard elements, e.g., valves and traps, are chosen from catalogues. The conductances and lengths of pipelines determine their diameters. If non-standard elements should be used, their equivalent schemes are designed on the basis of elements that have formulas for the calculation of conductance. After determining the sizes of pipelines, one should find their volume and check the accepted quasi-steady-state condition.

The size of ballast volume is determined from the condition of its work during the entire duration of steady-state mode, using Eq. (6.45):

$$V_b = \frac{Q t_s}{0.5 p_h - Q/(S_p K_e)}, \tag{7.32}$$

where p_h is the highest output pressure of the previous pump; S_p is the pumping speed and K_e is utilization coefficient of the subsequent pump. If the volume calculated using Eq. (7.32) is commensurable with the volume of the pipeline, one may not use a ballast volume. If this volume is large and commensurable with the size of the vacuum unit, then one may apply a ballast volume with adsorbent, the volume which is determined using Eq. (6.47), which can be written in the following form:

$$V_b = \frac{Qt_s}{[0.5p_h - Q/(S_pK_e)](K_T + \varepsilon)},$$

where K_T and ε are the adsorbability constant and the porosity of the adsorbent, respectively.

7.8   Example of Projecting Calculation of a Vacuum System in Steady-State Mode

We will choose for the non-computer aided calculation of a vacuum system for the production of ultrahigh vacuum. The initial data are the total rate of gas load $Q = 4 \times 10^{-6}$ m³ Pa/s, the working pressure in the vacuum chamber $p = 10^{-5}$ Pa, the operating time of the system in steady-state mode $t_s = 3600$ s, the sizes of the pumped volume $d = 500$ and $l = 1000$ mm, and the additional conditions are the use of magnetic discharge, vapor jet, and rotary pumps.

1. *Choice of the vacuum scheme.* We choose the vacuum scheme shown in Fig. 6.9.

2. *Choice of vacuum pumps.*

2a *Choice of ultrahigh-vacuum pump.* Pursuant to the additional condition for ultrahigh-vacuum pumping magnetic discharge pumps of the NMD series are selected with the limiting pressure $p_{lim} = 7 \times 10^{-8}$ Pa and the range of pumping speeds from 6×10^{-3} to 1.2   m³/s. The effective pumping speed in the pumped volume is $S_{e1} = Q/p_1 = 5 \times 10^{-6}/10^{-5} = 5 \times 10^{-1}$ m³/c.

We find the utilization coefficient of a magnetic discharge pump. At $n = 3$ we find from Fig. 6.11 for $S_e = 10^{-1}$ m³/s the optimum utilization coefficient $K_{e1} = 0.35$. We use Eq. (7.23) for the rated pumping speed:

$$S_{m1} = \frac{Q}{K_{e1}p_1 - p_{lim1}} = \frac{4 \times 10^{-6}}{0.35 \times 10^{-5} - 7 \times 10^{-8}} = 1.17 \text{ m}^3/\text{s}.$$

The closest by pumping speed magnetic discharge pump NMDO-1 has the following characteristics:

Rated pumping speed, m³/s	1.20
Inlet pipe diameter, mm	250
Maximum working pressure, Pa	10^{-1}
Highest start-up pressure, Pa	10^0
Limiting pressure, Pa	7×10^{-8}

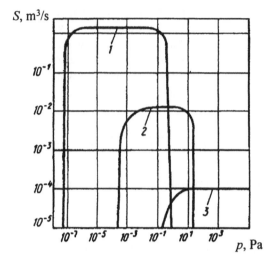

Figure 7.9 Speed of vacuum pumps: (*1*) NMD-1, (*2*) N-0.15S, and (*3*) VN-01 at various inlet pressures.

The dependence of the pumping speed of an NMDO-1 pump on the inlet pressure is shown in Fig. 7.9.

2b. *Choice of a high-vacuum pump.* Pursuant to the additional conditions we choose the series of vapor jet pumps of the N type with the limiting pressure $p_{lim} = 3 \times 10^{-4} - 4 \times 10^{-5}$ Pa and the range of pumping speeds from 1.5×10^{-2} to 30 m³/s.

The working pressure of an oil vapor pump is chosen from the highest working pressure of a magnetic discharge pump with a safety margin of 2. Then $p_2 = 10^{-1}/2 = 5 \times 10^{-2}$ Pa, which corresponds to the effective pumping speed $S_{e2} = Q/p_2 = 4 \times 10^{-6}/(5 \times 10^{-2}) = 8 \times 10^{-5}$ m³/s.

From Fig. 6.8 at $n = 5$ we find at $S_e = 8 \times 10^{-5}$ m³/s the optimum coefficient $K_{e2} = 0.13$. The rated pumping speed of a vapor jet pump is

$$S_{m2} = \frac{Q}{K_{e2}p_2 - p_{lim2}} = \frac{4 \times 10^{-6}}{0.13 \cdot 5 \times 10^{-2} - 3 \times 10^{-4}} = 6.5 \times 10^{-4} \text{ m}^3/\text{s}.$$

The closest by pumping speed vapor jet pump N-0.15S of the N series has the following characteristics:

Rated pumping speed, m³/s	1.5×10^{-2}
Inlet pipe diameter, mm	46
Outlet pipe diameter, mm	8
Highest output pressure, Pa	53
Limiting pressure, Pa	3×10^{-4}
Limiting pressure with a trap, Pa	3×10^{-5}

The dependence of the pumping speed of an N-0.15S pump on the inlet pressure is shown in Fig. 7.9.

2c. *Choice of a pump for medium and low vacuum.*

Pursuant to the additional condition we choose the VN series of rotary pumps with the limiting air pressure (with a trap) of $4\times10^{-1}-10^{-2}$ Pa and the range of pumping speeds from 10^{-4} to 1.5×10^{-1} m³/s.

We choose the working pressure of a mechanical pump from the maximum output pressure of the oil vapor pump with the safety margin of 2. Then $p_3 = 53/2 = 27$ Pa, which corresponds the effective pumping speed.

$$S_{e3} = \frac{Q}{p_3} = \frac{4\times10^{-6}}{27} = 1.5\times10^{-7} \text{ m}^3/\text{s}.$$

From Fig. 6.3 we find at $S_e = 1.5\times10^{-7}$ m³/s and $n = 3$ the optimum $K_e = 0.6$.
The rated pumping speed of a mechanical pump is

$$S_{m3} = \frac{Q}{K_{e3}p_3 - p_{\lim3}} = \frac{4\times10^{-6}}{0.6 \cdot 27 - 4\times10^{-1}} = 2.6\times10^{-7} \text{ m}^3/\text{s}.$$

The closest by pumping speed mechanical pump VN-01 has the following characteristics:

Rated pumping speed, m³/s	10^{-4}
Inlet pipe diameter, mm	8
Limiting pressure, Pa	4
Limiting pressure with a trap, Pa	4×10^{-1}
Maximum output pressure, Pa	10^5

The dependence of the pumping speed of a VN-01 pump on the inlet pressure is shown in Fig. 7.9.

3. *Determination of pipeline sizes and choice of vacuum system elements.*

3a. *The ultrahigh-vacuum system.*

We find the overall conductance of a section of a vacuum system from the magnetic discharge pump to the vacuum chamber using Eq. (7.25):

$$U_{o1} = S_{m1}\frac{K_{e1}}{1-K_{e1}} = 1.2\frac{0.35}{1-0.35} = 0.65 \text{ m}^3/\text{s},$$

where S_{m1} is the pumping speed of the pump, which is chosen from the catalogue. Figure 7.10 shows inner sizes of the pumped volume and lengths of the pipelines. The section of the vacuum system consists of three elements: pipelines 1 and 2 and valve 3.

We determine the conductance of the elements and the diameters of the pipelines. We consider, to a first approximation, that all the elements have the same conductance. Then $U_{ij} = 3U_{o1} = 3\times0.65 = 2$ m³/s. The mode of gas flow in a pipeline will be determined from the working pressure $p_1 = 10^{-5}$ Pa and the diameter of the inlet pipe of pump $d_{in} = 0.25$ m. The Knudsen criterion from Eqs. (2.1) and (1.29) is

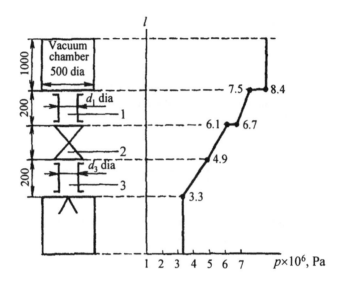

Figure 7.10 Schematic of the ultrahigh-vacuum section: (*1*) and (*3*) pipelines and (*2*) valve.

$$Kn = \frac{L}{d_{ef}} = \frac{L_1}{p_1 d_{in}} = \frac{7 \times 10^{-3}}{10^{-5} 0.25} = 3 \times 10^3 > 1.5,$$

i.e., the gas flow mode is molecular. The diameter of the first element can be calculated from the condition of sequential connection of the inlet orifice and the pipeline:

$$\frac{1}{U_{11}} = \frac{1 - d_1^2/(0.5)^2}{91 d_1^2} + \frac{0.2}{121 d_1^3} = \frac{1}{2},$$

whence we obtain $d_1 = 0.19$ m. From Table A.13 we choose the nominal pipeline diameter $d_1 = 0.2$ m. Then the conductance of the first stage $U_{11} = 2.29$ m³/s, the orifice conductance is 4.33 m³/s, and the pipeline conductance is 4.84 m³/s.

We choose a valve with the conditional passage diameter $d_n = 160$ mm (Table A.13) and a conductance of 3.34 m³/s in molecular gas flow mode. In view of the inlet resistance, the valve conductance will be found from the equality

$$\frac{1}{U_{12}} = \frac{1 - d_n^2/d_1^2}{91 d_n^2} + \frac{1}{3.34}.$$

Thus, $U_{12} = 2.2$ m³/s, the inlet orifice conductance is 6.47 m³/s, and the valve conductance is 3.34 m³/s.

The pipeline 3 diameter will be chosen from the condition $U_{13} = 2$ m³/s. Then, in view of the size of the previous element,

Table 7.3 Pressure distribution in the stage of the vacuum system from the magnetic discharge pump to the pumped volume.

Names of elements	Conductance, m^3/s	Difference of pressure, Pa	Inlet pressure, Pa	Outlet pressure, Pa
Element 3	2.48	1.6×10^{-6}	4.9×10^{-6}	3.3×10^{-6}
Valve (length)	3.34	1.2×10^{-6}	6.1×10^{-6}	4.9×10^{-6}
Valve (inlet)	6.47	6.0×10^{-7}	6.7×10^{-6}	6.1×10^{-6}
Element 1 (length)	4.84	8.3×10^{-7}	7.5×10^{-6}	6.7×10^{-6}
Element 1 (inlet)	4.33	0.9×10^{-6}	8.4×10^{-6}	7.5×10^{-6}

$$\frac{1}{U_{13}} = \frac{1 - d_3^2/d_2^2}{91 d_3^2} + \frac{0.2}{121 d_3^3} = \frac{1}{2}.$$

At $d_2 = 0.16$ m, the latter ratio yields $d_3 = 0,153$ m. According to the recommended sequence of diameters, we choose $d_3 = 0.160$ m. Thus, $U_{13} = 2,48$ m^3/s, and the overall conductance of the section, in view of the fact that the inlet conductance of the pump is infinity, is

$$\frac{1}{U_{o1}} = \frac{1}{U_{11}} + \frac{1}{U_{12}} + \frac{1}{U_{13}} = \frac{1}{2.29} + \frac{1}{2.2} + \frac{1}{2.48} = \frac{1}{0.77}.$$

The conductance of the chosen stage of the vacuum system is therefore 0.77 m^3/s, which is somewhat greater than the required value 0.65 m^3/s. The utilization coefficient of the magnetic discharge pump is

$$K_{e1} = \frac{U_{o1}}{S_{m1} + U_{o1}} = \frac{0.77}{1.2 + 0.77} = 0.39.$$

The efficiency 0.39 is close to the optimum value 0.35. We will calculate the pressure distribution over the length of the stage of the vacuum system from the magnetic discharge pump to the pumped volume. The results of calculation are listed in Table 7.3.

The pressure in the inlet section of the pump is, according to Eq. (6.68),

$$p_{p1} = p_{lim1} + \frac{Q}{S_{m1}} = 7 \times 10^{-8} + \frac{4 \times 10^{-6}}{1.2} = 3.3 \times 10^{-6} \text{ Pa.}$$

The difference of pressure at the element 3 is

$$\Delta p_3 = \frac{Q}{U_{13}} = \frac{4 \times 10^{-6}}{2.48} = 1.6 \times 10^{-6} \text{ Pa.}$$

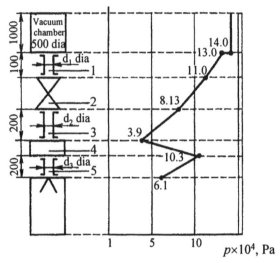

Figure 7.11 Schematic of the high-vacuum stage: (*1*), (*3*), and (*5*) pipelines, (*2*) valve, and (*4*) trap.

Similarly we find the differences of pressure at the other elements, calculate the pressure for the inlet and the outlet of each element and, on the basis of the results, plot the pressure distribution as shown in Fig. 7.10.

3b. *The high-vacuum system.*

We find the overall conductance of the section of the vacuum system from the vapor jet pump to the vacuum chamber using Eq. (7.25):

$$U_{o2} = S_{m2}\frac{K_{e2}}{1-K_{e2}} = 1.5 \times 10^{-2}\frac{0.13}{1-0.13} = 2.24 \times 10^{-3} \text{ m}^3/\text{s},$$

where S_{m2} is the pumping speed of the vapor jet pump and K_{e2} was chosen before.

We design the schematic of the considered vacuum system part. Figure 7.11 shows the inner sizes of the pumped chamber and the pipelines. The section of the vacuum system consists of five elements: pipelines 1, 3, 5, valve 2, and trap 4.

We determine the conductance of the elements and the diameters of the pipelines. We assume that all the elements have similar conductances. Then

$$U_{2j} = 5U_{o2} = 5 \cdot 2.24 \times 10^{-3} = 1.12 \times 10^{-2} \text{ m}^3/\text{s}.$$

The mode of gas flow in the pipeline will be determined from the working pressure $p_2 = 5 \times 10^{-2}$ Pa and the diameter of the inlet pipe of the pump is $d_{in} = 0.046$ m. The Knudsen criterion is

$$Kn = \frac{L}{d_{ef}} = \frac{L_1}{p_2 d_{in}} = \frac{7 \times 10^{-3}}{5 \times 10^{-2} \cdot 0.046} = 3 > 1.5 \,.$$

The mode of gas flow is therefore molecular. The diameter of the element 1 can be calculated from the condition of sequential connection of the inlet orifice and the pipeline

$$\frac{1}{U_{21}} = \frac{1 - d_1^2/(0.5)^2}{91\,d_1^2} + \frac{0.1}{121\,d_1^3} = \frac{1}{1.12 \times 10^{-2}}.$$

From this latter equation we find $d_1 = 0.023$ m. From Table A.13 we choose the nominal pipeline diameter $d_1 = 0.025$ m. Then the conductance of the first element is $U_{21} = 1.42 \times 10^{-2}$ m^3/s, the inlet conductance is 5.7×10^{-2} m^3/s, and the conductance of the pipeline is 1.89×10^{-2} m^3/s. We select as a next element the VEP-25 valve from Table A.13 with a nominal diameter of 25 mm and a molecular flow mode conductance of 0.014 m^3/s. Thus, $U_{22} = 0.014$ m^3/s, and the conductance of the inlet hole is infinity.

The pipeline 3 diameter will be chosen from the condition $U_{23} = 1.12 \times 10^{-2}$ m^3/s, whence, in view of the previous element size, and $l = 0.2$ m

$$d_3 = \sqrt[3]{\frac{0.2\,U_{23}}{121}} = 2.6 \times 10^{-2} \text{ m.}$$

We select $d_3 = 0.025$ m, whence $U_{23} = 9.45 \times 10^{-3}$ m^3/s. We choose a trap $d_n = 25$ mm and the conductance $U_{24} = 1.12 \times 10^{-2}$ m^3/s. The fifth element has the same sizes as the third element, i.e., $U_{25} = 9.45 \times 10^{-3}$ m^3/s and $d_5 = 25$ mm. The inlet conductance of the pump is infinity. The overall conductance is found from the following expression:

$$\frac{1}{U_{o2}} = \frac{1}{U_{21}} + \frac{1}{U_{22}} + \frac{1}{U_{23}} + \frac{1}{U_{24}} + \frac{1}{U_{25}} = 442.8 \text{ s/m}^3, \quad U_{o2} = 2.26 \times 10^{-3} \text{ s/m}^3.$$

The overall conductance of the chosen stage of the vacuum system is 2.26×10^{-3} m^3/s, which is somewhat greater than the required value 2.24×10^{-3} m^3/s. The utilization coefficient of the vapor jet pump is

$$K_{e2} = \frac{U_{o2}}{S_{m2} + U_{o2}} = \frac{2.26 \times 10^{-3}}{1.5 \times 10^{-2} + 2.26 \times 10^{-3}} = 0.13.$$

The coefficient $K_{e2} = 0.13$ is equal to the optimum value. We will calculate the pressure distribution over the length of this section of the vacuum system from the vapor jet pump to the pumped chamber. The pressure at the inlet section of the pump is, according to Eq. (6.68),

$$p_{p2} = p_{lim2} + \frac{Q}{S_{m2}} = 3 \times 10^{-4} + \frac{4 \times 10^{-6}}{1.5 \times 10^{-2}} = 6 \times 10^{-4} \text{ Pa.}$$

The difference of pressure at the element 5 is

Table 7.4 Pressure distribution in the stage of the vacuum system from the vapor jet pump to the pumped chamber.

Names of elements	Conductance, m³/s	Limiting pressure, Pa	Difference of pressure, Pa	Inlet pressure, Pa	Outlet pressure, Pa
Pipeline 5	9.45×10^{-3}	3×10^{-4}	4.23×10^{-4}	1.00×10^{-3}	6.00×10^{-4}
Trap 4	1.12×10^{-2}	3×10^{-5}	3.57×10^{-4}	3.90×10^{-4}	1.00×10^{-3}
Pipeline 3	9.45×10^{-3}	3×10^{-5}	4.23×10^{-4}	8.13×10^{-4}	3.90×10^{-4}
Valve 2	1.40×10^{-2}	3×10^{-5}	2.90×10^{-4}	1.10×10^{-3}	8.13×10^{-4}
Pipeline 1 (length)	1.89×10^{-2}	3×10^{-5}	2.10×10^{-4}	1.30×10^{-3}	1.10×10^{-3}
Pipeline 1 (inlet)	5.70×10^{-2}	3×10^{-5}	7.00×10^{-5}	1.40×10^{-3}	1.30×10^{-3}

$$\Delta p_5 = \frac{Q}{U_{25}} = \frac{4 \times 10^{-6}}{9.45 \times 10^{-3}} = 4.23 \times 10^{-4} \text{ Pa.}$$

Similarly we find differences of pressure on the other elements, calculating the inlet and outlet pressure at each of them. The results are summarized in Table 7.4, and the pressure distribution is plotted in Fig. 7.11.

3c. *The low-vacuum system.*

We find the overall conductance of the section of the vacuum system from the vapor jet pump to the mechanical pump using Eq. (7.25):

$$U_{o3} = S_{m3} \frac{K_{e3}}{1 - K_{e3}} = 1 \times 10^{-4} \frac{0.6}{1 - 0.6} = 2 \times 10^{-4} \text{ m}^3/\text{s,}$$

where S_{m3} is the pumping speed of the mechanical pump, chosen on catalogue and K_{e3} was chosen before. We will design the schematic of the considered section of the vacuum system. Figure 7.12 shows lengths of the pipelines, diameter of the outlet pipe of the vapor jet pump (8 mm), and diameter of the inlet pipe of the mechanical pump, 8 mm. This section of the vacuum system consists from 7 elements: four pipelines 1, 3, 5, and 7, valves 2 and 6, and trap 4.

We determine the conductance of the elements and the diameters of the pipelines in the assumption that all the elements have similar conductances:

$$U_{3j} = 7U_{o3} = 7 \cdot 2 \times 10^{-4} = 1.4 \times 10^{-3} \text{ m}^3/\text{s.}$$

The mode of gas flow in the pipeline will be determined from the working pressure $p_3 = 27$ Pa, and the diameter of the inlet pipe of the mechanical pump is $d_{in} = 0.008$ m. The Knudsen criterion is

$$Kn = \frac{L}{d_{ef}} = \frac{L_1}{p_3 d_{in}} = \frac{7 \times 10^{-3}}{27 \cdot 8 \times 10^{-3}} = 0.03 < 1.5,$$

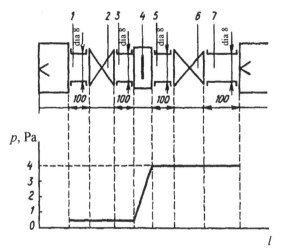

Figure 7.12 Schematic of the low-vacuum stage: (*1*), (*3*), (*5*), and (*7*) pipelines, (*2*) and (*6*) valves, and (*4*) trap.

but $0.03 > 5 \times 10^{-3}$, i.e., the flow mode is molecular-viscous. The resistance of the orifice at small differences of pressure, which is typical of steady-state gas flow mode at $K_{e3} = 0.6$, can be neglected in molecular-viscous flow mode. The diameter of the first pipeline can be calculated, at average pressure in the pipeline $p_{av} = p_3 = 27$ Pa, from the conductance, Eq. (2.59)

$$U_{31} = 121 \frac{d_1^3}{l_1} 0.9 + 1.36 \times 10^3 \frac{d_1^4}{l_1} p_{av} = 1.4 \times 10^{-3} \text{ m}^3/\text{s},$$

whence for $l_1 = 0.1$ m

$$d_1^4 + 3 \times 10^{-3} d_1^3 - 4.4 \times 10^{-9} = 0.$$

Solving this equation, we obtain $d_1 = 7.2 \times 10^{-3}$ m. From Table A.13, we choose $d_{31} = 8 \times 10^{-3}$ m, which corresponds to $U_{31} = 2.1 \times 10^{-3}$ m^3/s. Then we obtain $d_{33} = d_{35} = d_{37} = 8 \times 10^{-3}$ m.

For the second and the sixth parts of the system, from Table A.13 we choose the valve types KMU1-10 with the nominal diameter $d_n = 10$ mm and a conductance of 0.0014 m^3/s. The conductance of a valve in molecular-viscous mode is somewhat than in molecular mode. The difference in the conductances is neglected in this calculation.

We choose a trap with $d_n = 10$ mm and the conductance $U_{34} = 1.6 \times 10^{-3}$ m^3/s and find overall conductance from the following expressions

$$\frac{1}{U_{o3}} = \sum_{i=1}^{7} \frac{1}{U_{3i}},$$

Table 7.5 Pressure distribution in the stage of the vacuum system from the mechanical pump to the vapor jet one.

Names of elements	Conductance, m³/s	Limiting pressure, Pa	Difference of pressure, Pa	Inlet pressure, Pa	Outlet pressure, Pa
Pipeline 7	2.1×10^{-3}	4.0	1.9×10^{-3}	4.00	4.00
Valve 6	1.4×10^{-2}	4.0	2.9×10^{-3}	4.00	4.00
Pipeline 5	2.1×10^{-3}	4.0	1.9×10^{-3}	4.00	4.00
Trap 4	1.6×10^{-3}	0.4	2.5×10^{-3}	0.40	4.00
Pipeline 3	2.1×10^{-3}	0.4	1.9×10^{-3}	0.41	0.40
Valve 2	1.4×10^{-3}	0.4	2.9×10^{-3}	0.41	0.40
Pipeline 1	2.1×10^{-3}	0.4	1.9×10^{-3}	0.41	0.41

whence $U_{o3} = 2.53 \times 10^{-4}$ m³/s. The overall conductance of the chosen section of the vacuum system U_{o3} is somewhat greater than the required value 2.3×10^{-4} m³/s. The utilization coefficient of the mechanical pump of the system is

$$K_{e3} = \frac{U_{o3}}{S_{m3} + U_{o3}} = \frac{2.53 \times 10^{-4}}{1 \times 10^{-4} + 2.53 \times 10^{-4}} = 0.72 \,.$$

The coefficient $K_{e3} = 0.72$ is close to the optimum value 0.6. We calculate the pressure distribution over the length of this stage of the vacuum system from the mechanical pump to the vapor jet one. The pressure at the inlet section of the pump is, according to Eq. (6.68),

$$p_{p3} = p_{lim3} + \frac{Q}{S_{m3}} = 4 + \frac{4 \times 10^{-6}}{1 \times 10^{-4}} = 4 \text{ Pa.}$$

The difference of pressure on the element 7 is

$$\Delta p_7 = \frac{Q}{U_{37}} = \frac{4 \times 10^{-6}}{2.1 \times 10^{-3}} = 1.9 \times 10^{-3} \text{ Pa.}$$

Similarly we find the other differences of pressure by calculating the pressures at the inlet and the outlet of each element. The results are listed in Table 7.5 and the pressure distribution is plotted (Fig. 7.12). For the elements 1, 2, 3, and 4, we accept the limiting pressure of the mechanical pump with a trap to be 0.4 Pa.

7.9   Test Questions

7.1   What kind of information on a vacuum system can be stored in the automated database?

7.2   Which are the methods of automated structural designing of vacuum systems?

7.3 How can one use the methods of expert evaluations for the choice of weight factors for setting the goal function in the structural design of vacuum systems?

7.4 What are the tasks of the parametric synthesis of vacuum systems?

7.5 What are the advantages and the drawbacks of designing vacuum systems on the basis of the optimum vacuum pump utilization coefficients?

7.6 Which optimization criteria can be formulated for designing of vacuum systems?

7.7 Which types of restrictions are set in the parametric design of a vacuum system?

7.8 What is the main functional restriction of the optimum parametric design of a vacuum system?

7.9 For what purpose the connection scheme of vacuum system elements is developed?

7.10 Which main stages are included in the projecting calculation of a vacuum system?

Chapter 8

Construction of Vacuum Systems

8.1 Materials of Vacuum Engineering

The design of vacuum systems is in many respects determined by the properties of the materials used. In addition to the usual requirements of durability, adaptability to manufacture, low weight, etc., the materials used in vacuum engineering are required to provide:

1. the vapor pressure of the material at working temperature should be much lower than the working pressure;
2. the gas evolution of the material in operating conditions should be minimum;
3. vacuum tightness at small thickness;
4. corrosion resistance;
5. the absence of creep at temperatures from 500 to 600°C;
6. non-magnetic property.

If the vapor pressure of the material at the working temperature is equal to or higher than the working pressure, it causes its intensive evaporation and the formation of undesired coatings on the surfaces of various parts, for example, insulators of electric lead-in contacts.

Table C.9 shows the vapor pressures of some structural materials at two working temperatures typical of vacuum systems.

If brass is heated in high vacuum to 600°C, zinc contained in it evaporates and forms porous gas-untight material. The presence of lubrication and oils in a vacuum system limits the lowest achievable pressure.

Gas evolution of the materials at the working pressures and temperatures is determined by the presence of dissolved gases in the volume of the material and adsorbed gases on the surface. To remove the gases dissolved in metals, they are remelted in vacuum. The surfaces of vacuum materials should be carefully cleaned of contamination which is additional source of outgasing.

To improve the conditions for cleaning the internal surfaces of vacuum system elements, it is desirable to have a microroughness of 5 to 10 μm for highvacuum and 0.5–1 μm for ultrahigh vacuum systems to their medium height. Gas evolution of structural vacuum materials (Tables C.19 and C.20) depends on the method of preliminary treatment. One effective method that allows one to reduce gas evolution is high-temperature degassing to reduce the concentration of gases dissolved in the volume of a material. Reduction of hydrogen evolution from stainless steel is possible by deposition of oxide films or coatings with aluminum, silver, copper, etc. The presence of a superficial film hampers the transition of dissolved atoms from the crystal lattice to the surface and, at constant concentration of dissolved gases, it considerably reduces gas evolution.

Gas permeation is typical of many materials, but in some cases it is especially high. Silver is permeable for oxygen, iron, nickel, platinum, and palladium for hydrogen, glass for helium and hydrogen, and rubber for helium, hydrogen, and nitrogen. The permeabilities of some vacuum materials are listed in Table C.21.

Vacuum materials should be tight at small thickness. Cast materials often do not meet these requirements because they are porous. Sheet- and section-rolled products has unequal vacuum tightness in different directions. Slag inclusions form fibers in the strain direction of the processed material. Leaking of these fibers often shows itself only after heating in vacuum. Repar of the parts in which leakage is found is almost impossible, because soldering does not moisten slag inclusions, and heating during welding causes gas evolution and pore formation. When designing thin-walled parts it is necessary to control that the slag fibers be not directed across the wall, e.g., for ends of cylinder, the replacement of sheet-rolled products by section-rolled ones is undesirable. Vacuum-remelted metals have the best vacuum tightness.

Corrosion resistance is necessary for vacuum materials because corrosion increases gas evolution of materials, reduces the durability of thin-walled parts, and causes leakage.

Requirements to the corrosion resistance of vacuum materials are especially strict for ultrahigh vacuum systems, which are regularly heated to 400–500°C. For example, copper at these temperatures corrodes rapidly in air and, hence, it cannot be used for manufacturing of frequently heated parts contacting with atmosphere.

Sterssed parts of heated vacuum systems should not have appreciable creep at maximum working temperatures of 500–600°C. The creep of materials from which the parts of detachable flange connections are made causes appearance of leaksafter multiple heating cycles.

Non-magnetic propertyis a specific requirement to some parts of vacuum systems through which magnetic fluxes enter the vacuum chamber. These parts are available in the designs of magnetic lead-in contacts in vacuum, magnetic discharge pumps, and manometric transducers.

In vacuum engineering, such structural materials as cast iron, steel, copper, refractory metals, special alloys, glass, ceramics, plastics, rubber, oil, sealants, glue, etc. are widely utilized.

Cast iron is used for bodies of parts working in oil at low vacuum. Especially dense, fine-grained cast iron is used there. Among other cast alloys, bronzes are used that do not contain zinc, cadmium, phosphorus, and aluminum alloys.

High-quality structural low-carbon steel 20 (σ_t = 320–440 MPa) is easily soldered and welded and can be used for manufacturing non-heated parts of vacuum systems for low and medium vacuum. Steel 45 (σ_t = 640 MPa) is much less weldable and is not recommended for welded vacuum connections, but it can be used for manufacturing non-heated threaded parts, shafts, and other stressed parts. For parts of heated high-vacuum systems stainless steels with chromium contents higher than 13% are recommended that are not subjected to intergranular corrosion at high temperatures. Stainless steel 12Kh18N10T is acid-proof, non-magnetic, weldable, and soldered with special fluxes, and it is widely used in vacuum engineering. Cold-hardened stainless steels 1Kh21N5T and Kh17G9AN4 are stronger than steel 12Kh18N10T and can be used for manufacturing of heavily stressed parts such as bolts, pins, etc. Steel N36KhTU (EI702) retains good elastic properties at 600°C and can be used for manufacturing of spring equalizers in flange connections. Structure and basic properties of these steels are listed in Table C.22.

Copper is widely used in vacuum engineering for manufacturing of lining, internal fixture, and casings of soldered devices. The ultimate strength of soft copper ranges from 220 to 240 MPa, for solid copper, it is about 450 MPa, the creep strength is 70 MPa at 200°C and only 14 MPa at 400°C. It is recommended to use pure copper Grades MB (oxygen-free), MO and M1. The presence of oxygen in copper is detrimental for welding and soldering or annealing in hydrogen. Welded seams then become porous, and hydrogenation results in the restoration of copper oxide with the formation of water vapor that produces microscopic areas of great pressure and thus causes cracking effect ("hydrogen illness"): $Cu_2O + H_2 = 2Cu + H_2O$.

Brasses Lb2 (62% Cu and 38% Zn) and LS59-1 (59% Cu, 1% Pb, 40% Zn) are used for manufacturing of heated parts. Aluminum Grades AD1M and AMTS are applied for manufacturing of lining, pipelines of oil pumps, cryogenic screens, etc. Their linear expansion coefficient at temperatures from 20 to 300°C is 25.5×10^{-6}, and the ultimate strength is 120 MPa. Aluminum is easily-welded in helium arc and forms vacuum-tight seams. Duralumin D1 or D16 with an ultimate strength of 380 to 430 MPa does not provide tight vacuum welded seams.

Special alloys with certain physical properties that are necessary for some units of vacuum equipment are widely used in vacuum engineering. Among them there are Kovar (N29K18A) with a linear expansion coefficient of $(4.7–6.4) \times 10^{-6}$ for joining with glasses of molybdenum group C-47, C-49; Kh18TFM steel and N47D5 alloy with linear expansion coefficients of 10^{-5} and 8×10^{-6}, respectively, for joining with glasses C-87, C-89, C-90; Special alloys Grades N42, N45, N50 for soldering with various groups of glasses; N33K17 for joining with steatite ceramics; Invar (N36, EN3b) with low linear expansion coefficient and heat conductivity.

Refractory metals, e.g., tungsten, molybdenum, tantalum, and niobium are used for manufacturing of heaters, thermal screens, current lead-in contacts, etc.

Titanium (BT1, VT3, T3, and T4) is used to manufacture cathodes and getters for ion-sorption pumps.

C-47 and C-87 glasses are widely used in vacuum engineering to manufacture pipelines, cranes, traps, bodies, devices, pumps, manometers, insulators of electrical lead-in contacts, etc. In the glass Grade designations, the letter C is followed by figures corresponding to the linear expansion coefficient multiplied by 10^7. Physical properties of glasses are listed in Table C.23.

Ceramics are used in vacuum engineering instead of glass for manufacturing of high-temperature insulators. The following vacuum-dense ceramic are widely used: steatite, alundum, forsterite, and zircon. Alundum is the most termostable ceramics with 70–96% Al_2O_3, a softening temperature of 1900°C and a compression strength of 2000 MPa. Alundum ceramic is easily-soldered by the metallic coating method or active solders. Steatite and forsterite are produced on the basis of talc with addition of magnesium oxide, carbonic barium, and high-quality clay. Joining of steatite with metals is complicated, and forsterite is joined with titanium and forms good joint seams. Zircon has good thermal conductivity, but it is very hard and cannot be processed after annealing.

Plastics are used in vacuum engineering to produce many parts, i.e., seals, membranes, insulators, flexible pipelines, etc.

Fluoroplastic-4 is widely used, which is easily fabricated, and formed by pressure under condition of slow deformation. At temperatures above 200°C, fluoroplastic starts to evolve fluorine compounds. At temperatures lower than −70°C it becomes brittle. Gas evolution from fluoroplastic is less intense than from rubber, and fluoroplastic has good electric insulation properties and low friction coefficient.

Fluoroplastic-4 has the following physical properties:

Density, kg/m^3	$(2.1–2.3)\times10^3$
Tensile strength, MPa	14
Compression yield strength, MPa	3
Maximum working temperature, °C	200
Coefficient of heat conductivity, W/(m °C)	0.006
Coefficient of linear expansion	$(55–210)\times10^{-6}$
Dielectric permeability	9–2.2
Electrical strength, kV/mm	26
Gas evolution, m^3 Pa/(m^2 s)	10^{-4}
Nitrogen and oxygen gas permeability (at 20°C), m^3 Pa mm/(m^2 s Pa)	10^{-9}

Polyethylene has low gas evolution, but it may be applied only at room temperature because of insufficient thermal stability.

Vacuum rubber is widely used, especially in low vacuum engineering. Its gas evolution is much lower than for ordinary rubber, but much greater than for fluoroplastic. Moreover, the gas permeability of rubber is higher than that of fluoroplastic. The thermal stability of rubber is low, but the working temperature of the most termal stable rubbers is about 300°C. Due to the excellent elastic properties, rubber is a perfect material for vacuum sealings. Vacuum rubbers are subdivided into ordinary and termal stable, oil-resistant and non-oil-resistant.

Non-oil-resistant white vacuum rubber 7889 is very elastic and has low gas permeability. Its working temperature range is from +90 to −100°C. Oil-resistant black rubbers (9024 and IRP-1015) have poorer elasticity and greater gas evolution, but lower gas permeability in comparison with rubber 7889. IRP-1015 rubber is more oil-resistant, but softer. Before use, termal stable and oil-resistant rubbers (IRP-1368 and IRP-2043) with working temperatures of 250°C should be degassed in vacuum for 24 h at the maximum working temperature. Comparative characteristics of various vacuum rubbers are shown in Table C.24.

Glued connections are often used in low-vacuum systems and repair. L-4 epoxyde glue is applied to steel, glass, and ceramics with working temperatures of not more than 140°C, and it loses durability after long exposure to warm water. IP-9, KT-9, VKT-3, and VS-10T silicon organic glues withstand short-term heating to 300–350°C or long operation at 150–180°C.

For good hermetic sealing of low-vacuum connections, Ramsay paste (mixture of rubber, vaseline oil and paraffin) is added. Temporary repair of leaks and tightness test are performed using seals, such as Vakoplast modeling clay. Picein is made on the basis of wax and colophony and melts at 60–90°C.

8.2   Non-Detachable Connections

Vacuum-tight soldering is of great importance for vacuum mechanical engineering and instrument making, where it allows manufacturing of units from various materials: steel, ceramics, glass, and molybdenum.

In comparison with welding, tight soldering of metals allows considerably lower heating temperature of connected metals. Some metals and alloys can be soldered that do not provide vacuum tight weld junctions, such as steel with brass, aluminum with nickel, etc. The solder should have high mechanical durability, ductility, corrosion resistance, wettability and fluidity at the melting point. This property allows it to penetrate in gaps between connected parts.

The allowed gap size in the seam area during melting is 0.05–0.12 mm to ensure suction of the solder into the gap by the surface tension forces:

$$h = D_s - D_h; \quad D_s = D'_s(1 + \alpha_s \Delta t); \quad D_h = D'_h(1 + \alpha_h \Delta t),$$

where D'_s and D'_h are the diameters of the shaft and the opening at room temperature and α_s and α_h are the linear thermal expansion coefficients of the shaft and aperture materials.

The melting point of the solder is chosen by 100°C higher than the maximum degassing temperature of the unit, but by 100°C lower than the melting point of the connected materials.

Soldering is subdivided by the relative refractoriness of the solder into solid and soft. Solders that melt at temperatures lower than 300°C are used for soft soldering, and solders with melting points above 300°C, for solid soldering. The basic marks, structure and melting points of soft solders are summarized in Table C.25, and for solid solders, in Table C.26.

The manufacture of vacuum equipment units may involve soft soldering for elements with working temperatures of not more than 120°C, and solid soldering, for max. 450°C.

For the production of tight connections it is important that the crystallization temperature range of the solder material be not higher than 50°C. This value being greater, solders tend to liquifaction and do not provide tight connection. The same drawback is inherent to solders with higher quantity of dissolved gases and organic inclusions. Such solders boil during vacuum melting, and the seams become porous.

Soldering can be performed in the air or in a protective environment. During air soldering, fluxes are used for oxidation protection of surfaces. For soft soldering, fluxes such as mixture of equal fractions of $ZnCl_2$ and HCl or colophony solution in alkohol are used. Solid soldering of stainless steel is performed with flux that contains 40% potassium fluoride and 60% boric acid. Solid soldering of structural steels, copper, brass, and bronze can be performed with water-free drill flux.

The parts do not become soiled by fluxes and are degassed during soldering in protective environment, hydrogen or vacuum.

Vacuum hydrogen treatment in furnace for soldering provides a maximum working space temperature of 1150°C. Hydrogen soldering of copper Grades M1, M2, and M3 is not practic because of 'hydrogen illness', while soldering of stainless steel parts can be performed in hydrogen dried to about the dew point at temperatures not higher than –60°C,

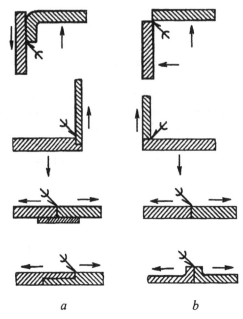

Figure 8.1 Butt and corner solder junctions: (*a*) regular and (*b*) irregular.

because of possible solidification of the stable superficial chromium oxides. For producing tight seams, the uniformity of assembly unit heating and cooling are of great importance.

Welded parts should be assembled and installed in special holders ensuring their mutual immobility, because soldering does not provide high mechanical integrity during tension. At a pressure of 2.5 MPa the connection of parts from steel 20 with solder POS-40 fails in 5000 hr, and at a pressure of 9.4 MPa, in 85 hr. The strength limit of solid solders is 200–400 MPa. Soldered connections based on shear are the most durable.

Regular and irregular joint shapes and angular weld junctions are shown in Figs. 8.1 *a* and *b*.

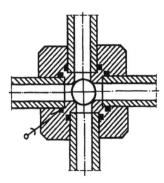

Figure 8.2 Vacuum element with solder jucntions made simultaneously in a vacuum-hydrogen furnace.

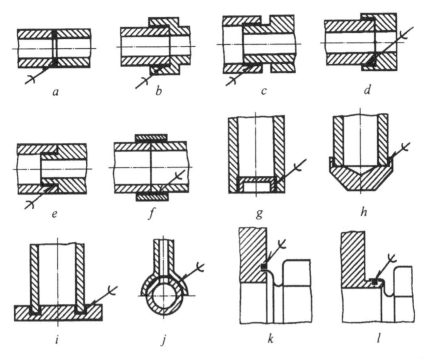

Figure 8.3 Vacuum solded connections: (*a*, *b*, *c*, *d*, *e*) connections of pipes, (*g*, *h*, *i*) end cups, flanges, bottoms, (*j*) branch pipes, and (*k*, *l*) bellows with flanges.

During manufacturing of complex solded units with a large number of seams, which can be soldered in hydrogen vacuum furnace, one heating cycle of the whole soldered unit with a diameter of 1 or 1.5 mm solder is performed soldering in special internal grooves (Fig. 8.2). Large difficulties are met in manufacturing of complex soldered units, which cannot be carried out with heating of the whole unit. In this case, solders with various melting temperatures are used.

The examples of soldered connections of pipes, flanges, end cups, bottoms, and bellows are shown in Fig. 8.3.

In the design of vacuum equipment elements and devices, it is often necessary to connect metallic parts to glass, or ceramic parts. Good alloying of metal with glass can be obtained only if the surface of metal soldered with glass has fine, but dense oxide layers. Metal oxides, as well as glass, have ionic structure and are well soluble in glass to form vacuum-tight connections. Seams of glass with metal may be of three types: coordinated, unmatched and with metal solder.

In seams (Fig. 8.4) glass and metal have close coefficients of linear expansion over the whole range of working temperatures. Many grades of technical glass have linear expansion coefficient of $(3-10) \times 10^{-6}$ and the linear expansion coefficient of metals and alloys ranges from 4.4×10^{-6} for tungsten to 17.8×10^{-6} for copper. In coordinated seams on their cooling from temperatures of glass softening down to room temperature the strength of the connected materials decreases. Because complete coincidence of the temperature

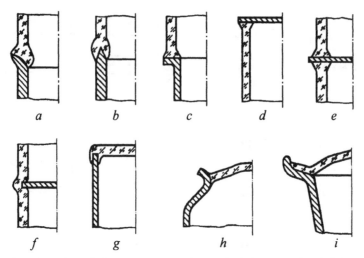

Figure 8.4 Coordinated metal-glass seams: (*a*, *b*, *c*) pipe-shaped seams, (*d*, *e*, *f*) disk-shaped seams, and (*g*, *h*, *i*) window-shaped seams.

dependences of linear expansion coefficients for both connected materials is unavailable, attempt is made to have them coincide at room temperature and the annealing temperature (on 50°C below than temperature softening of glass).

In unmatched seams, the linear expansion coefficients of connected materials may considerably differ from each other, and the dangerous stress appearing under these conditions is prevented by using flexible volume elements, metal parts of small diameter, metals with low yield stress, transition coordinated seams, and designs, in which the increased durability of glass due to compression is used. Figure 8.5 shows examples of unmatched seams. In constructions shown in Fig. 8.5, *a*, *b*, *d* the pointed pipe end serves as a flexible element, in Fig. 8.5, *c*, the same role is played by a thin cylindrical environment, in designs Fig. 8.5, *e*, *f*, the increased durability of glass due to compression is used. The linear expansion coefficient of metal in these seams should be higher than that of glass. Cooling of this parts causes internal compression stress in glass.

In seams with metal solder, silver layers are applied onto the surface of glass from a mixture with subsequent annealing. Such a seam is a metal cartridge with a glass tube, the gap between which is filled with a low-melting-point solder that well wets the metal and glass

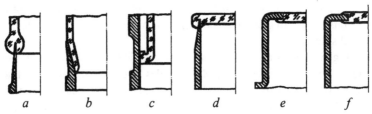

Figure 8.5 Uncoordinated metal-glass seams: (*a*, *b*, *c*) pipe-shaped seams and (*d*, *e*, *f*) window-shaped seams.

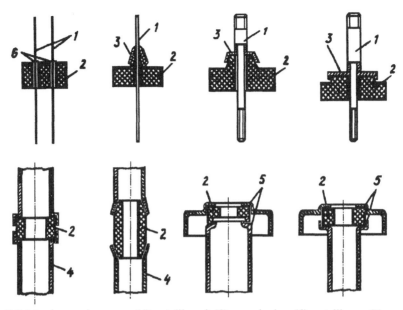

Figure 8.6 Metal-ceramics seams: (*1*) metallic rod, (*2*) ceramic ring, (*3*) metallic cup (Kovar, Nickel), (*4*) bushing (Kovar, nickel), (*5*) ring (nickel), and (*6*) seam.

and compensates their thermal stress. This metal-glass unit can be soldered to other metal parts.

Seams of ceramics with metals are widely applied in the elements of ultrahigh-vacuum systems, because they have higher thermal stability than the seams of glass with metals. Usually, they are made with active solders that form chemical compounds with ceramics during melting. Figure 8.6 shows examples of constructive elements in which seams of ceramics with metals are used. Seams of ceramics with glass can be received by direct connection of soldered materials.

For tight connection of parts in vacuum engineering the following types of welding may be used: (*a*) gas acetylene; (*b*) electric arc; (*c*) gas arc in protective environment: (*d*) cold plastic deformation; (*e*) diffusion and friction welding; (*f*) electron beam welding.

Gas acetylene welding is used for low-carbon steels with a wall thickness in the place of welding of max. 2 mm. Tight connections are obtained by welding with beading; joint welding of parts of ultrahigh-vacuum systems is not recommended.

Electric arc welding can be used for connection of parts of low-vacuum systems with a wall thickness of more than 2 mm. The best results can be obtained by automatic welding under a flux. It is not recommended, however, for ultrahigh-vacuum systems, because of the inadequate hermeticity.

Gas arc welding in protective environment with consumable and nonconsumable electrode can be used for connection of various metals in all types of vacuum systems. Stainless steel, copper, and aluminum with a thickness in the welding place of 0.1 to 2 mm are welded in argon or helium environments with nonconsumable tungsten electrodes. The best results are obtained by automatic welding in chambers filled with inert gas after removal of air.

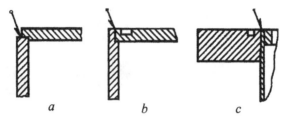

Figure 8.7 Corner weld junctions: (*a*) without beading, (*b*) with beading, and (*c*) connection of parts with different thicknesses.

Cold welding by the method of plastic deformation is used for connection of small parts of ductile materials (e.g., copper and aluminium) and requires complex pressing equipment.

Diffusion welding in vacuum and friction welding are used for connection of diverse materials, such as copper with ceramics, etc.

Electron beam welding is used for connection of chemically active and refractory materials, steel parts, and copper and aluminium alloys. This welding is conducted in vacuum chambers at pressures of not higher than 10^{-3} Pa.

All types of vacuum-tight welding should meet special requirements:

1. For producing of tight connections, welding should be conducted at a constant speed; breaks and backing seams are often the places of microcracks that cause leakage;
2. Welding is desirable from the side exposed while in service in vacuum, in order to reduce the number of cracks, pockets and roughnesses on the back side of the seam;
3. A concave seam in joint and corner connections turning out is allowed for welding without added material;
4. After manufacturing, welded seams should be checked for tightness using vacuum leak detectors.

The design and technology of vacuum parts prepared for welding should meet additional requirements:

1. parts before welding should be carefully cleaned and degreased;
2. the interface between welded parts should be set with a sliding fit.

Parts for corner seams can be prepared according to one of the variants shown in Fig. 8.7, *a–c*. The variant in Fig. 8.7, *a* is recommended for non-circular, and the variant in

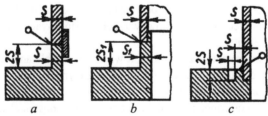

Figure 8.8 Welding of flanges with shell: (*a*) with lining, (*b*) without lining, and (*c*) with flange turning.

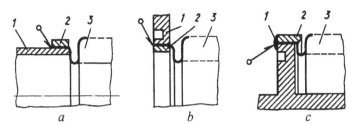

Figure 8.9 Welding of bellows: (*a*) with a pipe: (*1*) pipe, (*2*) protection ring, and (*3*) bellows, (*b*) with a flange: (*1*) flange, (*2*) protection ring, and (*3*) bellows, and (*c*) with a shaft: (*1*) shaft, (*2*) protection ring, and (*3*) bellows.

Fig. 8.7, *b*, for round parts. Connection of a thin-walled pipe with a thick plate is shown in Fig. 8.7, *c*. The peculiarity of this welding is the additional ring that evens the thickness of the welded parts. Without this ring, welding would be impossible because of the fusion of a thin-walled part.

Welding of shelled flanges shown in Fig. 8.8, *a–c*. In all cases the flanges do not require allowance for additional processing after welding, because the thin-walled elements have the same thickness, and a massive flange has not enough time to heat up during welding.

Stainless steel bellows that are widely used in high-vacuum engineering, may have a wall thickness from 0.05 to 0.25 mm. Connections of the bellows with pipes, flanges, and shafts are shown in Fig. 8.9.

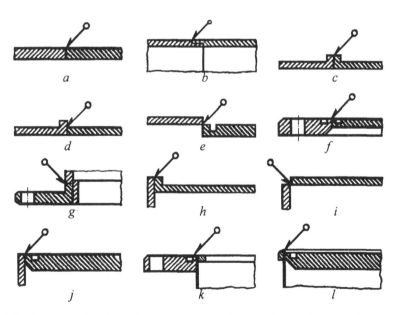

Figure 8.10 Examples of designs of vacuum weld junctions: (*a*) butt-weld without beading for flat parts, (*b*) butt-weld without beading for cylindrical parts, (*c*)–(*g*) butt-weld with beading for flat parts, (*h, j*) corner-weld with beading for flat parts, (*i*) corner-weld without beading for flat parts, (*k*) flange with a thin shell, and (*l*) bottom with a thin shell.

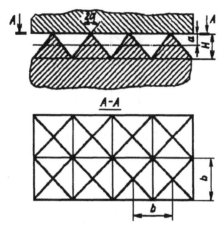

Figure 8.11 Calculation model of sealing surfaces.

Examples of flat, cylindrical, and corner welded vacuum connections with beading and without it are shown in Fig. 8.10.

8.3   Detachable Vacuum Connections

In detachable vacuum connections, one should provide a tight joint of two connected parts, the tightness of which should be close to that of a solid material. In the place of contact of two parts, manufacture always leaves microroughnesses that complicate the production of vacuum tight connections.

The tightness can be improved if the gap between the connected materials is filled with a sealer the viscosity of which is sufficient to fill the roughnesses at a contact pressure that is considerably lower than the elasticity limit of the main connected materials. For filling or isolation of microroughnesses, one may plastically deform at least one of the two connected parts. The sealings may be greases, rubber, fluoroplastic, and metals.

We determine the force of deformation of microridges that is necessary for maintenance of the desired leakage. We make the following assumptions:

1. the tight surface is perfectly even;
2. the microridges of the sealing are in the form of pyramids with the vertex corner 2α = 150 arc deg, and the basis of the pyramid is a square (Fig. 8.11);
3. the heights of the microridges is constant and is equal to H;
4. the probability of narrow and deep notches is small;
5. the bases of microridges lay in the same plane.

Deformation of microridges causes their hardening. The dependence between pressure and deformation in the superficial layer will be written as a power function:

$$\sigma = \sigma_s \left(\frac{\varepsilon}{\varepsilon_s}\right)^n = B\varepsilon^n,\qquad (8.1)$$

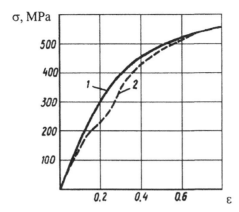

Figure 8.12 Contact deformation stress of a pyramid with square base and a vertex angle of 150 arc deg: (*1*) calculation and (*2*) experiment.

where

$$B = \frac{\sigma_s}{\varepsilon_s^n} = E\varepsilon_s^{1-n},$$

where ε is the generalized relative deformation of the microridges; σ is the pressure in the contact zone; σ_s is the yield stress of the sealing; ε_s is the relative deformation of the microridges corresponding to the pressure σ_s, n is the exponent ($0 \le n \le 1$); for perfectly ductile materials $n = 0$, and for perfectly elastic ones $n = 1$; E is the elasticity modulus of the sealing.

Figure 8.12 shows the dependence of contact pressure on the relative deformation of the height of square pyramids with a vertex corner of 150 arc deg for specimens of annealed copper MB. Curve 2 is obtained experimentally, and curve 1 is approximation by Eq. (8.1) at $B = 580$ MPa and $n = 0.29$.

Contact pressures for the chosen sealing model at specific pressure q is determines using the formula

$$\sigma = \frac{q}{\varepsilon^2}, \tag{8.2}$$

where $\varepsilon = a/H$ is the approachement of the contacting surfaces. Solving in common (8.1) and (8.2), we find the correlation between specific pressure and relative deformation of microridges:

$$\varepsilon = \left(\frac{q}{B}\right)^{\frac{1}{n+2}}. \tag{8.3}$$

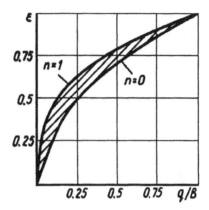

Figure 8.13 Relative deformation of microroughnesses as a function of specific pressure for ($n = 1$) absolutely elastic and ($n = 0$) absolutely plastic sealing materials.

The dependence of relative deformation of microridges on specific pressure, according to Eq. (8.3), for absolutely elastic and perfectly ductile sealings is shown in Fig. 8.13.

Calculation of the conductance of an elementary groove $b \times b$ (Fig. 8.11) in molecular flow mode using the Knudsen equation (2.55) gives the following expression at $\varepsilon \to 1$:

$$U_s = 0.22 v_{ar} H^2 (1 - \varepsilon)^3 . \tag{8.4}$$

By assumption of the absence of waviness on the surface of real sealing with the length l and width h, one may determine the total conductance of parallel and sequentially connected elementary grooves:

$$U_0 = \frac{l}{h} U_s . \tag{8.5}$$

Substituting expression for U_s from Eq. (8.4) and for ε from Eq. (8.3) into Eq. (8.5), we obtain the equation for the determination of the conductance of a perfect joint of smooth and rough surfaces in molecular gas flow mode:

$$U_0 = \frac{0.22 v_{ar} H^2 l}{h} \left[1 - \left(\frac{q}{B} \right)^{\frac{1}{n+2}} \right]^3 . \tag{8.6}$$

The flow of gas through a sealing per unit length is

$$Q_1 = U_0 \Delta p = K_g K_s K_f , \tag{8.7}$$

where K_g, K_s, K_f are the coefficients of gas, shape and force:

$$K_g = \Delta p v_{ar}; \quad K_s = \frac{0.22 H^2 l}{h}; \quad K_f = \left[1 - \left(\frac{q}{B} \right)^{\frac{1}{n+2}} \right]^3.$$

From Eq. (8.7) one may calculate the specific pressure that is necessary for obtaining the desired specific leakage Q_1:

$$q = B \left[1 - \left(\frac{Q_1}{K_g K_s} \right)^{1/3} \right]^{n+2}. \tag{8.8}$$

The efficiency of various sealings is often compared using not the specific pressure, but the force per unit of length of sealing:

$$F = qh = Bh \left[1 - \left(\frac{Q_1}{K_g K_s} \right)^{1/3} \right]^{n+2}. \tag{8.9}$$

From the latter formula it can be seen that complete tightness of a connection on the given model occurs at $F = Bh$. In real seals, it is practically impossible to achieve complete tightness, because of deep notches.

Detachable vacuum connections should meet the following requirements: minimum leakage and gas evolution; mechanical durability; thermal resistance to repeated heating without infringement of tightness; corrosion resistance; maximum number of cycles of disassembly and assembly with preservation of tightness; convenience of repair and adaptability to manufacture; possibility of easy check for tightness.

In vacuum engineering of systems with heating to 300°C, rubber seals are widely used. Rubber has good elastic properties, and requires only a small effort for the production of a vacuum-tight connection to a steel surface with $R_A = 10$ μm. For linings with a width of 4 mm specific pressure of sealing is 4–8 N/mm, which corresponds to a specific pressure of 1–2 MPa. Rubber sealing allow practically unlimited number of disassemblies and assemblies, they are simple in manufacturing, and seldom require repair. A disadvantage of rubber seals is the high gas evolution and gas permeability in comparison with metal or glass parts. The usual shapes of rubber seals are round or square. The diameter or side of square are chosen from constructive requirements, usually 3–5 mm.

The seal shown in Fig. 8.14, *a*, between two flat flanges is the simplest in manufacturing and repair, but it does not provide exact fixing of the tightened parts. The maximum pressure on lining in this sealing is not limited, which may result in overloading of the lining during assembly, accompanied by plastic deformation. For fixing of connected parts and restriction of the sealing pressure one may use additional constructive elements. The connection in Fig. 8.14, *b*, limits the maximum compression pressure of the lining and provides axial fixing at the expense of the contact of connected parts. The shape of the seal can be round or square. For producing the necessary pressure of hermetic sealing, the degree of deformation of the lining should be 30%. These connections about are frequently used in various vacuum equipment. Their basic sizes are shown in Fig. A.15 and A.16 of the application.

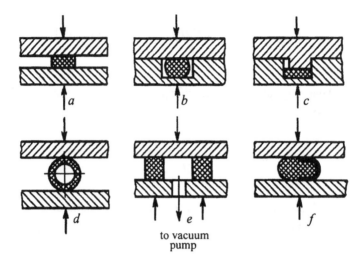

Figure 8.14 Constructive schemes of detachable vacuum connections with polymer sealing.

The connection in Fig. 8.14, *d*, with increased pressure inwards the sealing is recommended for sealing of large diameter flanges. It has high deformation of the sealing element and may compensate waviness of the flange surfaces.

The connection in Fig. 8.14, *e*, has double sealing with intermediate pumping. It is much more complex than other connections with rubber lining, but it provides high reliability and low gas permeability of the connection.

The connection in Fig. 8.14, *f*, differs by metal foil armoring of the sealing element from the side of vacuum, which reduces gas evolution and gas permeability of the connection by reducing the contact surface of the sealing with vacuum.

As a seal one may apply fluoroplastic, which has low gas evolution and gas permeability. A disadvantage of flouroplastic is the very low elasticity limit, for which reason seal requires intense bulk compression. The constructive scheme shown in Fig. 8.14, *c*, is the best for providing such conditions. It provides restriction of pressure on the lining and fixes connected parts in the axial and in the radial direction. The shape of this sealing is square. Fluoroplastic does not leak in 0.1 mm gaps even at very high specific pressures. The basic sizes of connection asymmetrical flanges and fluoroplastic sealing are shown in Fig. A.17 of the application.

In ultrahigh vacuum, metal sealing is of great importance because it allows heating to 450–500°C. Microroughnesses are filled at the expense of plastic deformation of the lining material. The fluidity of metals is much lower than for rubber, and, hence, the creation of sealing requires high specific pressure and best cleanliness of the surface. Gas evolution of metal lining is 10 times lower than for rubber, but the connection with metal linings is more complex in manufacturing and allows a limited number of heating and assembly cycles. Schematics of the most widely used metal sealing are shown in Fig. 8.15, *a–g*.

Connections with round sealing (Fig. 8.15, *a*) are the simplest in manufacturing, repair and work reliably with copper and gold lining. The pressure of hermetic sealing in such connections with a 0.8 mm gold wire is 350 N/mm. The pressure of hermetic sealing (Fig.

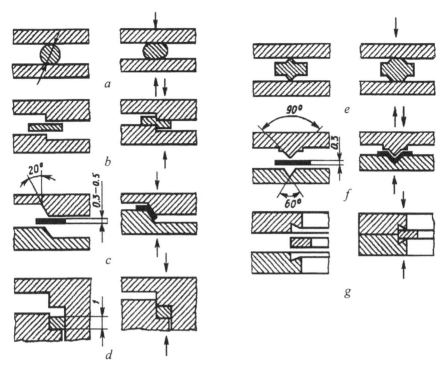

Figure 8.15 Schemes of sealing with metallic lining: (*a*) wire lining, (*b*) flat cutting lining, (*c*) conical, (*d*) with volume compression, (*e*) knife-like, (*f*) groove-wedge, and (*g*) Conflat.

8.15, *b*) is very sensitive to the coaxiality of the cutting edges. For lining with a thickness of 4 mm the minimum pressure of hermetic sealing for unannealed copper is 470 N/mm, and for aluminium it is 340 N/mm.

Conical connection (Fig. 8.15, *c*) requires, in comparison with flat one, somewhat lower pressure of hermetic sealing, but it has large dimensions and works poorly during non-uniform heating, because of the gap appearing at a radial temperature gradient. Sealing with volume compression (Fig. 8.15, *e*) is poorly demountable because of the penetration of lining material into the gaps between adjacent parts. Knife sealing (Fig. 8.15, *e*) has the lowest pressure of hermetic sealing among metal seals and can be used instead of rubber sealing, but its pressure of hermetic sealing increases after repeated assembly. Groove-wedge connection (Fig. 8.15, *d*) with a 0.5 mm thick flat copper lining requires a pressure of 280 N/mm for hermetic sealing. Its disadvantage is the presence of a closed gas volume and asymmetry of the flanges. Conflat-type sealing (Fig. 8.15, *g*) is widely used in heated ultrahigh-vacuum systems. Flange connection using this seal is shown in Fig. A.18.

During degassing of detachable connections, in order to avoid appearance of leaks after cooling, it is necessary to ensure uniformity of heating and to choose materials with identical linear expansion coefficients over the entire range of working temperatures. Copper and stainless steel have close linear expansion coefficients, and therefore copper is widely used as a lining material between stainless steel parts. To increase the reliability of sealing

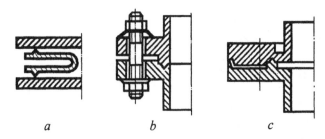

a   *b*   *c*

Figure 8.16 Elastic elements for compensation of thermal deformation in detachable connections: (*a*) elastic lining, (*b*) elastic bolts, and (*c*) elastic flanges.

gold lining or gold coating of copper lining can be used. This prevents corrosion of lining at repeated heating to 450–500°C. To compensate the small difference in the linear expansion coefficients of elastic lining, thin bolts, plate-like springs, disks facilitating turning in flanges, etc. can be used (Fig. 8.16).

8.4   Vacuum Pipelines

Elements of vacuum systems are placed in machines and equipment according to technological process requirements, convenience of operation, repair, etc.. They are connected using connecting elements, i.e., of pipelines. The nominal inner diameter in a pipe is referred to as the passage diameter and is designated as D_c. Recommended diameters are given in Table A.13.

Pipelines may be flexible or rigid. Flexible pipelines are more complex than rigid ones and are used for connection of elements that do not have a common structural basis. The allowance for their installation may be several millimeters in this case.

Constructive schemes of pipelines are shown in Fig. 8.17. The length of a pipeline is determined from constructive reasons associated with the convenience of elements positioning in the skeleton of a vacuum system. The diameter of a pipeline is determined during projecting calculation, on the basis of the requirements to its conductance.

Vacuum pipelines should maintain atmospheric pressure without destruction or loss of stability. The thickness S of pipeline walls derived from the condition of its durability for thin walls $S/D_c < 0.05$ is recommended to be determined using the formula

$$S = \frac{p_{atm}D_c}{2\sigma_{all}} + C, \tag{8.10}$$

where p_{atm} is the atmospheric pressure, C is the admittance for corrosion and technological allowance, and σ_{all} is the allowed stress.

The allowed stress is accepted equal to the least of the following three values:

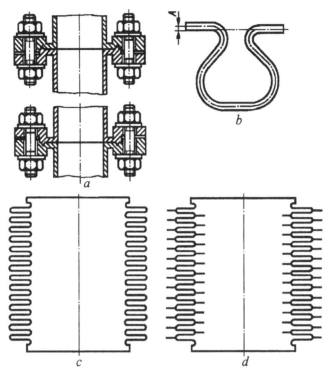

Figure 8.17 Shapes of pipelines: (*a*) rigid pipeline, (*b*) bent pipeline, (*c*) hydraulically shaped bellows, and (*d*) weld membrane bellows.

$$\frac{\sigma_p^t}{2.6} ; \ \frac{\sigma_t^t}{1.5} ; \ \frac{\sigma_{ad}^t}{1.5} ,$$

σ_b^t, σ_t^t, σ_{ad}^t which are the ultimate strength, yield stress, and limit of long-term durability of pipe material at their working temperature.

The admittance C is $(0.05–0.18)S$. If the calculation of S using the Eq. (8.10) yields that $S/D_c \geq 0.05$, the calculation should be repeared using the formulas for thick-walled cylinders. The distribution of the tangential σ_t and normal σ_n stress in a thick-walled pipe can be determined using the formulas

$$\sigma_t = \frac{p_2 r_2^2 - p_1 r_1^2}{r_1^2 - r_2^2} + \frac{(p_1 - p_2) r_1^2 r_2^2}{r^2(r_1^2 - r_2^2)} ,$$

$$\sigma_n = \frac{p_2 r_2^2 - p_1 r_1^2}{r_1^2 - r_2^2} - \frac{(p_1 - p_2) r_1^2 r_2^2}{r^2(r_1^2 - r_2^2)} \tag{8.11}$$

where p_1 and p_2 are the pressures outside and inside the cylinder, respectively, and r_1 and r_2 are the outer and inner radii of the cylinder. If $p_2 = 0$, the most dangerous is the stressed condition of external fibers of the pipe. The main stresses at $r = r_1$ are

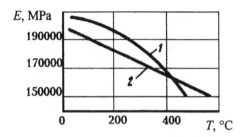

Figure 8.18 Elasticity modulus of pipe materials as a function of temperature: (*1*) carbon steels and (*2*) alloyed austenitic steels.

$$\sigma_1 = \sigma_t = \frac{2p_1 r_1^2}{r_1^2 - r_2^2}; \quad \sigma_2 = \sigma_n = -p_1; \quad \sigma_3 = -\frac{p_1 r_1^2}{r_1^2 - r_2^2}.$$

Using the theory of elasticity, we prove, that the condition

$$\sigma_{all} \geq \sqrt{0.5[(\sigma_1 - \sigma_2)^2 + (\sigma_2 - \sigma_3)^2 + (\sigma_3 - \sigma_1)^2]} \tag{8.12}$$

is satisfied.

The wall thickness of cylindrical pipelines, especially for large diameters, should be necessarily checked for the condition of stability

$$S = 1.25 D_c \left(\frac{pl}{ED_c}\right)^{0.4} + C, \tag{8.13}$$

where p and E are the pressure and elasticity modulus of the pipe material, Pa; D_c and l are the passage diameter and of the pipeline length, m; and S and C are the wall thickness and allowance, m. The temperature dependence of elasticity modulus for typical structural materials of pipelines is shown in Fig. 8.18.

8.5   Devices for Transmitting Movement to Vacuum

The necessity of transmitting movement to vacuum occurs because the improvement of working conditions requires the driving mechanisms should be placed outside the vacuum chamber. Devices for transmitting movement to vacuum are subdivided into three groups:

1. for transmitting reciprocating movement;
2. for transmitting lateral movement;
3. for transmitting rotary movement.

Each group is subdivided accoding to the limiting pressure, transmitted force, movement speed, stroke, etc.

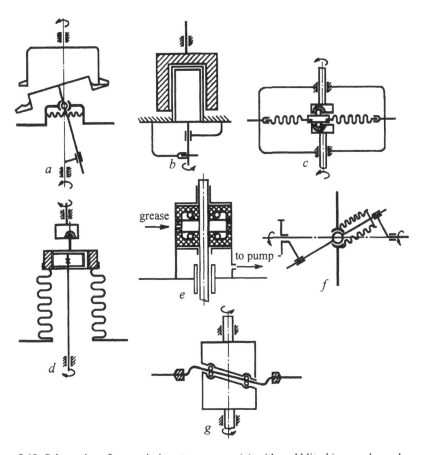

Figure 8.19 Schematics of transmissions to vacuum: (*a*) with wobblity hinge and membrane, (*b*) magnetic, (*c*) with plane-parallel moving part and membrane, (*d*) with plane-parallel moving part and bellows, (*e*) molecular sealing with additional pumping, (*f*) with wobbling hinge and bellows, and (*g*) wave-like transmission.

Transmissions to vacuum should be designed so the sealing element be not affected by the transmitted force, and the driving elements and supports be located, whenever possible, outside the vacuum chamber. For friction pairs placed in vacuum, it is necessary to use special means to prevent binding of friction materials. For this purpose it is possible to use as lubricant molybdenum disulfide MoS_2, sulfidize the friction surface or to use materials, with greatly different physical properties, for example, metals and ceramics.

Transmission of reciprocating movement for low and medium vacuum are made usually with rubber and fluoroplastic seals. Rubber and steel have high friction coefficient, and rubber sealing of mobile connections always requires lubrication. Fluoroplastic can be used without greasing, but its wear in operating time should be compensated by additional elastic elements. The sealing for reciprocating movement transmission in high and ultrahigh vacuum are made completely from metal and can be heated for degassing to 400–500°C. The transmission of rotary movement in vacuum differs by a large variety of the constructive solutions, kinematic circuits of which are shown in Fig. 8.19, *a–g*.

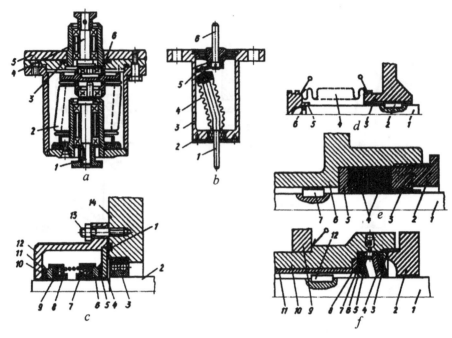

Figure 8.20 Construction of movement transmissions to vacuum: (*a*) bellows transmission with a step shaft: (*1*) in shaft, (*2*) bellows, (*3*) rubber seal, (*4*) out shaft, (*5*) vacuum chamber, and (*6*) intermediate coupling; (*b*) bellows transmission with bent shaft: (*1*) in shaft, (*2*) lid, (*3*) cup, (*4*) bellows, (*5*) carrier coupling, and (*6*) out shaft; (*c*) transmission with ground disks: (*1*) sealing, (*2*) shaft, (*3*) bearing, (*4*) lid, (*5*) and (*11*) stationary rongs, (*6*) and (*10*) rotating rings, (*7*) and (*9*) seals, (*8*) spring, (*12*) cup, (*13*) fixture, and (*14*) casing, (*d*) bellows transmission of reciprocating movement, (*e*) gland transmission, and (*f*) rubber seal.

Transmission of rotation to vacuum with a spatial joint and sealing in the form of a flexible element performing oscillating movement provides kinematically rigid transmision of large momentum with low rotation frequency. The maximum rotation frequency is limited by the fatigue durability of the flexible element, for example, bellows or a membrane, and the serviceability of the friction pair that operates in vacuum. This transmission withstands heating to 450–500°C, which allows it to be applied in ultrahigh-vacuum systems.

Magnetic transmission of rotation to vacuum is used in high-vacuum systems for high-frequency rotary movement with small twisting momentum. It can be made heated, but does not provide kinematically rigid transmission.

Molecular transmission of rotation with additional pumping is used to transmit large twisting moment to high vacuum at a high rotation frequency. The sealing between atmosphere and intermediate vacuum during baking of a vacuum system requires forced cooling. The pressure in the intermediate chamber is chosen within 10^{-3}–10^1 Pa. The sealing between the vacuum chamber and intermediate vacuum is provided by resistance of the gap between the shaft and the device casing.

Wave-like transmission of rotation to vacuum provides kinematically rigid transfer of rotary movement with low rotation frequency. It can be used in high-vacuum systems.

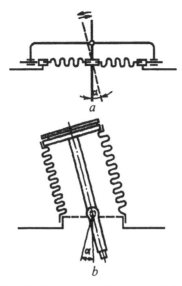

Figure 8.21 Transmissions of oscillating movement to vacuum: (*a*) membrane and (*b*) bellows.

The design of transmission of rotation with a part performing plane-parallel movement shown in Fig. 8.20, *a, b* is applied in backed high-vacuum systems for transmission of small twisting moment at small rotation frequency. Figure 8.20, *c* shows the design of transmission of rotation with ground disks used for work with lubrication in rotary pumps.

The transmission of reciprocating movement for low and medium vacuum (Fig. 8.20, *e, f*) are usually made with rubber or fluoroplastic sealing. Glanded seals (Fig. 8.20, *e*) consist of rod 1, sealing nut 2, metal disk 3, rubber lining 4, fluoroplastic cartridge 5, case 6, and pin 7. The rubber lining is an elastic element for the thin fluoroplastic cartridge, the wall thickness of which is 0.5 mm. This transmission is suitable for reciprocating motion with large stroke, but small velocity of not more than 0,1 m/s.

Seals with rubber packing glands (Fig. 8.20, *f*) consist of rod 1, sealing nut 2, metal disks 3 and 8, rubber packing glands 4 and 7, lubricating valves 5, oil distributor ring 6, vacuum chamber wall 9, casing 10, sliding bearing 11, and pin 12.

Sealing for the transmission of reciprocating movement in high and ultrahigh vacuum (Fig. 8.20, *d*) is made completely of metal and can be heated for degassing to 450–500°C. It consists of rod 1, pins 2, case 3, bellows 4, head 5, and pins 6. Stainless steel bellows 4 are welded to the head 5 and the case 3 by electric beam or argon arc welding.

Schematics of transmission of oscillating movement to vacuum shown in Fig. 8.21 has the oscillation angle α in membrane inputs not more than 10 arc deg, and in bellows not more than 30 arc deg. Producing of large oscillation angles requires accelerating transmissions or rotary movement ones. One may also use reciprocating transmissions with subsequent transformation of the reciprocating movement to oscillating.

In the determination of the serviceability of rotary, reciprocating, and oscillation movement transmissions, a certain difficulty is faced when calculating bellows. Bellows can be considered as a spring with rigidity

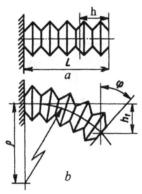

Figure 8.22 Schemes of bellows deformation: (*a*) axial and (*b*) bending flexure.

$$C_0 = \frac{Q}{\lambda} = \frac{2.5ED_o(1 + 0.013t^2)\delta^{2.43}}{n(D_o - D_i)^3},$$ (8.14)

where Q is the elasticity force of the bellows; λ is the elastic stroke of the bellows; D_o and D_i are the outer and inner diameters of the bellows, mm, t is the corrugation pitch, mm, δ is the wall thickness of the bellows, mm, n is the number of the corrugations, and E is the elasticity modulus of the bellows material. For copper alloy L80 this modulus is $E = 116,000$ MPa, and for 12Kh18N10T stainless steel it is $202,000$ MPa. The difference in the outer and inner diameters and thickness of the bellows wall shows the greatest effect on its rigidity. The rigidity of bellows should be taken into account when calculating precision and lightly-loaded mechanisms. The durability of bellows can be calculated using the caclulation scheme in which the bellows is considered as a system of ring-shaped plates alternately connected by their external and internal contours. Figure 8.22 shows two basic deformation schemes of bellows for transmission of movement to vacuum. The axial deformation of bellows is used in transmission of reciprocating movement, and their bending flexure, in transmission of rotary movement.

Axial deformation of bellows under the scheme in Fig. 8.22, *a*, by the value h produces stress

$$\sigma_0 = \frac{4hE\delta K_2}{D_i^2 n(1 - \mu^2)},$$ (8.15)

where μ is the Poisson ratio, and K_2 is the coefficient determined by $\alpha = R_o/R_i$. For the edges of bellows, we have for the outer edge

$$K_{2o} = \frac{\alpha^2 - 1 - 2\ln\alpha}{(\alpha^2 - 1)^2 - 4\alpha^2\ln^2\alpha},$$

for the inner edge

$$K_{2i} = \frac{2\alpha^2\ln\alpha - (\alpha^2 - 1)}{(\alpha^2 - 1)^2 - 4\alpha^2\ln^2\alpha}.$$

During bending flexure deformation of bellows under the scheme in Fig. 8.22, b the stress pressure depends on bellows end face rotation angle φ. The nominal length of bellows does not change. The calculation formula for the stress at the outer and inner contours of bellows is

$$\sigma_o = \frac{\varphi E\delta K_3}{n(1 - \mu^2)D_i},\tag{8.16}$$

where for calculation of the stress at the outer edge

$$K_{3o} = \frac{\alpha^2 - 1}{\alpha(\alpha^2 + 1)\ln\alpha - \alpha^2 + 1},$$

and for the inner edge

$$K_{3i} = \frac{\alpha^2 - 1}{(\alpha^2 + 1)\ln\alpha - \alpha^2 + 1}.$$

From Eqs. (8.15) and (8.16) it can be seen that the stress at the inner edge is greater than at the outer one.

In transmissions of movement to vacuum, bellows undergo cycled loading. Approximated calculation of the durability of bellows N, expressed as the number of loading cycles that a bellows can withstand before destruction is performed using the dependence such as

$$N = \frac{K}{\sigma^x},\tag{8.17}$$

where $K = 11\times10^{11}$ and $x = 4.42$ for stainless steel 12Kh18N10T (σ is the maximum stress in the bellows under one of the loading schemes, MPa). The required durability of bellows depends on their working conditions and may range from 10,000 to 1,000,000 cycles. In especially heavy cases the durability may be as low as only 1500 cycles.

8.6 Electrical Vacuum Contacts

Feeding of electric current to various devices working inside a vacuum chamber requires hermetic electrical contacts isolated from the casing of the vacuum chamber. Depending on their purpose, electrical contacts may be low- or high-voltage, small- or strong-current, low- or high-frequency, low- or high-vacuum.

For feeding electric power to heating elements, low-frequency low-voltage strong-current contacts are widely used. Example of the design of such input for unheated vacuum

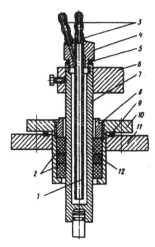

Figure 8.23 Water-cooled electric contact for low-vacuum systems.

systems is shown in Fig. 8.23. In chamber casing 11, cup 10 is sealed by lining 9. Nut 8 fizes gland 12 at the casing 7 of contact 6. Head 4 is sealed by lining 5. The material for the linings is rubber, and insulators 2 can be made of any grades of vacuum ceramics. During the operation the system is cooled by water through union 3 and pipe 1.

High-voltage contact for ultrahigh-vacuum systems (Fig. 8.24, a) is welded to basis 7 by bottom cap 6. Ceramics 5 is connected to caps 6 and 4 by vacuum-tight soldering. Head 2 and jack 3 are simultaneously soldered to the top of cap 4. Central core 1 is screwed into jack 3.

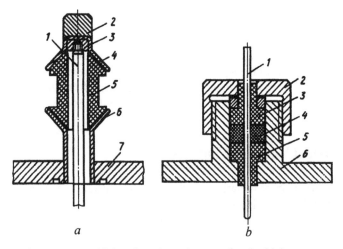

a   b

Figure 8.24 Electric contacts: (a) high-voltage heated contact for ultrahigh-vacuum systems and (b) contact for thermocouples.

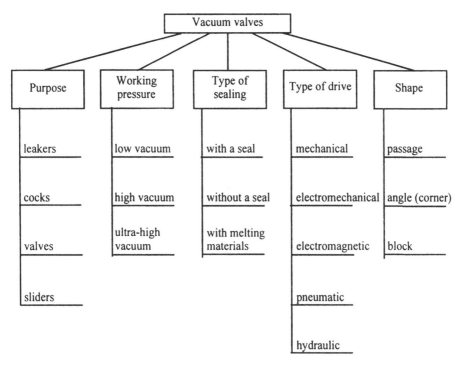

Figure 8.25 Classification of vacuum valves.

During precision measurement of temperature at the contact by a thermocouple, one should provide the absence of interrupted contacts. The contact shown in Fig. 8.24, *b* prevents interrupted contacts and can be used in unheated vacuum systems. In casing 6, insulating cartridges 5 and 3 with lining 4 seal one or both wires of thermocouple 1 with the help of nut 2. The material of lining may be rubber, and the insulating cartridges can be made of fluoroplastic.

8.7 Vacuum Valves

In vacuum machines and systems, connection of various parts is often made using vacuum valves, the classification of which is summarized in Fig. 8.25.

Leaks can be made of porous materials, some metals that are permeable for particular gases, for example, platinum for hydrogen, silver for oxygen, etc. Microscopic movement of the needle in the hole, and change of the gap between materials with various linear expansion coefficients can be used for creating and adjusting large vacuum resistances.

Vacuum valves are used in systems with small gas evolution, at rough vacuum ranges, and when it is not required to obtain large conductance. In metal vacuum valves and closures working at room temperature, rubber and fluoroplastic sealing can be used (Fig. 8.26, *a*).

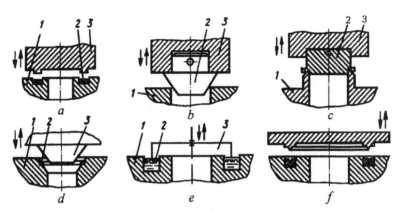

Figure 8.26 Sealing elements of vacuum valves: (*1*) seat, (*2*) seal, and (*3*) sealing element.

In heated devices, hermetic sealing is obtained at the expense of plastic deformation of the material of one of contacting surfaces. The design with a conical sealing element (Fig. 8.26, *b*) made of copper or aluminium is widely used. A disadvantage of this design is the gradual increase of the area of the sealing surface, and, hence, the effort of closing the valve. In the design shown in Fig. 8.26, *c* the sealing element works by cutting, and the effort of hermetic sealing does not depend on the number of operations, but the stroke for closure increases.

In the design shown in Fig. 8.26, *d*, the pressure of hermetic sealing and the stroke of the sealing element do not depend on the number of operation cycles, but a clean surface in the place of contact of the sealing elements is not achieved. Reliable operation of this device is provided by sealing made of noble metals.

The pressure of hermetic sealing per 1 mm of sealing length for rubber is 2.5–5 N, for aluminium it is 100–150 N, and for copper, 200–300 N. In any design of vacuum valves, the drive should ensure constant sealing pressure. Exceeding of the optimum sealing pressure reduces the number of operation cycles.

Sealing with melted metals, usually gallium, indium, tin, or lead (Fig. 8.26, *e, f*) does not require large effort for clothing. In the design shown in Fig. 8.26, *e*, seal 2 should be in melted condition at the moment of motion of sealing element 3. In the design shown in Fig. 8.26, *f*, the seal is melted only from time to time for renewing the sealing surface. Valves with melted seal have short service life because of the fast oxidation, evaporation, and mechanical removal of the sealing material.

Choice of the mechanism of vacuum valve drive is determined by the pessure of hermetic sealing and the stroke of the sealing element. The mechanism of combined hand-operated and electromechanical drive (Fig. 8.27) is often used in practice. In case 1 with two connecting pipes 10 and 12, sealing element 11 that is connected with the casing of bellows 2 moves in reciprocating mode. The sealing pressure is provided by the screw pair between shaft 3 and casing 1. The hand-operated drive is wheel 7, which is rigidly connected with shaft 3, the disks of friction coupling 4 being separated by the finger driven in the corner groove of half buching 6. The effort of the hand-operated drive is not adjusted. The electromechanical drive from electric motor 9, through worm gears 8 and 5, friction coupling

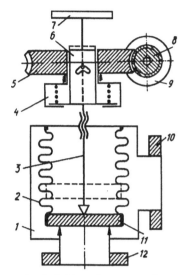

Figure 8.27 Scheme of electromechanical drive of vacuum valves.

4, and screw pair on shaft 3, produces the sealing pressure, which is limited by the maximum twisting moment that can be transmitted by the coupling 4.

The electromagnetic drive is convenient for remote control of vacuum valves. The sealing pressure is usually produced by a spring. The valve is opened by turning of electric current. To facilitate the working conditions of the electric magnet, valves with large sealing pressure are equipped with a hand-operated drive and an electromagnetic release.

To increase the sealing pressure and obtain more convenient configuration of valves, hinge mechanisms are used (Fig. 8.28, *a*), which, at the moment of valve closing, are close

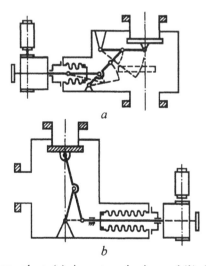

Figure 8.28 Drive of vacuum valves: (*a*) six-part mechanism and (*b*) ellipsographic.

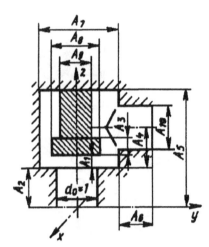

Figure 8.29 Scheme of angle valve: A_1–A_{10} are relative dimensions.

to extremal position, in which the greatest amplification of transmitted force is obtained. The ellipsographic mechanism (Fig. 8.28, b) has additional advantage, because at the point of sealing the component of the locking force perpendicular to the motion of the sealing element is absent. This prevents the occurrence of friction during sealing and allows one to use complex shapes of sealing elements.

To reduce the overall system size, vacuum valves for large diameter pipelines are possibly made flat using the parallelogram mechanism with a short crankshaft.

When designing vacuum valves, one should calculate their conductance in molecular mode of gas flow. For analytical calculation of valve conductance one should construct its equivalent scheme from of simple elements, for which analytical dependences are available. These simple elements are considered earlier the long or short pipes, holes, concentric pipes, etc. Construction of equivalent circuits faces the difficulty during the determination of inlet apperture resistance.

The analytical formulas for conductance of holes, are true for their connection to infinite or semi-infinite volumes. For arbitrarily located surfaces before the inlet apperture, analytical expressions are absent. For determining apperture conductance in this case one should use the formula for the connection of two pipelines with various crosssections.

One may estimate the error of this approximation. It is well-known that the maximum error of pipeline calculation without account of inlet resistance (12%) occurs at pipeline length $l = 1.2d$, where d is the pipeline diameter. For example, an angle valve has not less than two inlet resistances connected to a limited volume; thus, the error may be 24% of the total valve conductance. The accuracy of calculation can be increased by using the statistical method.

By way of example, we consider calculation of the angle vacuum valve, the scheme of which is represented in Fig. 8.29. We construct the equivalent scheme of this valve. The difficulties of constructing the scheme (Fig. 8.30) arise in sections B and G for direction of flow change from axis to right. It is impossible to analytically take into account the

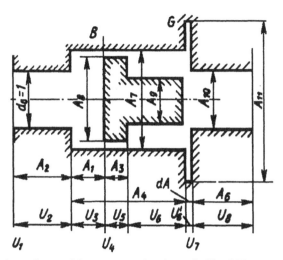

Figure 8.30 Equivalent scheme of the vacuum valve shown in Fig. 8.29.

influence of the A_7/d_0 ratio and the size A_1 on the conductance of the inlet apperture in section B; in section G the size A_{11} is preset rather arbitrarily, the conductance U'_6 of a portion with length $dA \to 0$ being equal to infinity.

The equation of the conductance of this equivalent scheme, which is considered as a series of sequentially connected elements, can be written in the following form:

$$\frac{1}{U_\Sigma} = \frac{1}{U_1} + \frac{1}{U_2} + \dots + \frac{1}{U_8} \tag{8.18}$$

Considering that the calculation will be performed for air at room temperature, the conductance of the long tube and the apperture in molecular gas flow mode, we obtain

$$\frac{1}{U_\Sigma} = \frac{1}{91d_0^2} + \frac{A_2}{121d_0^3} + \frac{A_1}{121A_7^3} + \frac{A_8^2}{91A_7^2(A_7^2 - A_8^2)} + \frac{A_3A_7}{170A_8[(A_7 - A_8)^2(A_7 + A_8)]}$$

$$+ \frac{A_7[A_4 - (A_1 + A_3)]}{170A_9[(A_7 - A_9)^2(A_7 + A_9)]} + \frac{A_{11}^2}{91A_{10}^2(A_{11}^2 - A_{10}^2)} + \frac{A_6}{121A_{10}^3}. \tag{8.19}$$

In Eq. (8.19), U_Σ is the total conductance of the valve, m^3/s, and the sizes A_1–A_{11} and d are expressed in meters. We perform several transformations in Eq. (8.19), taking into account the following conditions and restrictions:

1. we choose A_{11} from a condition $\pi A_7 A_{10} = \pi A_{11}^2/4$, whence $A_{11} = 2\sqrt{A_7 A_{10}}$ and
2. at negative lengths of portions, their conductance is taken to be infinity.

Thus, Eq. (8.19) is transformed to the following:

$$\frac{1}{U_\Sigma} = \frac{1}{91d_0^2} + \frac{A_2}{121d_0^3} + \frac{A_1}{121A_7^3} + \frac{A_8^2}{91A_7^2(A_7^2-A_8^2)} + \frac{A_7Bf(A_4-A_1)}{170[(A_7-A_8)^2(A_7+A_8)]A_8}$$

$$+ \frac{A_7[A_4-(A_1+A_3)]f(A_4-A_1-A_3)}{170[(A_7-A_9)^2(A_7+A_9)]A_9} + \frac{A_7}{22.8A_{10}^2(4A_7-A_{10})} + \frac{A_6}{121A_{10}^3}. \quad (8.20)$$

For convenience of determining valve conductance at various sealing element positions, location of the lateral pipe, etc., when separate portions of the equivalent scheme may be absent, two functions are introduced into Eq. (8.20):

$$B = \begin{cases} A_3, \text{ if } (A_4-A_1) \geq A_3; \\ A_4-A_1, \text{ if } (A_4-A_1) < A_3; \end{cases}$$

$$f(x) = \begin{cases} 1, \text{ if } \text{sign}(x) = 1; \\ 0, \text{ if } \text{sign}(x) = -1; \end{cases}$$

We compare calculations using Eqs. (8.19) and (8.20) with those performed using mathematical modeling by the method of statistical tests from the probability of passage of molecules, entering the apperture of the bottom entrance flange, through the valve: $P_{1\rightarrow 2} = U_\Sigma/(91d_0^2)$. The results show that for valve stroke $A_1 > 0.245$ the relative error of analytical calculations does not exceed 25%.

8.8 Test Questions

8.1 Which are the special requirements to materials of vacuum systems?

8.2 Which seams of metal with glass are referred to as coordinated?

8.3 Which is the way to reduce gas evolution from metallic vacuum systems?

8.4 In which cases is it expedient to use oxygen-free copper grades?

8.5 For which purposes Kovar alloy is used in vacuum engineering?

8.6 Which type of detachable vacuum connections provides higher accuracy of the axial dimensions in assembly?

8.7 In which case it is expedient to use double vacuum sealing with intermediate pumping?

8.8 Which type of transmitting rotation movement with small twisting moment and large rotation frequency is expedient for heated vacuum systems?

8.9 Which type of load is especially dangerous for bellows connections?

8.10 Why is it necessary to reduce the number of intermediate contacts in thermocouple connections?

Chapter 9

Problems

9.1 Properties of Gases at Low Temperatures

1.1 How will the pressure of gas change if one half of nitrogen molecules is replaced by hydrogen at constant volume and temperature?

1.2 The volume of gas increaess by 2 times due to heating by 10 K. In which range of temperatures is the gas heated?

1.3 By what factor will the mean free path of nitrogen molecules change if temperature increases from 700 to 350 K?

1.4 The pressure of hydrogen in a 10 l volume at 17°C is 10^{-2} Pa. After heating the volume to 7°C, the pressure reached the same value in 1 h. Find the gas flow into the volume.

1.5 Find the pressure in a 100 l vacuum chamber that contains 10^5 nitrogen molecules at 298 K ($m = 4.65 \times 10^{-2\,6}$kg).

1.6 A pressure of 10^{-9} Pa is obtained in a 1 m³ vacuum chamber at 293 K. Find the number of gas molecules that remained in the chamber.

1.7 Find the pumping speed of nitrogen by a vacuum pump with an inlet hole diameter of 0.01 m at 300 K, given the molecule capturing coefficient in the inlet hole is 0.5.

1.8 By what factor will the adsorption time of water molecules on stainless steel change due to a change in temperature from 300 to 77 K, given the adsorption heat is 80 MJ/kmole?

1.9 What pressure of nitrogen establishes in a 4 l volume one half of which is filled with 1 kg CaA zeolite, due to cooling from 293 to 77 K, given the adsorbability coefficients of CaA for nitrogen are 10^3 for $T = 298$ K and 10^6 for $T = 77$ K?

1.10 Find the diameter of a spherical vacuum chamber the pressure of water vapor in which increases twofold due to an increase in temperature from 300 to 500 K, given the probability of condensation at the free surface is 1, the Henry law is valid for adsorption, and the adsorption heat is 80 MJ/kmole.

1.11 Calculate the time of establishing adsorption equilibrium in a 1 l volume after rapid heating of a 0.1 mm diam 10 mm long wire, given the adsorption heat is 12 MJ/kmole.

1.12 Calculate the room temperature regeneration time of a nitrogen trap with a cooled surface area of 10^4 cm^2 and a volume of 10 l after blowing by residual gases with an adsorption heat of 12 MJ/kmole, given the pump efficiency is 10^{-4} (m^3 Pa)/s and the initial pressure is 10^{-2} Pa.

1.13 How will the pressure of nitrogen change in a 1 l vacuum tight chamber after heating to 100 K, given the initial pressure is 10^{-6} Pa, if the initial temperature is 300 K, and the adsorption heat is 12 MJ/kmole?

1.14 What is the surface coverage degree of stainless steel by water molecules at 10^{-3} Pa and 298 K, given the adsorption heat is 80 MJ/kmole?

1.15 What is the minimum pressure that can be obtained by degassing a spherical chamber 1 m in diameter by a perfect pump with a pumping speed of 1 m^3/s, given the specific gas evolution is 10^{-6} (m^3 Pa)/(m^2 s)?

1.16 What is the minimum thickness of a stainless steel vacuum chamber wall that is required for the gas evolution and gas penetration flows through it be equal?

1.17 What is the scatter of the thickness of a coating with an area of 0.01 m^2 placed symmetrically at a distance of 0.25 m over a point evaporation source?

9.2   Theory of Vacuum Engineering

2.1 Determine the degree of vacuum in a spherical chamber 1 m in diameter filled with air at $T = 293$ K and a pressure of 10^{-4} Pa.

2.2 Find the pressure of nitrogen that corresponds to the transition from high to medium vacuum in a vacuum chamber formed by two concentric cylinders 1 m in length and diameters of 1 and 0.5 m at 293 K.

2.3 By what factor will differ the pressures of transition between different degrees of vacuum for a spherical chamber 1 m in diameter and a cubic chamber with a side length of 1 m?

2.4 Calculate the friction force at low vacuum for a plate with an area of 2 m^2 that moves with a speed of 100 m/s in hydrogen at a distance of 0.1 m from a stationary flat surface.

2.5 What amount of liquid nitrogen evaporates from an oil vapor trap with a 100 cm^2 surface made of stainless steel, given the evaporation heat of liquid nitrogen is 200 J/g, the residual gas at 10^{-2} Pa is nitrogen, and the trap wall temperature is 300 K?

2.6 Find the ratio of pressures indicated by manometers installed on a vacuum furnace with a temperature of 1273 K and on a pipeline with a temperature of 293 K in high vacuum and the absence of gas flows.

2.7 Find the ratio of pressures in two pipelines at room temperature, between which a trap with a temperature of 77 K is placed.

2.8 A vacuum system with nitrogen at 10 Pa consists of three volumes connected by pipelines with diameters of 0.4 and 0.01 m. The temperatures of the volumes are 293, 500, and 1000 K. Determine the ratio of pressures that establish in the volumes in the absence of gas flows.

2.9 Two volumes with temperatures of 300 and 600 K and pressures of 10^{-4} and 10^{-6} Pa are connected under the conditions of high vacuum. Find the pressures in the volumes after their connection, given the volumes are 1 and 2 l, respectively.

2.10 A vacuum device with a volume of 0.01 m³ without pumping means is to be stored for one year (3.5×10⁷ s) in a hydrogen atmosphere at a pressure difference of 2×10⁵ Pa. The admissible increase in pressure is 10⁻² Pa. What are the requirements to the sensitivity of the leak detector with helium as a test gas at a pressure difference of 10⁵ Pa?

2.11 Find the flow of air through a hole 0.2 m in diameter the pressure to one side of which is 10 Pa and to the other side 8 Pa, the temperature being 298 K?

2.12 By what factor will the gas flow through a pipeline 0.1 m in diameter and 1 m in length change after changing the mean pressure in the pipeline from 1 to 100 Pa at 500 K?

2.13 Find the conductance of a 1 m long round conical pipeline for air, given the minimum pipeline diameter is 0.1 m and the maxima diameter is 0.3 m, $T = 300$ K, and the gas flow mode is molecular.

2.14 Determine the required length of a pipe 0.001 m in diameter one end of which is in contact with air and the second is connected to a pump with an efficiency of 5×10³ (m³ Pa)/s?

2.15 To which pressure will correspond the minimum breakdown voltage of a 30 mm gap at $T = 300$ K for $M = 28$ kg/kmole?

2.16 Find the distance of the outlet slot from the inlet slot of a mass-spectrometer in order for nitrogen ions to get into it in crossed electric and magnetic fields, given the voltage before the inlet slot is 1000 V, the magnetic field induction is 0.1 T, and the electron charge is 1.6×10⁻¹⁹ C?

2.17 To which temperature should an electron-emitting surface be heated in order for the electrons to have the same energy as due to acceleration in a 1 V electric field?

2.18 Find the number of nitrogen ions that are formed by an electron in a 20 mm gap with a linear potential distribution from 0 to 40 V at a pressure of 10⁻² Pa, given the ionization efficiency is 0.5 Pa⁻¹ m⁻¹ and the ionization potential is 14.5 V.

2.19 Find the path of an α-particle with an initial energy of 4.75×10⁶ eV in nitrogen at a pressure of 10⁵ Pa and a temperature of 300 K until the particle loses excess energy, given the ionization potential is 14.5 eV.

2.20 Determine the pressure of nitrogen, at which the free path length for cyclotrone resonance is equal the ion trajectory until incidence upon the collector at a distance of 20 mm from the axis of the ionizing radiation, the induction being $B = 0.1$ T.

2.21 Find the distance between an oxygen ion and a nitrogen one after passing a distance of 1 m, the initial energies being 100 eV.

2.22 A flat capacitor is in hydrogen with a pressure of 10² Pa. Determine whether hydrogen atoms will be ionized by electrons in the capacitor, given the voltage is 400 V, the distance between the capacitor plates is 25 mm, and the hydrogen ionization potential is 13.54 eV.

2.23 Which temperature of a tungsten cathode is required for electrons having the mean kinetic energy corresponding to this temperature to overcome a work function of 4.54 V?

2.24 A cathode in the form of a filament with a diameter of 0.16 mm and a length of 50 mm emits 1.5×10¹⁷ electrons per second from one squared cemtimeter. Find the current of positive ions at a nitrogen pressure 10⁻² Pa.

9.3   Calculation of Vacuum Systems

3.1 Determine the conductance of a vacuum system between the pump ($S = 2$ m^3/s and $p_{lim} = 10^{-5}$ Pa) and the vacuum chamber ($d = 1$ m and $l = 0.5$ m) at a specific gas evolution of the chamber material of 10^{-4} (m^3 Pa)/(m^2 s) that is required to obtain a pressure of 2×10^{-4} Pa in the vacuum chamber.

3.2 Determine the steady state pressure established in a vacuum chamber with a diameter of 1 m and a length of 2 m, given the specific gas evolution of the chamber material is 10^{-4} (m^3 Pa)/(m^2 s) and the chamber is connected to a pump ($S = 0.1$ m^3/s and $p_{lim} = 10^{-6}$ Pa) through a pipeline 160 mm in diameter and 0.2 m in length.

3.3 Determine the required speed of a pump connected to a pipeline 0.1 m in diameter and 1 m in length in order for the effective pumping speed for the vacuum chamber at $T = 293$ K to be 50 l/s.

3.4 Calculate the pressures in the beginning and the end of a pipeline with a radius of 10 mm and a length of 2 m pumped by a vacuum pump with $S = 0.1$ m^3/s and $p_{lim} = 10^{-5}$ Pa, given the inlet gas flow is 10^{-5} (m^3 Pa)/s and the specific gas evolution is 10^{-4} (m^3 Pa)/(m^2 s).

3.5 Determine the pressure in the middle of a round pipeline 5 mm in diameter and 1 m in length pumped from one side by a vacuum pump $S = 0.1$ m^3/s and $p_{lim} = 10^{-5}$ Pa, given the inlet gas flow is 10^{-5} (m^3 Pa)/s, and the specific gas evolution of the pipeline material is 10^{-4} (m^3 Pa)/(m^2 s).

3.6 A pumped chamber 500 mm in diameter is connected through a pipeline 200 mm in diameter to a pump with a speed of 0.1 m^3/s and a limiting pressure of 10^{-5} Pa. Find the pressure establishing in the pumped chamber, given the air evolution flow is 10^{-5} (m^3 Pa)/s at 298 K and the pipeline length is 500 mm.

3.7 Find the working time of a 300 l ballast cylinder at a gas flow of 10^{-3} (m^3 Pa)/s from the high-vacuum pump with a maximum outlet pressure of 10 Pa, if a roughing pump with a speed of 2 l/s and a limiting pressure of 10^{-1} Pa is connected to the ballast cylinder through a pipeline 32 mm in diameter and 1 m in length.

3.8 Calculate the time of pumping a vacuum object in the form of a pipe 0.1 m in diameter and 0.5 m in length required to reduce the maximum pressure from 10^{-3} to 10^{-5} Pa. The pumping is performed by a pump with a speed of 10^3 l/s and a limiting pressure of 10^{-7} Pa.

3.9 Determine the probability of gas molecules to pass through a pipeline 0.1 m in diameter and 0.5 m in length in the middle of which a diaphragm 0.05 m in diameter is installed. The gas flow mode is molecular.

3.10 Determine the time of pumping a 300 l vacuum chamber from atmospheric pressure to 0.1 Pa through a pipeline 0.05 m in diameter and 2 m in length by a pump with a speed of 5 l/s and a limiting pressure of 10^{-2} Pa.

Reference Tables

Table C.1 Relation of pressure units.

Pressure units	1 Pa	1 mm Hg	1 dyn/cm^2	1 phys/at.	1 kgf/cm^2	1cal/m^3
1 Pa	1	7.50×10^{-3}	10	9.87×10^{-6}	1.02×10^{-5}	2.39×10^{-1}
1 mm Hg	1.33×10^2	1	1.33×10^3	1.32×10^{-3}	1.36×10^{-3}	3.18×10^2
1 dyn/cm^2	1.00×10^{-1}	7.50×10^{-4}	1	9.87×10^{-7}	1.02×10^{-6}	2.39×10^{-2}
1 phys. at.	1.01×10^5	7.60×10^2	1.01×10^6	1	1.03×10^0	2.39×10^4
1 kgf/cm^2	9.81×10^4	7.36×10^2	9.81×10^5	9.68×10^{-1}	1	2.34×10^4
1 cal/m^3	4.19×10^0	3.14×10^{-2}	4.19×10^1	4.13×10^{-5}	4.27×10^5	1

Table C.2 Composition of dry atmospheric air.

Gas	Content, %	Partial pressure, Pa (the total is 10^5 Pa)	Gas	Content, %	Partial pressure, Pa (the total is 10^5 Pa)
N_2	78.1	7.8×10^4	CH_4	2.0×10^{-4}	2.0×10^{-1}
O_2	21	2.1×10^4	Kr	1.0×10^{-4}	1.0×10^{-1}
Ar	0.9	9×10^2	N_2O	5.0×10^{-5}	5.0×10^{-2}
CO_2	0.03	3×10^1	H_2	5.0×10^{-5}	5.0×10^{-2}
Ne	1.8×10^{-3}	1.8×10^0	Xe	9.0×10^{-6}	9.0×10^{-3}
He	5.2×10^{-4}	5.2×10^{-1}			

Table C.3 Arithmetical mean velocities of molecules of some gases at different temperatures.

Gas	v_{am}, m/s				Gas	v_{am}, m/s			
	4.2 K	77 K	293 K	600 K		4.2 K	77 K	293 K	600 K
N_2	56	245	470	672	Xe	26	110	220	311
Ar	47	200	395	563	H_2O	70	300	590	839
O_2	53	225	440	629	Air	54	235	460	661
CO_2	45	195	375	536	CO	56	245	470	672
Ne	67	285	555	792	He	150	640	1250	1785
Kr	33	140	270	325	CH_4	75	320	625	889
H_2	210	906	1770	2524					

Table C.4 Molecular characteristics of some gases.

Gas	$m \times 10^{26}$, kg	$d_m \times 10^{10}$, m	C, K	$d_T \times 10^{10}$, m at T, K			
				600	293	77	4.2
N_2	4.65	2.74	116	2.99	3.24	4.34	14.7
O_2	5.31	3.01	125	3.31	3.60	4.88	16.7
Ar	6.63	3.00	142	3.34	3.66	5.06	17.7
CO_2	7.31	3.36	254	4.01	4.59	6.97	26.3
Ne	3.35	2.35	56	2.46	2.56	3.09	8.90
Kr	19.9	3.17	188	3.63	4.06	5.88	21.4
H_2	0.33	2.41	84	2.57	2.73	3.48	11.0
Xe	21.8	3.53	252	4.21	4.81	7.30	27.6
H_2O	2.99	2.53	659	3.66	4.56	7.82	31.8
Air	4.81	3.13	112	3.41	3.68	4.90	16.5
He	0.66	1.94	80	2.07	2.19	2.77	8.69

Note: Molecular diameters d_m are determined from gas viscosities.

Table C.5 Effects of interaction of two similar molecules.

Gas	Inductional $\times 10^{19}$		Orientational $\times 10^{19}$		Dispersional $\times 10^{19}$		Total $\times 10^{19}$
	$J \times m^6$	%	$J \times m^6$	%	$J \times m^6$	%	$J \times m^6$
H_2O	10	4	190	77	47.0	19.0	247
CO	0.06	0.1	0.003	0.005	65.7	99.9	67.6
NH_3	10	5.35	84	45	93.0	49.6	187
HJ	1.68	0.44	0.4	0.1	382	99.5	384
N_2	–	–	–	–	57.2	100	57.2
O_2	–	–	–	–	39.8	100	39.8
H_2	–	–	–	–	11.4	100	11.4
He	–	–	–	–	1.49	100	1.49
Ne	–	–	–	–	7.97	100	7.97
Ar	–	–	–	–	69.5	100	69.5

Table C.6 Physical adsorption heats Q_a and condensation heats E of some gases on carbon.

Gas	Q_a, MJ/kmol	E, MJ/kmol
Vacuum oil vapors	90–100	96
H_2O	92	45.3
CO_2	28–33	25.3
Air	12–20	5.7
N_2	11–19	5.6
O_2	13–20	6.8
CH_4	18	9.2
Ar	14.7	6.5
CO	12.6	6.3
H_2	9	0,8
Ne	4.2	1.8
He	2	0.008

Table C.7 Chemisorption heats Q_{ch} of some substances on carbon.

Substance	C	H	N	O	S	F	Cl	Br	J
Q_{ch}, MJ/kmol	344	415	292	350	259	441	238	276	240

Table C.8 Critical parameters of some gases.

Gas	H_2O	Kr	CO_2	N_2	O_2	Ar	H_2	Ne
T_{cr}, K	647	209	304	126	155	151	33.2	44.4
T_b, K	273	116	217	63.2	54.4	83.8	13.9	24.5
p_T, Pa	560	7.7×10^4	4.5×10^5	1.2×10^4	146	1.5×10^4	7.2×10^3	4.3×10^4

Table C.9 Saturation pressures of vacuum materials at various temperatures, Pa.

Material	Temperature	
	20°C	500°C
Transformer oil	10^{-1}	–
Mercury	10^{-1}	10^6
VM-4 vacuum oil	5×10^{-3}	–
Ramsay paste	10^{-3}	–
VM-1 vacuum oil	5×10^{-6}	–
Piceine	10^{-6}	–
Zinc	10^{-12}	10^2
Indium	10^{-18}	10^{-6}
Silver	–	10^{-7}
Tin	–	10^{-8}
Aluminum	–	10^{-8}
Copper	–	10^{-9}
Gold	–	10^{-9}

Table C.10 Evaporation rates of some substances at 1.33 Pa.

Substance	Cu	Ae	Zn	Ni	Fe	Cr
T, K	1545	1480	615	1670	1740	1665
Evaporation rate, kg/(m^2 s)	1.2×10^{-3}	7.9×10^{-4}	1.9×10^{-3}	1.1×10^{-3}	1.1×10^{-3}	10^{-3}

Table C.11 Solubility of gases in solids.

Gas	Absorbent	n	S_o, $(m^3\,Pa)/(Pa^{1/n}\,m^3)$	Q_S, MJ/kmol	S, $(m^3\,Pa)/m^3$ $(p = 10^5\,Pa)$ 293 K	673 K
H_2	Fe	2	1×10^3	56	3×10^0	2×10^3
	Ni	2	8×10^2	25	1×10^3	3×10^4
	1Kh18N9T stainless steel	2	5×10^2	20	3×10^3	3×10^4
	Cu	2	2×10^3	77	1×10^{-1}	6×10^2
	Mo	2	3×10^2	59	5×10^{-1}	5×10^2
	Al	2	Almost insoluble			
	Pd	2	2×10^3	10	8×10^4	3×10^5
	Pt	2	1×10^4	145	4×10^{-7}	7×10^0
	Ag	2	2×10^2	50	2×10^0	4×10^2
	Resin	1	–	–	4×10^3	–
	Fluoroplastic	1	–	–	3×10^3	–
N_2	Cu	2	Almost insoluble			
	Ag	2	Almost insoluble			
	Mo	2	2×10^4	160	3×10^{-8}	4×10^0
	W	2	2×10^4	310	1×10^{-21}	6×10^{-6}
	Resin	1	–	–	5×10^3	–
	Fluoroplastic	1	–	–	3×10^3	–
O_2	Fe	2	2×10^3	18	2×10^4	1×10^5
	Cu	2	1×10^3	34	3×10^2	2×10^4
	Resin	1	–	–	1×10^4	–
	Fluoroplastic	1	–	–	3×10^3	–

Note: S are calculated using Eqs. (1.64) and (1.66).

Table C.12 Diffusion coefficients of gases in solids.

Gas	Absorbent	n	D_o, m²/s	Q_D, MJ/kmol	D, m²/s 293 K	D, m²/s 673 K
	Fe	2	1×10^{-6}	8	2×10^{-7}	5×10^{-7}
	Kh18N9T stainless steel	2	1×10^{-7}	49	5×10^{-12}	1×10^{-9}
	Ni	2	2×10^{-7}	73	6×10^{-14}	3×10^{-10}
H_2	Cu	2	1.1×10^{-6}	153	2×10^{-20}	1×10^{-12}
	Mo	2	1×10^{-7}	174	3×10^{-21}	2×10^{-14}
	Pd	2	6.6×10^{-5}	65	1×10^{-10}	2×10^{-7}
	SiO_2	1	6×10^{-10}	36	2×10^{-16}	1×10^{-11}
N_2	Fe	2	1.1×10^{-5}	285	4×10^{-31}	1×10^{-16}
	Kh18N9T stainless steel	2	1.0×10^{-5}	566	3×10^{-56}	1×10^{-27}
	Fe	2	4×10^{-12}	84	1×10^{-19}	2×10^{-15}
O_2	Ni	2	2×10^{-9}	679	6×10^{-70}	8×10^{-36}
	Cu	2	7.8×10^{-6}	238	4×10^{-27}	4×10^{-15}
	Ti	2	5×10^{-8}	218	2×10^{-27}	2×10^{-16}
	Fe	2	1.3×10^{-5}	163	4×10^{-20}	6×10^{-12}
CO	Kh18N9T stainless steel	2	1×10^{-5}	566	3×10^{-56}	1×10^{-27}
	Ni	2	5.4×10^{-7}	197	1×10^{-24}	1×10^{-14}
He	SiO_2	1	3×10^{-8}	23	2×10^{-12}	5×10^{-10}

Note: D are calculated using Eq. (1.67).

Table C.13 Ionization energies of gases.

Gas	N_2	He	Ne	Ar	CO	O_2	H_2
Ionization energy, eV	14.5	24.6	21.6	15.8	14.1	12.6	13.6

Table C.14 Dynamic viscosity coefficients.

Gas	H_2	He	CH_4	Ar	Ne	N_2	CO	O_2	CO_2	Air
$\eta\times10^5$, N s/m²	0.88	1.90	1.10	2.10	3.00	1.75	1.70	2.02	1.40	1.80

Table C.15 Specific heat capacities and heat conductivities of gases.

Gas	c_p, kJ/(kg K)	c_V, kJ/(kg K)	$\lambda_H \times 10^2$, W/(m K)	$q = K_{ta}/K_{tg}$
Air	1.01	0.72	2.4	1.00
N_2	1.03	0.73	2.4	1.06
O_2	0.92	0.66	2.4	1.06
CO_2	0.85	0.66	1.45	0.98
Ar	0.52	0.31	1.6	1.80
H_2O	1.95	1.47	1.6	0.63
He	5.36	3.13	14.2	0.57
Ne	1.05	0.68	4.65	1.27
H_2	13.8	10.2	16.8	0.27

Table C.16 Accommodation coefficients at 300 K.

Gas	W	Pt	Ni	SiO_2	Gas	W	Pt	Ni	SiO_2
Air	0.87	0.90	0.80	0.8	CO_2	0.85	0.80	0.8	0.80
N_2	0.87	0.80	0.80	0.8	Ne	0.78	0.70	0.8	0.70
O_2	0.85	0.83	0.85	0.8	He	0.40	0.50	0.4	0.38
Ar	0.85	0.89	0.95	0.9	H_2	0.35	0.25	0.3	0.37

Table C.17 Interdiffusion coefficients for air with different gases at 273 K and 10^5 Pa.

Gas	O_2	CO_2	CH_4	H_2O	H_2	C_2H_2
$D_o \times 10^4$, m^2/s	0.178	0.138	0.196	0.220	0.661	0.194

Table C.18 Formulas for calculating the conductances of apertures and pipelines for air at 293 K.

Element	Flow mode	
	viscous	molecular
Round orifice with diameter d, m	$U = 160d^2$ for $\dfrac{p_2}{p_1} \leq 0.1$	$U = 91d^2$
Arbitrary-shape orifice with area A, m^2	$U = 200A$ for $\dfrac{p_2}{p_1} \leq 0.1$	$U = 116A$
Pipeline with diameter d and length l, m	$U = 1.36 \times 10^3 \dfrac{d^4}{l} p_m$	$U = 121 \dfrac{d^3}{l}$
Rectangular section pipeline $a \geq b$ m	$U = 865f \dfrac{ab^3}{l} p_m$	$U = 308\varphi \dfrac{a^2 b^2}{l(a+b)}$
Pipeline with equal-side triangle section; a is the triangle side, m	$U = 299f \dfrac{a^4}{l} p_m$	$U = 48.1 \dfrac{d^3}{l}$
Elliptical pipeline; a is the long axis and b is the small axis, m	$U = 2.72 \times 10^3 \dfrac{a^3 b^3}{(a^2 + b^2)l} p_m$	$U = 171 \dfrac{a^2 b^2}{l\sqrt{a^2 + b^2}}$
Pipeline with diameter d_1 and coaxial rod with diameter d_2, m	$U = 1.36 \times 10^8 \dfrac{p_m}{l} \left[d_1^4 - d_2^4 - \dfrac{(d_1^2 - d_2^2)^2}{\ln(d_1/d_2)} \right]$	$U = 121 \dfrac{(d_1 - d_2)^2(d_1 + d_2)}{l}$

a/b	1	2	5	10	100	∞
f	2.3	3.7	4.7	5.0	5.3	5.3
φ	1.1	1.2	1.3	1.4	–	–

Note: U is in m^3/s; p in Pa; $p_m = (p_1 + p_2)/2$.

Table C.19 Specific gas evolution of structural materials at room temperature.

Material	Processing	Specific gas evolution one hour after degassing, q, m^3 Pa/(m^2 s)	Coefficients	
			A	B
Structural steel	Unprocessed	4×10^{-4}	-3.2	4.2×10^{-5}
	Chromium plating	1.3×10^{-7}	$-$	$-$
	Vacuum, 450°C, 15 h	$10^{-8} - 10^{-10}$	$-$	$-$
Stainless steel	Unprocessed	4×10^{-5}	-4.1	8.3×10^{-5}
	Vacuum, 450°C, 15 h	$10^{-8} - 10^{-10}$	$-$	$-$
Copper	Unprocessed	2×10^{-4}	-3.5	4.0×10^{-5}
	Vacuum, 450°C, 15 h	$10^{-8} - 10^{-10}$	$-$	$-$
Brass	without treatment	3×10^{-4}	-3.4	3.1×10^{-5}
Aluminum	Ditto	6×10^{-6}	$-$	$-$
Nickel	Ditto	7×10^{-6}	$-$	$-$
Vacuum rubber	Ditto	10^{-2}	$-$	$-$
Polyethylene	Ditto	10^{-4}	$-$	$-$
Fluoroplastic	Ditto	3×10^{-4}	$-$	$-$

Note: $\log q = A - Bt$, where t is time, s.

Table C.20 The effect of processing on gas evolution from stainless steel.

Processing	Gas evolution rate, m^3 Pa/(s m^3) Time after start of degassing, h			
	1	5	10	20
Unprocessed	4×10^{-5}	2×10^{-5}	4×10^{-6}	2×10^{-6}
Mechanical polishing	4×10^{-6}	$-$	2×10^{-7}	$-$
Ultrasonic cleaning	4×10^{-6}	2×10^{-6}	7×10^{-7}	4×10^{-7}
Glass beading	$-$	2×10^{-7}	1×10^{-7}	5×10^{-8}
Chemical cleaning	1×10^{-5}	4×10^{-6}	3×10^{-6}	2×10^{-6}
Chemical polishing	2×10^{-6}	8×10^{-7}	5×10^{-7}	3×10^{-7}
Electrolytical polishing	1×10^{-5}	2×10^{-6}	8×10^{-7}	3×10^{-7}
Vacuum degassing at 300°C for 2 h	1×10^{-6}	$-$	4×10^{-7}	2×10^{-7}
High-temperature vacuum degassing	10^{-9}	$-$	$-$	$-$
Oxidation in air at $T = 450$°C	7×10^{-10}	$-$	$-$	$-$
Aluminum plating	5×10^{-10}	$-$	$-$	$-$

Table C.21 Gas permeability of vacuum materials.

Gas	Material	n	K_0, (m³ Pa) m/(Pa$^{1/n}$ m²)	Q_p, MJ/kmol	q, (m³ Pa) m/(m² s) 293 K $\Delta p = 10^5$ Pa	673 K $\Delta p = 10^5$ Pa
	Fe	2	1×10^{-3}	64	6×10^{-7}	1×10^{-3}
	Ni	2	2×10^{-4}	98	5×10^{-9}	9×10^{-6}
	1Kh18N9T stainless steel	2	5×10^{-5}	69	2×10^{-8}	3×10^{-5}
	Cu	2	2×10^{-3}	230	2×10^{-21}	6×10^{-10}
H_2	Al	2	almost impermeable			
	Mo	2	3×10^{-5}	238	2×10^{-21}	1×10^{-11}
	Pd	2	1×10^{-1}	75	8×10^{-6}	6×10^{-2}
	Resin	1	–	–	1×10^{-6}	–
	Polyethylene	1	–	–	2×10^{-7}	–
	Fluoroplastic	1	–	–	1×10^{-9}	–
	Mo	2	8×10^{-3}	380	3×10^{-34}	4×10^{-15}
	Cu, Ag	2	almost impermeable			
N_2	Resin					
	7889	1	–	–	1×10^{-9}	–
	9024	1	–	–	1×10^{-10}	–
	Fluoroplastic	1	–	–	1×10^{-10}	–
	Fe	2	3×10^{-2}	181	8×10^{-16}	6×10^{-7}
O_2	Cu	2	8×10^{-3}	272	1×10^{-25}	8×10^{-11}
	Fluoroplastic	1	–	–	3×10^{-10}	–
CO	Fe	2	1×10^{-4}	156	4×10^{-16}	3×10^{-8}
	Resin	1	–	–	–	–
He	Quartz	1	–	–	–	7×10^{-9}
	Fluoroplastic	1	–	–	3×10^{-9}	–
H_2O	Polyethylene	1	–	–	1×10^{-7}	–
	Resin	1	–	–	1×10^{-6}	–

Note: q are calculated using Eq. (1.68).

Table C.22 Stainless steels used in vacuum engineering.

Steel grades	σ_l, MPa	σ_s, MPa	Composition, %							
			C	Si	Mn	Cr	Ni	Ti	Fe	N_2
12Kh18N10T	540	200	0.12	0.8	1.5	18	10	0.7	Balance	–
1Kh21N5T	600	350	0.14	0.8	0.8	21	5.5	0.8	»	–
Kh17G9TN4	700	350	0.12	0.8	9	17	4	–	»	0.2
N39KhTYu	750	–	0.05	0.6	0.8	15	35	0.9	»	–

Table C.23 Physical properties of glasses.

Glass Grade	$\alpha \times 10^7$ (at 20–200°C)	σ_p, MPa		Heat conductivity λ, W/(m K)	Softening temperature T_s, °C	Heat resistance t, °C
		Compression	Tension			
Quartz	5.5–6.1	1800	95.0	0.096	1500	–
C-37 (tungsten)	36–39	1100	84.0	0.072	796–816	185
C-47 (molybdenum)	46–48	1090	90.5	0.050	580–600	200
C-87 (lead)	86–90	850	50.0	0,038	480–500	100

Table C.24 Properties of vacuum rubbers.

Characteristics	Rubber trademarks				
	7889	9024	IRP-1015	IRP-1368	IRP-2043
Tension strength limit, MPa	17	10	9	3	10
Relative tension, %	550	350	400	150	–
Maximum working temperature, °C	70	70	70	250	250
Residual strain after 45% vertical compression at 70°C for 100 h, %	10	15	15	–	–
Swelling in vaseline oil at 70°C, %	90	6	4.5	18	5
Frost resistance (impact brittleness), °C	–50	–40	–30	–57	–30
Sealing pressure, MPa	1.6–1.8	2.0–2.2	2.0–2.2	0.4–0.7	–

Table C.25 Chemical compositions and melting points of soft solders for vacuum tight junctions.

Solder trademark	Chemical composition, %						Melting point, °C	
	Sn	Pb	Cu	Ag	Sb	Zn	onset	end
POS-30	30	Balance	0.15	–	2	–	183	256
POS-40	40	–	0.10	–	2	–	183	235
POS-61	61	–	0.10	–	0.9	–	183	183

Table C.26 Chemical compositions and melting points of hard solders for vacuum tight junctions.

Solder trademark	Chemical composition, %					Melting point, °C	
	Cu	Ag	Zn	Au	Si	onset	end
PSr-45	30	45	Balance	–	–	660	725
PSr-72	28	72	–	–	–	779	779
PZlM80	20	–	–	80	–	889	889
PMK 4	96	–	–	–	4	910	1000
MB copper	100	–	–	–	–	1083	1083

Table C.27 Gas concentration in vacuum materials.

Material	Gas	s, $(m^3\ Pa)/m^3$
12Kh18N10T Stainless steel	H_2	7×10^4
	N_2	4×10^4
	O_2	2×10^4
Steel 20	H_2	2×10^4
	N_2	3×10^4
	O_2	2×10^4
Iron	H_2	9×10^4
	N_2	3×10^4
	O_2	6×10^4
Vacuum melted sheet steel	H_2	0.4
	O_2	1×10^4

Appendix

A.1 Derivation of the Maxwell–Boltzmann Function (Velocity Distribution of Gas Molecules)

We will consider a deduction of the distribution function, Eq. (1.16), the existence of which is postulated in the vacuum physics. The number of molecules, the velocity of which are in the range from v_x to $v_x + dv_x$, is proportional to the total number of molecules n and velocity increment dv_x, and is determined by the distribution function. The similar relationships can be written for the y and z coordinate axes. Thus,

$$
\begin{aligned}
dn_{v_x} &= nf(v_x)dv_x, \\
dn_{v_y} &= nf(v_y)dv_y, \\
dn_{v_z} &= nf(v_z)dv_z
\end{aligned}
\tag{A1.1}
$$

where $f(v_x)$, $f(v_y)$, $f(v_z)$ – distribution functions.

The probability that the velocity vectors, the ends of which are inside the parallelepiped with the sides dv_x, dv_y, dv_z, with account of the independence of the coordinates, is determined from the probability theory as the product of the partial probabilities by the formula

$$
dn_v = nf(v_x)f(v_y)f(v_z)dv_xdv_ydv_z.
\tag{A1.2}
$$

The distribution function does not depend on direction and is determined only by the module of speed v, i.e.,

$$
f(v) = f(v_x)f(v_y)f(v_z).
\tag{A1.3}
$$

The following functions satisfy this equation :

$$
\begin{aligned}
f(v_x) &= A\exp(-Bv_x^2), \\
f(v_y) &= A\exp(-Bv_y^2), \\
f(v_z) &= A\exp(-Bv_z^2),
\end{aligned}
\tag{A1.4}
$$

that one may verify by substitution of Eq. (A1.4) into Eq. (A1.3):

$$f(v) = A^3 \exp(-B(v_x^2 + v_y^2 + v_z^2)) = A^3 \exp(-Bv^2). \tag{A1.5}$$

Therefore, Eq. (A1.1) with account of Eq. (A1.4) can be represented as

$$dn_{v_x} = An \exp(-Bv_x^2)dv_x,$$
$$dn_{v_y} = An \exp(-Bv_y^2)dv_y,$$
$$dn_{v_z} = An \exp(-Bv_z^2)dv_z, \tag{A1.6}$$

We rewrite Eq. (A1.2), using Eqs. (A1.3) and (A1.4):

$$dn_v = A^3 n \exp(-Bv^2)dv_x dv_y dv_z, \tag{A1.7}$$

Since the space of gas molecules is isotropic and the concentration of the particles that have the velocity v is identical in the entire space of velocities, we can transit from the Cartesian coordinates to the spherical ones, i.e.,

$$dv_x dv_y dv_z = 4\pi v^2 dv,$$

Then we have

$$dn_v = A^3 4\pi n \exp(-Bv^2)v^2 dv. \tag{A1.8}$$

To determine the constants A and B, we integrate Eq. (A.1.8) over the possible range of velocities:

$$n = \int_{v=0}^{v=\infty} dn_v = 4\pi nA^3 \int_0^\infty \exp(-Bv^2)v^2 dv. \tag{A1.9}$$

It is well-known that

$$\int_0^\infty \exp(-ax^2)x^{2k}dx = \frac{\sqrt{\pi} \cdot 1 \cdot 3 \cdot \ldots \cdot (2k-1)}{2^{(k+1)}a^{(k+1/2)}}. \tag{A1.10}$$

Calculating the integral in Eq. (A1.9) according to Eq. (A1.10), we have

$$A = \sqrt{\frac{B}{\pi}}. \tag{A1.11}$$

Equation (A1.11) establishes relationship between the above-mentioned constants. To determine their value, we use expression for root-mean-square speed from Eq. (1.11).

$$v_s = \sqrt{\frac{3kT}{m}} = \sqrt{\frac{1}{n}\int_0^\infty v^2 dn_v} . \tag{A1.12}$$

Substituting Eq. (A1.8) into Eq. (A1.12), we have

$$v_s = \sqrt{4\pi A^3 \int_0^\infty \exp(-Bv^2)v^4 dv} . \tag{A1.13}$$

We determine the integral in Eq. (A1.13) according to Eq. (A1.10):

$$\int_0^\infty \exp(-Bv^2)v^4 dv = \frac{3\sqrt{\pi}}{8B^{5/2}} . \tag{A1.14}$$

Then, taking into account Eqs. (A1.11) and (A1.13), we obtain

$$v_s = \sqrt{\frac{3}{2B}} = \sqrt{\frac{3kT}{m}} ,$$

whence

$$B = \frac{m}{2kT} \text{ and } A = \left(\frac{m}{2\pi kT}\right)^{1/2} .$$

Using the coefficients A and B, we rewrite the distribution function (A1.5) as

$$f(v) = \left(\frac{m}{2\pi kT}\right)^{3/2} \exp\left(-\frac{mv^2}{2kT}\right) .$$

Therefore, the number of molecules that have velocities in the range from v to v + dv, according to Eq. (A1.8), is

$$dn_v = \left(\frac{m}{2\pi kT}\right)^{3/2} 4\pi n \exp\left(-\frac{mv^2}{2kT}\right) v^2 dv ,$$

which coincides with Eq. (1.16).

A.2 Average Velocity

By definition, with account of Eq. (1.16), the average speed is

$$V_a = \frac{1}{n}\int_0^\infty v\,dn_v = \int_0^\infty 4\pi v^3\left(\frac{m}{2\pi kT}\right)^{3/2}\exp\left(-\frac{mv^2}{2kT}\right)dv = 4\pi\left(\frac{m}{2\pi kT}\right)^{3/2}\int_0^\infty v^3\exp\left(-\frac{mv^2}{2kT}\right)dv.$$

We denote $m/(2kT) = a$, then

$$V_a = 4\pi\left(\frac{a}{\pi}\right)^{3/2}\int_0^\infty v^3\exp(-av^2)dv.$$

The integral can be calculated integrating by parts:

$$\int_0^\infty v^3\exp(-av^2)dv = -\frac{1}{2a}\int_0^\infty v^2 d(\exp(-av^2)) = -\frac{1}{2a}\left[v^2 e^{-av^2}\big|_0^\infty - \int_0^\infty \exp(-av^2)d(v^2)\right].$$

$$= -\frac{1}{2a}\left[0 + \frac{1}{a}\exp(-av^2)\big|_0^\infty\right] = \frac{1}{2a^2}.$$

Then

$$V_a = 4\pi\left(\frac{a}{\pi}\right)^{3/2}\frac{1}{2a^2} = \sqrt{\frac{8kT}{\pi m}},$$

which coincides with Eq. (1.18).

A.3 The Equation of Polymolecular Adsorption

We will consider the derivation of Eq. (1.56). We write the condition of adsorption balance for multilayer adsorption as the equality of condensation speeds in each previous adsorption layer and evaporation in the subsequent layer:

$$\mu_i S_i = v_{i+1} S_{i+1}, \tag{A3.1}$$

where μ_i and v_i are the specific condensation and evaporation speeds; S_i and S_{i+1} are a free surfaces of the previous and subsequent layer.

    The specific condensation speed of molecules in the ith layer is

$$\mu_i = f_i N_q, \tag{A3.2}$$

where f_i is the probability of condensation of molecules in the ith layer, N_q is the number of molecules incident onto unit surface in unit time. Using Eqs. (1.35) and (1.18), we rewrite Eq. (A3.2) as

$$\mu_i = \frac{f_i p}{\sqrt{2\pi mkT}} . \tag{A3.3}$$

The specific evaporation speed of molecules is

$$\nu_i = \frac{a_i}{\tau_{ai}}, \tag{A3.4}$$

where a_i is the number of molecules adsorbed at unit surface and τ_{ai} is the time of molecules adsorption in the ith layer.

With account of the expression for adsorption time (1.42) we rewrite Eq. (A3.4) as

$$\nu_i = \frac{a_i}{\tau_{0i}} \exp\left(-\frac{Q_{ai}}{RT}\right) . \tag{A3.5}$$

According to the assumptions of the polymolecular adsorption theory,

$$\tau_0 = \tau_{01} = \tau_{02} = ... = \tau_{0i} = ... = \tau_{0N},$$
$$f = f_1 = f_2 = ... = f_i = ... = f_N,$$

then condition (A3.1) with account of Eqs. (A3.3) and (A3.5) can be represented as a set of equations:

$$gpS_0 = \exp\left[-\frac{Q_1}{RT}\right]S_1 ,$$

$$gpS_1 = \exp\left[-\frac{Q_1}{RT}\right]S_2$$

$$..............................$$

$$gpS_{i-1} = \exp\left[-\frac{Q_i}{RT}\right]S_i$$

$$.............................. \tag{A3.6}$$

Here p is the gas pressure, Q_i is the adsorption heat in the ith layer, $S_0, S_1, ..., S_i$ are the free surfaces of adsorbent in the layer appropriate to the index, and g is constant.

According to the above assumptions, $Q_1 = Q_a$; $Q_2 = Q_3 = ... = Q_i = ... = E$, where E is the condensation heat. With these simplifications, we write set (A3.6) as:

$$S_1 = yS_0 \, ; \, y = \frac{\ell}{g}\exp\left[-\frac{Q_1}{RT}\right] \, ;$$

$$S_2 = xS_1 \, ; \, x = \frac{\ell}{g}\exp\left[-\frac{E}{RT}\right] \, ;$$

$$S_3 = xS_2 = x^2 S_1 \, ;$$

$$\cdots\cdots\cdots\cdots\cdots\cdots\cdots\cdots$$

$$S_i = xS_{i-1} = Cx^i S_0 \, ;$$

$$\cdots\cdots\cdots\cdots\cdots\cdots\cdots\cdots \qquad\qquad \text{(A3.7)}$$

where

$$C = \frac{y}{x} = \exp\left[\frac{Q_a - E}{RT}\right].$$

The entire adsorbent surface is

$$A = \sum_{i=0}^{\infty} S_i \, ,$$

and the quantity of adsorbed matter is

$$G = a_m \sum_{i=0}^{\infty} iS_i \, .$$

The specific quantity of absorbed gas is determined as

$$a = \frac{G}{A} = \frac{a_m \sum_{i=0}^{\infty} iS_i}{\sum_{i=0}^{\infty} S_i} \, . \qquad\qquad \text{(A3.8)}$$

Taking into account that

$$\sum_{i=0}^{\infty} S_i = S_0\left(1 + C\sum_{i=0}^{\infty} x^i\right) = S_0\left(1 + \frac{Cx}{1-x}\right) \, ;$$

$$\sum_{i=0}^{\infty} iS_i = CS_0 \sum_{i=0}^{\infty} ix^i = CS_0 x\frac{d}{dx}\sum_{i=0}^{\infty} x^i = \frac{xCS_0}{(1-x)^2} \, ,$$

Eq. (A3.8) can be transformed to

$$\frac{a}{a_m} = \frac{Cx}{(1-x)(1-x+Cx)}.$$ (A3.9)

At pressure p that is equal to the saturation pressure p_T, an indefinitely great number of layers can be formed on a free surface.

$$\frac{p_T}{g}\exp\left(\frac{E}{RT}\right) = 1.$$

Then $x = p/p_T$. Substituting the resultant expression for x into Eq. (A3.9) we obtain the equation of polymolecular adsorption

$$\frac{a}{a_m} = \frac{C(p/p_T)}{(1-p/p_T)(1+(C-1)p/p_T)},$$

which coincides with Eq. (1.56).

A.4 The Equation of Gas Flow Through an Aperture in Viscous Flow Mode

At low vacuum and viscous mode of gas flow, the law of conservation of energy for adiabatic gas flow can be written as the equality of the increase in gas kinetic energy and the change of its enthalpy:

$$\frac{Gw_{r2}^2}{2} = G(I_1 - I_2),$$ (A4.1)

where G is the gas flow, w_{r2} is the velocity of gas flow at the outlet of the aperture, and I_1 and I_2 are the enthalpies of gas before and after the passage of the aperture. Using the condition $I = c_pT$, we rewrite Eq. (A 4.1) in the following form:

$$\frac{w_{r2}^2}{2} = c_pT_1\left(1 - \frac{T_2}{T_1}\right).$$ (A4.2)

Introducing the specific volume of gas v that is measured in m^3/kg, we write the equation of gas condition (1.12) as $pv = RT/M$. It is well-known from technical thermodynamics that for an adiabatic process

$$\frac{v_1}{v_2} = \left(\frac{p_2}{p_1}\right)^{1/\gamma}, \; \gamma = \frac{c_p}{c_v}, \; c_p - c_v = \frac{R}{M}.$$

We transform the equation of conservation of energy (A4.2) to the following form:

$$w_{r2} = \sqrt{\frac{2\gamma}{\gamma - 1}p_1 v_1 \left[1 - \left(\frac{p_2}{p_1}\right)^{(\gamma - 1)/\gamma}\right]}. \qquad (A4.3)$$

Taking into account Eq. (A4.3) for w_{r2}, gas flow (kg/s) through the aperture is

$$P = \frac{w_{r2}A}{v_2} = \psi A \sqrt{\frac{p_1}{v_1}}, \qquad (A4.4)$$

where A is the aperture cross-section area,

$$\psi = r^{1/\gamma}\sqrt{\frac{2\gamma}{\gamma - 1}[1 - r^{(\gamma - 1)/\gamma}]}, \; r = \frac{p_2}{p_1}. \qquad (A4.5)$$

It follows from the equation of gas condition that $v_1 = RT_1/(Mp_1)$. Then Eq. (A4.4) can be rewritten as

$$P = \psi A p_1 \sqrt{\frac{M}{RT_1}}.$$

In conventional units (m^3 Pa)/s, the gas flow

$$Q = \frac{PRT_1}{M} = \psi A p_1 \sqrt{\frac{RT_1}{M}},$$

that coincides with Eq. (2.37).

Table A.1 Characteristics of vacuum gauges.

Type	Trademark	Range of working pressures	Measurement accuracy, %	Dimension (length × width × height), mm	Weight, kg	Power consumption, W	Type of transducer
Deformation	VDG-1	1×10^{0}–1×10^{3}	±5	–	–	–	PMGD-1
Thermal	VTSO-1	3×10^{-1}–1×10^{4}	±10	80×150×295	–	–	–
	VSB-1	1×10^{0}–4×10^{3}	±30	390×248×257	14	140	MT-6, MT-6-3
	VT-2A	1×10^{-1}–3×10^{2}	±30	226×248×257	9	110	PMT-2, PMT-4M, MT-8
	VT-3	1×10^{-1}–7×10^{2}	±30	320×185×150	4.5	35	PMT-2, PMT-4M, MT-8
	VTB-1	1×10^{0}–4×10^{3}	±40	200×158×319	6	–	MT-6
		1×10^{0}–4×10^{3}	-40 – +60	100×158×206	2.5	–	MT-6
Magnetic	VIM-2	1×10^{-11}–1×10^{-2}	±70	386×278×292	22	120	PMM-14
	VMB-3	1×10^{-5}–3×10^{0}	±170	380×290×240	19	150	PMM-13
	VMB-6	1×10^{-5}–1×10^{-1}	-50 – +80	400×200×390	16	100	PMM-32
	VMB-8	1×10^{-8}–1×10^{-1}	-55 – +130	240×158×335	35	75	PMM-32
	VMB-10	1×10^{-4}–1×10^{0}	±90	–	–	–	PMM-38
	VMB-11	1×10^{-8}–1×10^{-1}	-50 – +100	–	–	–	PMM-46
	VMB-12	1×10^{-4}–1×10^{0}	-60 – +130	–	–	–	PMM-44
	VMTsB-12	1×10^{-6}–1×10^{-1}	-50 – +100	–	–	–	PMM-32-1
	VMB-14	1×10^{-7}–1×10^{0}	-50 – +110	–	–	–	PMM-32-1
Ionization	VIO-1	7×10^{-9}–1×10^{-1}	±30	228×95×328	26	–	PMI-39
	VIO-2	1×10^{-4}–1×10^{-1}	±13	–	–	–	MI-29
	VMTsB-11	2×10^{-3}–1×10^{2}	–	100×158×230	3	–	IM-12, PMI-10-2
	VI-12	1×10^{-4}–1×10^{0}	±50	448×340×287	30	280	PMI-12, PMI-12-8
	VI-14	1×10^{-8}–1×10^{1}	55	480×220×360	20	140	IM-12, MI-12-8, PMI-12-8
Combined	VIT-2	1×10^{-5}–1×10^{1}	30	320×280×215	10	75	PMT-2, PMT-4M LM-3-2, PMI-2
	VIT-3	1×10^{-5}–1×10^{2}	35	230×360×485	15	75	PMT-2, PMT-4M PMI-2, PMI-10, PMI-51
	VTM-2	1×10^{-7}–1×10^{5}	-40 – +85	–	–	–	PMT-6-3, PMM-32-1M

Table A.2 Characteristics of partial pressure gauges.

Types of partial pressure gauge	Trademark	Resolution	Mass range, a.u.	Working pressure range, Pa
Static	MSD-1	60	2–150	$1 \times 10^{-3} - 1 \times 10^{-8}$
	MKh-1304	100	1–120	$1 \times 10^{5} - 1 \times 10^{-7}$
	MKh-1306	800	2–900	$1 \times 10^{5} - 1 \times 10^{-7}$
Linear resonance	APDP-2	20	2–200	$1 \times 10^{-3} - 1 \times 10^{-7}$
(farvitron)	MKh7301	200	1–200	$1 \times 10^{-2} - 1 \times 10^{-8}$
Cyclotrone	MKh4301	25	1–100	$1 \times 10^{-3} - 3 \times 10^{-8}$
(omegatron)	IPDO-1	20	2–100	$1 \times 10^{-3} - 5 \times 10^{-8}$
	IPDO-2	35	1–250	$1 \times 10^{-3} - 7 \times 10^{-9}$
Radio-frequency (topatron)	MKh-6401	50	1–56	$1 \times 10^{-2} - 7 \times 10^{-7}$
Pass-time	RMS-2M	45	2–200	$1 \times 10^{-2} - 1 \times 10^{-6}$
(chronotron)	MSKH-3A	30	1–250	$1 \times 10^{-3} - 1 \times 10^{-7}$
Mass-filter	APDM-1	400	1–400	$7 \times 10^{-2} - 1 \times 10^{-11}$
(monopolar)	EFM-1	50	1–100	$1 \times 10^{-1} - 1 \times 10^{-7}$
	KM-2	300	2–300	$1 \times 10^{-3} - 1 \times 10^{-7}$
Mass-filter (quadrupolar)	MSKh-2M	100	1–250	$1 \times 10^{-3} - 1 \times 10^{-7}$
	MSKh-4A	100	1–600	$1 \times 10^{-3} - 1 \times 10^{-2}$

Table A.3 Characteristics of leak detectors.

Principle	Trademark	Minimum flow, m^3 Pa/s	Gauge type	Test gas	Weight, kg	Power consumption, W	Dimensions, mm
Spark	IO43.009	7×10^{-6}	Vacuum glass	Air	2	60	250×210×200
Cataro-metric	TP7101	2×10^{-6}	Atmospheric	CO_2, freon, helium	3.5	–	164×136×62
Halogen	BGTI-7	5×10^{-7}	Atmospheric	Freon	9	35	89×304×330
	GTI-6	1×10^{-7}	Atmospheric	Freon	10	75	360×160×220
		7×10^{-9}	Vacuum				
Electron capturing	13TE-9-001	7×10^{-10}	Atmospheric	Elegas, freon	4	20	295×255×125
Mass-spectro-scopic	PTI-7	7×10^{-12}	Vacuum	Helium	250	1500	600×780×1250
	PTI-10	5×10^{-13}	Vacuum	Helium	215	1100	1470×670×620
	TI1-15	7×10^{-14}	Vacuum	Helium	95	750	398×667×470

Table A.4 Working liquids of vacuum pumps.

Type	Trademark	Technical Standard	Vapor pressure at 293 K, Pa	Kinematic viscosity at 323K $\times 10^6$ m²/s	Density at 293 K, g/cm³	Molar mass kg/kmole	Evaporation heat, MJ/kmole	Oxidation stability[*]
Mercury	R1, R2	–	3×10^{-1}	–	13.6	200	–	E
Mineral	VM-1	OST38-01402-86	1×10^{-6}	70	0.89	450	120	P
	VM-4		5×10^{-3}	50	0.87			
	VM-5		2×10^{-7}	70	0.89			S
	VM-6		4×10^{-5}	40	0.87			P
	VM-7		4×10^{-6}	90	0.89			S
	Apiezon A	–	1×10^{-5}	–	0.89	350	90	P
	Apiezon B	–	1×10^{-6}	–		350	90	P
	Apiezon C	–	5×10^{-7}	–		574	120	P
Silicon-organic	PES-V-1	GOST16 480-70	1×10^{-5}	25	0.97	700	100	G
	PES-V-2	GOST16 480-70	1×10^{-4}		0.98		90	G
	Octoil	–	1×10^{-5}	75	1.06	391	120	G
	PFMS-2	TU6-02-777-73	1×10^{-4}	10	1.10	571	110	G
	PFMS-13	TU6-02-274-74	3×10^{-5}	20	1.10	480	100	G
	MFT-1	TU6-02-934-73	1×10^{-6}	36	1.10	544	130	G
	DC-704	–	1×10^{-6}	47	1.56	484	100	G
	DC-705	–	1×10^{-7}	170	1.58	546	120	G
	FM-1	TU6-02-758-73	5×10^{-9}	36	1.10	547	120	G
	FM-2	TU6-02-286-73	1×10^{-10}	60	1.10	689	140	G
Ethers	NeovacSY	–	1×10^{-6}	250	1.20	405	–	S
	MVD	–	1×10^{-6}	20	1.20	3000	130	S
	Santovac-5	–	1×10^{-7}	–	–	447	–	S
	5F4E	TU609-4626-78	1×10^{-9}	120	1.20	446	130	S
	N-PFE	TU609-06-822-76	1×10^{-10}	130	1.20	450	145	S
Hydro-carbon	Alkaren-35	TU6-01-28-54-85	5×10^{-4}	100	0.94	–	–	G
	RZhM-130	TU6-02-3-354-87	4×10^{-4}	30	1.90	–	–	G
	Alkaren-24	–	5×10^{-8}	55	0.90	420	130	G

[*]E, excellent; G, good; S, satisfactory; P, poor.

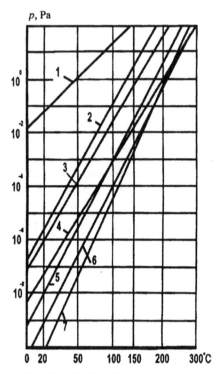

Figure A.4 Vapor pressures of working liquids of vacuum pumps. (*1*) mercury, (*2*) PFMS-2, (*3*) VM1, DC-704, (*4*) Santovac 5, (*5*) DC-705, (*6*) 5F4E, (*7*) FM-2.

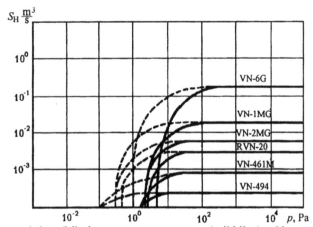

Figure A.5 Characteristics of displacement vacuum pumps (solid line), without a baffle; (dashed line), with a baffle.

Table A.5 Characteristics of mechanical vacuum pumps.

Main parameters	Piston-type				Vane-rotor			Multivane	
	VNK-0.5M	3VNP-3	2DVNP-6	VN-01	3VNR-1D	2VNR-5DM	RVN-6I	RVN-25	RVN-50
Working pressure range, Pa	1×10^4–1×10^5	2×10^3–1×10^5	2×10^2–1×10^5	3×10^0–1×10^5	4×10^{-1}–1×10^5	5×10^{-2}–1×10^5	2×10^4–1×10^5	2×10^4–1×10^5	2×10^4–1×10^5
Pumping speed, m³/s	0.0080	0.0630	0.1050	0.0001	0.0010	0.0050	0.1000	0.4000	0.8000
Limiting pressure: total with gas ballast, Pa	–	–	–	–	7×10^0	3×10^0	–	–	–
total without gas ballast, Pa	5×10^3	4×10^2	4×10^1	5×10^0	1×10^{-1}	7×10^{-1}	1×10^4	1×10^4	1×10^4
Partial for air, Pa	5×10^3	4×10^2	4×10^1	5×10^{-1}	7×10^{-2}	1×10^{-2}	1×10^4	1×10^4	1×10^4
Rotor speed, rpm	1500	750	1500	1400	2800	1430	1450	600	500
Oil charge, dm³	2.75	–	–	–	0.5	1.2	–	–	–
Cooling water consumption, dm³/s	–	0.04	0.05	–	–	–	0.08	0.18	0.36
Motor power, kW	3	5.5	11	0.12	0.25	0.55	15	55	75
Inlet hole diameter, mm	80	100	100	8	10	16	110	150	250
Dimension: length, mm	862	1430	1770	306	320	540	1500	2250	3000
width, mm	640	795	795	135	130	160	700	1000	1200
height, mm	1725	925	925	170	200	280	740	1100	1500
Weight, kg	540	750	1500	8.3	9.5	30	310	2250	4500

Table A.5. (cont.) Characteristics of mechanical vacuum pumps.

Main characteristics	Water-ring				Plunger-type			
	VVN1-0.75	VVN1-3	VVN1-12	VVN1-300	AVZ-20D	AVZ-90	AVZ-180	NVZ-500
Working pressure range, Pa	2×10^4–1×10^5	2×10^4–1×10^5	2×10^4–1×10^5	2×10^4–1×10^5	5×10^{-2}–1×10^5	4×10^0–1×10^5	4×10^0–1×10^5	5×10^0–1×10^5
Pumping speed, m³/s	0.0125	0.0500	0.2000	5.000	0.0200	0.0900	0.1850	0.5000
Limiting pressure: total with gas ballast, Pa	–	–	–	–	7×10^0	4×10^2	4×10^2	4×10^2
total without gas ballast, Pa	8×10^3	8×10^3	8×10^3	1×10^4	1×10^0	7×10^0	7×10^0	7×10^0
partial for air, Pa	8×10^3	8×10^3	8×10^3	1×10^4	1×10^{-2}	7×10^{-1}	7×10^{-1}	1×10^0
Rotor speed, rpm	1500	1500	1000	250	1500	1500	1500	1000
Oil charge, dm³	–	–	–	–	3.5	14	28	140
Cooling water consumption, dm³/s	0.05	0.13	0.38	12	–	0.20	0.40	2.00
Motor power, kW	2.2	7.5	22	630	2.2	11	15	55
Inlet hole diameter, mm	63	78	135	300	40	100	100	160
Dimension: length, mm	815	1195	1840	6600	650	1000	1070	2765
width, mm	332	385	710	3110	400	575	875	1760
height, mm	333	755	1220	2120	665	1035	1055	1355
Weight, kg	90	91	885	25000	185	600	880	4000

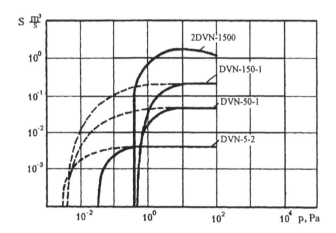

Figure A.6 Characteristics of two-rotor vacuum pumps (solid line), without a trap; (dashed line), with a trap.

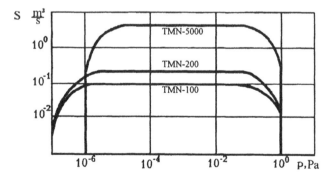

Figure A.7 Characteristics of turbomolecular pumps.

Table A.6 Characteristics of two-rotor vacuum pumps.

Characteristics	Trademarks							
	DVN-5-1	DVN-5-2	DVN-50-1	DVN-50-2	DVN-1 50-1	2DVN-500	2DVN-1500	DVN-5000
Working pressure range, Pa	$4{\times}10^0$–$2{\times}10^2$	$3{\times}10^{-2}$–$2{\times}10^1$	$4{\times}10^{-2}$–$2{\times}10^2$	$3{\times}10^{-1}$–$1{\times}10^2$	$4{\times}10^{-2}$–$1{\times}10^2$	$4{\times}10^0$–$2{\times}10^1$	$4{\times}10^0$–$2{\times}10^1$	$4{\times}10^0$–$2{\times}10^1$
Pumping speed, m³/s	0.005	0.007	0.045	0.050	0.135	0.500	1.500	5.000
Limiting pressure: total, Pa	$7{\times}10^{-1}$	$5{\times}10^{-2}$	$7{\times}10^{-1}$	$5{\times}10^{-2}$	$7{\times}10^{-1}$	$7{\times}10^{-1}$	$7{\times}10^{-1}$	$7{\times}10^{-1}$
partial for air, Pa	–	$5{\times}10^{-3}$	$7{\times}10^{-3}$	–	$7{\times}10^{-3}$	–	–	–
Maximum start-up pressure, Pa	$1{\times}10^3$	$1{\times}10^2$	$1{\times}10^3$	$5{\times}10^2$	$5{\times}10^2$	$1{\times}10^2$	$1{\times}10^2$	$1{\times}10^2$
Cooling water consumption, dm³/s	0.01	–	0.01	0.02	0.02	0.1	0.1	–
Motor power, kW	0.18	0.18	0.7	1.5	1.5	7.5	10	28
Dimensions:								
length, mm	458	261	415	645	623	1375	1835	2580
width, mm	158	186	240	360	240	600	580	890
height, mm	187	224	290	325	260	845	485	1145
Weight, kg	17	23	28	60	45	565	830	1900
Inlet pipe diameter, mm	40	40	85	85	100	175	250	–

Table A.7 Characteristics of turbomolecular pumps.

Main characteristics	Trademarks						
	TMN-100	TMN-200	TMN-500	TMN-1000	TMN-5000	TMN-10000	TMN-20000
Working pressure range, Pa	5×10^{-7}–1×10^{-2}	5×10^{-7}–1×10^{-2}	1×10^{-6}–1×10^{-2}	5×10^{-7}–1×10^{-2}	5×10^{-7}–1×10^{-2}	5×10^{-7}–1×10^{-2}	1×10^{-6}–1×10^{-2}
Air pumping speed, m³/s	0.10	0.25	0.50	1.00	6.30	11.0	18.0
Limiting residual pressure, Pa	10^{-7}	10^{-7}	8×10^{-7}	10^{-7}	10^{-7}	10^{-7}	2×10^{-6}
Maximum outlet pressure, Pa	10^{0}	10^{0}	10^{0}	10^{0}	10^{0}	10^{0}	10^{0}
Revolution speed, rps	300	300	300	400	100	100	83.3
Motor power, kW	0.3	0.3	0.8	0.25	2.0	7.0	7.0
Motor electric power source (trademark)	NVR-3	NVR-3	NV3	–	–	–	–
Cooling water consumption, dm³/s	0.01	0.01	0.03	–	0.06	–	–
Inlet pipe diameter, mm	125	160	260	–	500	–	–
Outlet pipe diameter, mm	32	50	50	–	100	–	–
Dimensions:							
length, mm	–	675	–	–	–	–	–
width, mm	–	310	–	–	–	–	–
height, mm	–	385	–	–	–	–	–
Weight, kg	110	205	210	190	1500	3500	3910

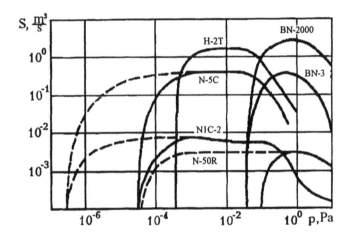

Figure A.8 Characteristics of vapor-jet pumps (solid line), without a trap; (dashed line), with a trap.

Table A.8 Characteristics of high-vacuum oil-vapor pumps.

Characteristics	Trademarks						
	N-0,025-2	NVD-0,15	NVDS-100	DFR-3000	DFR-8000	DFR-20000	DFR-50000
Working pressure range, Pa	10^{-4}–10	10^{-3}–10	10^{-4}–1	10^{-4}–1	10^{-4}–1	10^{-4}–1	10^{-4}–1
Pumping speed, m³/s	0.01	0.08	0.24	2.7	7.2	18	45
Limiting pressure, Pa	1×10^{-5}	3×10^{-4}	1×10^{-5}	3×10^{-5}	3×10^{-5}	3×10^{-5}	3×10^{-5}
Maximum outlet pressure, Pa	92	66	40	60	60	60	60
Heater power, kW	0.23	0.45	0.6	2.2	4.8	12	24
Water consumption, dm³/s	–	0.02	0.02	0.1	0.1	0.2	0.2
Working liquid charge, dm³	0.02	0.1	0.1	0.6	1.7	5.0	12
Inlet pipe diameter, mm	40	85	100	250	400	630	1000
Outlet pipe diameter, mm	14	20	25	50	63	100	160
Dimensions:							
length, mm	–	295	275	–	–	–	–
width, mm	–	200	180	–	–	–	–
height, mm	–	350	360	560	785	1130	1890
Weight, kg	–	9	8	23	69	173	420

Table A.9 Characteristics of booster oil-vapor pumps.

Characteristics	Trademarks			
	NVBM-0,5	NVBM-2,5	NVBM-5	NVBM-15
Working pressure range, Pa	5×10^{-3}–14	4×10^{-3}–40	4×10^{-3}–40	4×10^{-3}–40
Speed, m^3/s	0.80	2.60	5.70	17.5
Limiting pressure, Pa	1×10^{-3}	7×10^{-4}	7×10^{-4}	7×10^{-4}
Maximum inlet pressure, Pa	90	200	200	200
Inlet pipe diameter, mm	160	250	400	630
Outlet pipe diameter, mm	70	100	160	160
Oil loading, dm^3/s	7	23	57	120
Cooling water consumption, dm^3/s	0.04	0.1	0.2	0.5
Heater power, kW	2	6	12	36
Dimensions:	515	740	1150	1550
length, mm	465	590	910	1500
width, mm	1065	1710	2340	2700
height, mm				
Weight, kg	67	230	380	1550

Table A.10 Characteristics of cryoadsorption pumps.

Characteristics	Trademarks			
	TsVN-0.1-2	TsVN-0.3-2	TsVN-1-2	TsVN-1.5-3
Working pressure range, Pa	5×10^0–10^5	3×10^1–10^5	5×10^0–10^5	1×10^0–10^5
Pumping speed, m^3/s	0.002	0.004	0.006	0.010
Limiting pressure, Pa	1×10^0	7×10^0	1×10^2	3×10^0
Amount of nitrogen pumped, m^3 Pa	1×10^3	3×10^3	1×10^4	1×10^4
Initial liquid nitrogen consumption, dm^3	1.3	2	6	6
Established liquid nitrigen consumption, dm^3/h	0.25	0.35	0.25	0.5
Amount of adsorbent, kg	0.1	0.3	1	1.5
Regeneration time of CaA-4 zeolite at atmospheric pressure, h	3	2	3	3
Heater power, kW	0.4	0.14	0.83	0.35
Dimension:	192	–	124	–
length, mm	192	–	124	–
width, mm	390	–	390	–
height, mm				
Weight, kg	1.3	–	4.2	–

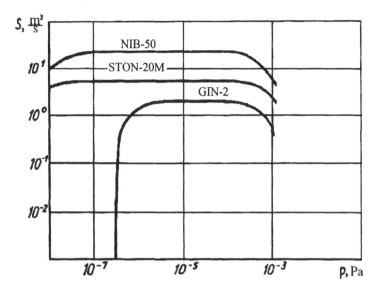

Figure A.11 Characteristics of gettering pumps.

Table A.11 Characteristics of getter-ion pumps.

Characteristics	Trademarks				
	EGIN-0.8/0.1	EGIN-1.0/0.1	EGIN-1.6/0.25	EGIN-5/1	EGIN-10/1.5
Working pressure range, Pa	4×10^{-7}–3×10^{0}	4×10^{-7}–3×10^{0}	4×10^{-7}–3×10^{0}	4×10^{-7}–3×10^{0}	4×10^{-7}–3×10^{0}
Pumping speed, m³/s	0.8	1.0	1.6	5.0	10.0
Limiting pressure, Pa	7×10^{-8}	7×10^{-8}	7×10^{-8}	7×10^{-8}	7×10^{-8}
Start-up pressure, Pa	13	13	13	13	13
Number of evaporators	2	2	2	6	6
Power consumption, kW	1.0	1.13	2.25	9.8	14.85
Dimensions:	430	530	630	760	1100
length, mm	612	370	430	870	1100
width, mm	580	525	540	1000	1000
height, mm					
Power supply control unit dimensions:		–	–	–	–
		–	–	–	–
length, mm	647	–	–	–	–
width, mm	712				
height, mm	1034				
Weight, kg	100	100	150	300	550

Table A.12 Characteristics of magnetic discharge pumps.

Main characteristics	Trademarks						
	NMD-0.0063	NMD-0.025	NMD-0.063	NMD-0.1	NMD-0.25	NMD-0.68	NMD-1
Working pressure range, Pa	4×10^{-7}–2×10^{-1}	4×10^{-7}–2×10^{-1}	4×10^{-7}–2×10^{-1}	4×10^{-7}–2×10^{-1}	4×10^{-7}–2×10^{-1}	4×10^{-7}–2×10^{-1}	4×10^{-7}–2×10^{-1}
Pumping speed, m^3/s	0.006	0.022	0.06	0.11	0.25	0.65	1.20
Limiting pressure, Pa	7×10^{-8}	7×10^{-8}	7×10^{-8}	7×10^{-8}	7×10^{-8}	7×10^{-8}	7×10^{-8}
Maximum start-up pressure, P$_a$	1×10^{0}	1×10^{0}	1×10^{0}	1×10^{0}	1×10^{0}	1×10^{0}	1×10^{0}
Inlet pipe diameter, mm	25	100	100	100	160	250	250
Dimensions:							
length, mm	85	157	320	320	320	500	554
width, mm	80	84	106	180	327	350	500
height, mm	160	220	320	320	340	580	580
Weight, kg	2.9	8.4	21	32	53	190	290
Power supply unit (trademark)	BP-0.0063	BP-0.025	BP-0.063	BP-0.1	BP-0.25	BP-0.63	2BP-0.63
Power supply unit dimension:							
length, mm	480	480	480	480	480	480	480
width, mm	300	300	320	320	320	320	320
height, mm	220	220	300	300	300	300	600
Power supply unit weight, kg	21	20	37	35	47	47	247

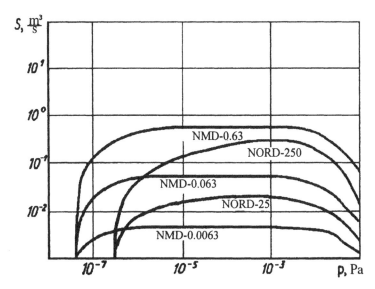

Figure A.12 Characteristics of magnetic discharge pumps.

Table A.13 Conductance of vacuum valves in molecular gas flow mode.

Fixture trademark	Conditional passage diameter, mm	Conductance, m³/s
ZVE-100	100	1.2
ZVE-160	160	3.34
ZVE-250	250	13.4
ZVE-400	400	46.25
VEP-25	25	0.014
VEP-63	63	0.148
VEP-100	100	0.470
VRP-25	25	0.011
VRP-63	63	0.102
VRP-100	100	0.332
CMU1-10	10	0.0014
CMU1-16	16	0.0040
CMU1-25	25	0.0140
CMU1-40	40	0.0400
CMU1-63	63	0.1480

The following nominal sizes of conditional passages of vacuum system elements are recommended (mm): 0.1; 0.25; 0.63; 1.0; 1.6; 2.5; 4.0; 6.3; 10; 16; 25; 40; 63; 100; 160; 250; 400; 630; 1000; 1600; 2500; 4000; 6300.

For flange and union junctions and pipeline elements, the following nominal sizes of conditional passages can be used (mm): 8; 12; 20; 32; 50; 80; 125; 200; 320; 500; 800; 1250; 2000; 3150; 5000 .

Table A.14 Notations of vacuum system elements.

Vacuum system element	Notation	
	Graphic	Text
Vacuum pump (gemeric notation)		N
Rotary pump		NL
Two-rotor pump		NZ
Turbomolecular pump		NR
Water-ring pump		NW
Ejector pump		NH
Diffusion pump (vapor-jet)		ND

Table A.14 (Cont.) Notations of vacuum system elements.

Vacuum system element	Notation	
	Graphic	Text
Adsorption pump		NA
Getter pump		NG
Cryosorption pump		NC
Ion-cryosorption pump		NE
Magnetic discharge pump		NM
Trap (generic notation)		B
Water-cooled flow trap		BW
Air-cooled trap		BA
Priming trap		BL

Table A.14 (Cont.) Notations of vacuum system elements.

Vacuum system element	Notation	
	Graphic	Text
Thermoelectric trap		BT
Sorption trap		BS
Ion trap		BE
Manometric transducer (generic notation)		P
Deformation transducer		PD
Liquid transducer		PL
Ionization transducer		PA
Magnetic transducer		PM
Thermal transducer		PT
Leak detector (generic notation)		G

Table A.14 (Cont.) Notations of vacuum system elements.

Vacuum system element	Notation	
	Graphic	Text
Mass-spectrometric leak detector	$3/4a$ ... S ... $3/4a$	S
Valve (generic notation)	a ... $a/2$	V
Hand-driven valve	$a/2$ ... $a/10$	VH
Remote controlled valve	$a/2$ ... $a/2$	VA
Electromagnetic drive valve	$a/4$	VE
Pneumetic/hydraulic drive valve		VP
Electric motor drive valve	M ... $a/2$	VM
Passage valve	$a/2$ ... a	VR
Angle valve		VN

Table A.14 (Cont.) Notations of vacuum system elements.

Vacuum system element	Notation	
	Graphic	Text
Three-way valve		VT
Leak		VF
Vacuum chamber (generic notation)	$2a$ a	CV
Heated part of vacuum chamber	$a/10$	CT
Pipeline (generic notation)		T
Straight pipeline		TR
Bend		TN
T-joint		TT
Four-way union		TK
Manifold		TG

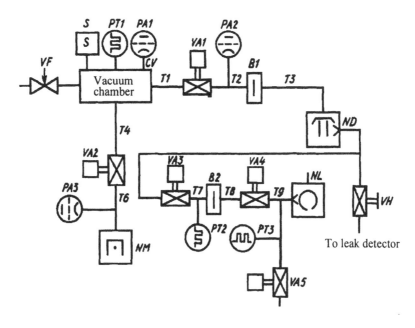

Figure A.14 Example of a principal vacuum scheme.

Notation	Name	Q-ty	Notes
VF	NK-20 Leaker	1	
VA5	Du 16 valve	1	U = 4 l/s
VA3, VA4, VH	Du 16 valve	3	U = 4 l/s
VA2	ZEPM-100 valve	1	U = 3340 l/s
VA1	KEU-63 valve	1	U = 170 l/s
T7–T9	DU10 pipeline	3	L = 50 mm
T4, T6	DU160 pipeline	2	L = 200 mm
T1–T3	DU 63 pipeline	3	L = 200 mm
S	Analyzer RMO-4C	1	
PT1–PT3	PMT-2 thermocouple transducer	3	
PA2–PA3	PMI-12 ionization transducer	2	
PA1	PMI-2 ionization transducer	1	
NM	NMD-0.25 pump	1	S = 250 l/s
NL	3NVR-1D pump	1	S =1 l/s
ND	N1S2 pump	1	S =100 l/s
CV	Vacuum chamber d = 500, L = 500	1	V = 100 l
B2	LS1T-25 trap	1	U = 2 l/s
B1	LS1T-63 trap	1	U = 10 l/s

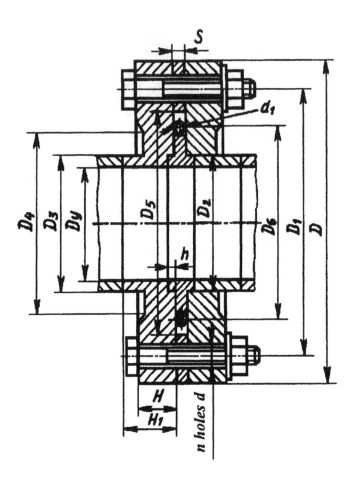

Table A.15 Detachable junction with symmetrical flanges and rubber sealing.

D_y	D	D_1	D_2	D_3	D_4	D_5	D_6	H	H_1	h	d	n	S	d_1
10	55	40	12.2	14	22	–	19	8	11	2.5	6.6	4	–	5.0
16	60	45	18.2	20	28	–	23	8	11	2.5	6.6	4	–	5.0
25	70	55	27.2	30	38	–	33	8	11	2.5	6.6	4	–	5.0
40	100	80	42.2	45	53	–	47	12	14	2.5	9	4	–	5.0
63	130	110	66.0	70	78	96	80	12	14	2.5	9	8	3.9	5.0
100	165	145	103	110	118	128	110	12	14	2.5	9	8	3.9	5.0
160	225	200	–	170	178	180	166	16	17	–	11	12	4.8	6.0
250	335	310	255	260	268	290	268	16	17	4.5	11	12	6.4	8.0
400	510	480	405	410	418	450	408	20	21	4.5	14	16	6.4	8.0

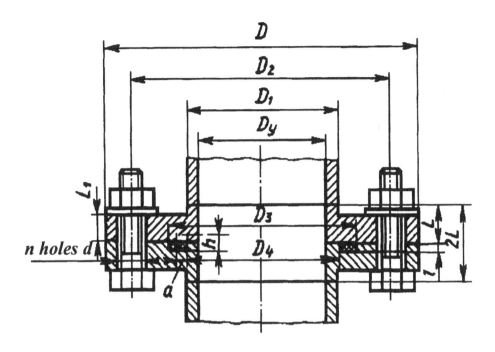

Table A.16 Detachable junction with asymmetrical flanges and rubber seals.

D_y	D	D_1	D_2	D_3	D_4	L	L_1	l	d	n	a	h
10	46	14	34	22.5	13.5	10	5	1.8	6	4	3	2.0
15	52	19	39	27.5	18.5	10	5	1.8	6	4	4	3.5
20	62	24	48	32.5	23.5	11	6	1.8	7	4	4	3.5
25	70	30	55	38.0	29.0	11	6	1.8	7	4	4	3.5
32	78	37	62	45.0	36.0	13	8	1.8	7	4	4	3.5
50	110	56	90	69.0	55.0	15	10	3.0	9	6	4	3.5
60	120	66	102	79.0	65.0	15	10	3.0	9	6	5	3.5
80	145	86	125	100	86.0	15	10	3.0	9	6	5	3.5
100	170	106	145	120	106	15	10	3.0	12	6	5	3.5
125	195	131	170	146	132	15	10	3.0	12	6	5	3.5
160	235	166	210	188	168	17	12	4.2	12	8	5	3.5
200	275	206	250	228	208	17	12	4.2	14	8	5	3.5
260	340	266	308	288	268	17	12	4.2	14	8	5	5.0
300	380	306	350	328	308	19	14	4.2	14	8	5	5.0
400	490	407	455	430	410	21	16	4.2	16	8	5	5.0
500	600	508	565	530	510	25	20	4.2	18	8	6	5.0

Note: All dimensions in mm.

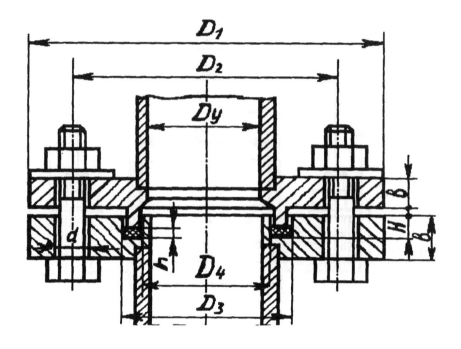

Table A.17 Detachable junction with fluoroplastic seals.

D_y	D_1	D_2	D_3	D_4	b	H	Number of bolts, n	d	h
10	46	34	19.5	13.5	8	4	4	6	2.0
15	52	39	24.5	18.5	8	4	4	6	2.0
20	62	48	31.5	23.5	10	4	4	7	2.0
25	70	55	37	29	10	4	4	7	2.0
32	78	62	44	36	10	4	4	7	2.0
40	85	70	52	44	10	4	4	7	2.0
50	110	90	70	55	12	6	4	9	3.5
60	120	102	80	70	12	6	4	9	3.5
80	145	125	100	90	12	6	4	9	3.5
100	170	145	120	110	14	6	4	9	3.5
125	195	170	145	135	14	6	8	12	3.5
160	235	210	180	170	14	6	8	12	3.5
200	275	250	220	210	18	6	8	12	3.5
260	340	308	280	270	18	6	8	14	5.0
300	380	350	320	310	20	8	8	14	5.0
380	460	430	400	390	20	8	8	18	5.0
500	600	565	525	515	24	8	8	18	5.0

Note: All dimensions in mm.

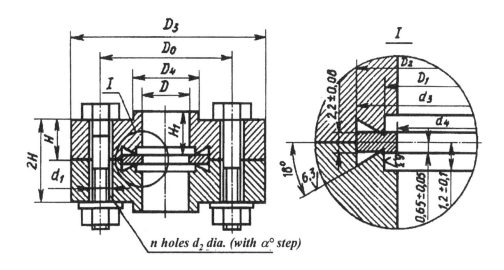

Table A.18 Conflat-type detachable junction.

D_y	D $H11$	D_0	$D_1 \pm$ 0.1	D_2 I_S9	D_3 $h11$	D_4 $h11$	H $h11$	H_1 $h11$	d_1	d_2 $H13$	n	α
16	16.0	27.0	18.3	21.4	34	18.5	8	13.0	M4	4.4	6	60°
40	35.0	58.7	41.8	48.3	70	38.5	13	18.0	M6	6.6	6	60°
63	59.5	92.1	77.0	82.6	114	66.5	18	23.0	M8	8.4	8	45°
100	100.4	130.2	115.2	120.7	152	104.5	21	26.0	M8	8.4	16	22°30′
160	150.0	181.0	166.0	171.5	202	160.6	22	27.0	M8	8.4	20	18°
200	200.0	231.8	216.8	222.3	253	206.0	25	30.0	M8	8.4	24	15°
250	250.0	282.6	267.5	273.0	305	256.0	25	30.0	M8	8.4	32	11°15′

	Gasket (copper)	
D_y	d_3	d_4
16	21.2±0.05	16.0±0.2
40	48.1±0.05	36.8±0.3
63	82.4±0.05	63.6±0.3
100	120.5 (+0.05 −0.1)	101.7±0.3
160	171.3 (+0.05 −0.1)	152.5±0.5
200	221.1 (+0.05 −0.1)	203.4±0.5
250	272.8 (+0.05 −0.1)	254.0±0.5

Note: All dimensions in mm.

Reference

1. E.S. Frolov, V.E. Minaichev, A.T. Aleksanrova *et al.*, *Vacuum Technique: A Handbook* (Mashinostroenie, Moscow, 1992).

2. V.V. Kuzmin, *Vacuum measurements* (Izdatelstvo standartov, Moscow, 1992).

3. B.S. Danilin and V.E. Minaichev, *The Design of Vacuum Systems* (Energiya, Moscow, 1971).

4. S. Deshman, *Scientific Basis of Vacuum Technique* (Mir, Moscow, 1964).

5. A.I. Pipko, V.Ya. Pliskovskii, and E.A. Penchko, *Design and Rating of Vacuum Systems* (Energiya, Moscow, 1979).

6. L.N. Rozanov, *Vacuum Equipment* (Mashinostroenie, Leningrad, 1975).

7. G.L. Saksaganskii, *Molecular Flow in Multistage Vacuum Systems* (Atomizdat, Moscow, 1980).

8. E.S. Frolov, *Turbomolecular Vacuum Pumps* (Mashinostroenie, Moscow, 1980).

9. A.B. Tzeitlin, *Vapour Jet Vacuum Pumps* (Energiya, Moscow-Leningrad, 1965)

Index

Printed and bound by CPI Group (UK) Ltd, Croydon, CR0 4YY

21/10/2024

01777095-0016